T0191892

Lecture Notes in Physics

Volume 949

Founding Editors

W. Beiglböck
J. Ehlers
K. Hepp
H. Weidenmüller

Editorial Board

M. Bartelmann, Heidelberg, Germany
P. Hänggi, Augsburg, Germany
M. Hjorth-Jensen, Oslo, Norway
R.A.L. Jones, Sheffield, UK
M. Lewenstein, Barcelona, Spain
H. von Löhneysen, Karlsruhe, Germany
A. Rubio, Hamburg, Germany
M. Salmhofer, Heidelberg, Germany
W. Schleich, Ulm, Germany
S. Theisen, Potsdam, Germany
D. Vollhardt, Augsburg, Germany
J.D. Wells, Ann Arbor, USA
G.P. Zank, Huntsville, USA

The Lecture Notes in Physics

The series Lecture Notes in Physics (LNP), founded in 1969, reports new developments in physics research and teaching-quickly and informally, but with a high quality and the explicit aim to summarize and communicate current knowledge in an accessible way. Books published in this series are conceived as bridging material between advanced graduate textbooks and the forefront of research and to serve three purposes:

- to be a compact and modern up-to-date source of reference on a well-defined topic
- to serve as an accessible introduction to the field to postgraduate students and nonspecialist researchers from related areas
- to be a source of advanced teaching material for specialized seminars, courses and schools

Both monographs and multi-author volumes will be considered for publication. Edited volumes should, however, consist of a very limited number of contributions only. Proceedings will not be considered for LNP.

Volumes published in LNP are disseminated both in print and in electronic formats, the electronic archive being available at springerlink.com. The series content is indexed, abstracted and referenced by many abstracting and information services, bibliographic networks, subscription agencies, library networks, and consortia.

Proposals should be sent to a member of the Editorial Board, or directly to the managing editor at Springer:

Christian Caron
Springer Heidelberg
Physics Editorial Department I
Tiergartenstrasse 17
69121 Heidelberg/Germany
christian.caron@springer.com

More information about this series at http://www.springer.com/series/5304

Claude Amsler

The Quark Structure of Hadrons

An Introduction to the Phenomenology and Spectroscopy

Claude Amsler
Stefan Meyer Institute
Austrian Academy of Sciences
Vienna, Austria

Physics Institute
University of Zurich
Zurich, Switzerland

ISSN 0075-8450 ISSN 1616-6361 (electronic)
Lecture Notes in Physics
ISBN 978-3-319-98526-8 ISBN 978-3-319-98527-5 (eBook)
https://doi.org/10.1007/978-3-319-98527-5

Library of Congress Control Number: 2018953170

© The Editor(s) (if applicable) and The Author(s) 2018, corrected publication 2018
This work is subject to copyright. All rights are reserved by the Publisher, whether the whole or part of the material is concerned, specifically the rights of translation, reprinting, reuse of illustrations, recitation, broadcasting, reproduction on microfilms or in any other physical way, and transmission or information storage and retrieval, electronic adaptation, computer software, or by similar or dissimilar methodology now known or hereafter developed.
The use of general descriptive names, registered names, trademarks, service marks, etc. in this publication does not imply, even in the absence of a specific statement, that such names are exempt from the relevant protective laws and regulations and therefore free for general use.
The publisher, the authors and the editors are safe to assume that the advice and information in this book are believed to be true and accurate at the date of publication. Neither the publisher nor the authors or the editors give a warranty, express or implied, with respect to the material contained herein or for any errors or omissions that may have been made. The publisher remains neutral with regard to jurisdictional claims in published maps and institutional affiliations.

This Springer imprint is published by the registered company Springer Nature Switzerland AG
The registered company address is: Gewerbestrasse 11, 6330 Cham, Switzerland

Foreword

This textbook is based on a series of lectures given by Prof. Claude Amsler at the Stefan Meyer Institute for Subatomic Physics of the Austrian Academy of Sciences in Vienna, 2016. The guest lecture was held within the framework of the graduate school (Doktoratskolleg) Particles & Interactions, DKPI, funded by the Austrian Science Fund FWF and encompassing all faculty in Vienna working in particle and nuclear physics at the University of Vienna, the Vienna University of Technology (TU Wien), the Institute for High Energy Physics HEPHY of the Austrian Academy of Sciences, and the Stefan Meyer Institute.

The author is one of the leading experts in the field of hadron spectroscopy. He is a member of the meson team of the Particle Data Group since 1995, and has contributed to several chapters on the quark model and hadron structure of the annual Review of Particle Properties. He participated in many experiments, most notably in the Crystal Barrel experiment at CERN/LEAR, studying mesons created by antiproton-proton annihilation, and wrote numerous review articles on this subject. Recently he also got involved in experiments at the Antiproton Decelerator of CERN.

The text is aimed at advanced students in hadron physics and describes in detail the experimental and phenomenological basis of hadron structure as well as the underlying quark model. It is a complete and detailed account of this subject and includes a description of virtually all known hadrons, including the most recently discovered exotic states and their potential interpretation as multi-quark states. The text addresses in particular experimentalists who would like to better understand the foundations of their research. It is very timely and welcome as an update and extension of the textbook by the same author on Nuclear and Particle Physics, published in 2015 as a broad introductory course of the whole field.

Speaker of the Graduate School DKPI, TU Wien, Vienna, Austria — Prof. Anton Rebhan

Director, Stefan Meyer Institute of the Austrian Academy of Sciences, Vienna, Austria — Prof. Eberhard Widmann

Contents

Chapter 1
Introduction

Particles which interact through the strong interaction are called hadrons (from the Greek word for "strong"). They are divided into two classes, baryons (for "heavy") with half-integer spins and mesons (for "medium" heavy) with integer spins. Baryons include their antimatter counterparts, the antibaryons. The total number of baryons minus that of antibaryons remains constant in any physical process, while mesons can be created or destroyed (provided that energy permits and electric charge is conserved).

Hadrons are not elementary (point-like) particles, but composite structures with finite dimensions, made of partons: quarks, antiquarks and gluons. The word "parton" was coined in 1969 by Richard Feynman in the context of deep inelastic scattering. Quarks were proposed in 1963 by Murray Gell-Mann, Yuval Ne'eman, André Petermann and George Zweig, as a bookkeeping tool to reduce the large number of known hadrons. In the words of Gell-Mann "such particles presumably are not real, but we may use them in our field theory anyway". Today we know, e.g. from the scaling law in deep inelastic scattering, the observation of particle jets at high energy colliders, and—as we shall see—from the observation of γ-transitions in quark-antiquark bound systems (quarkonia), that quarks and gluons are real entities that, however, remain confined in hadrons.

Baryons consist of three quarks (qqq), mesons of quark-antiquark pairs ($q\overline{q}$). The word "quark" was introduced by Gell-Mann, inspired by the sentence "Three quarks for Muster Mark" in James Joyce's novel Finnegans Wake (1939). Joyce was apparently thinking of the German word for cottage cheese, also meaning "rubbish" or "nonsense". Perhaps the origin of the word can be found even earlier in Mephisto's words "in jeden Quark begräbt er seine Nase" ("in every rubbish he lays his nose", Goethe's Faust, 1808) [1].

Quarks are spin-$\frac{1}{2}$ fermions and have by convention positive parity, while antiquarks have negative parity (Sect. 2.1). There are three families (or generations) of quarks, made of the up, down, charm, strange, top (or true) and bottom (or beauty)

© The Author(s) 2018

C. Amsler, *The Quark Structure of Hadrons*, Lecture Notes in Physics 949,
https://doi.org/10.1007/978-3-319-98527-5_1

Table 1.1 The quantum numbers of the quarks are additive

		u	d	c	s	t	b
Q	Electric charge [e]	$\frac{2}{3}$	$-\frac{1}{3}$	$\frac{2}{3}$	$-\frac{1}{3}$	$\frac{2}{3}$	$-\frac{1}{3}$
i_3	Isospin projection	$\frac{1}{2}$	$-\frac{1}{2}$	0	0	0	0
B	Baryon number	$\frac{1}{3}$	$\frac{1}{3}$	$\frac{1}{3}$	$\frac{1}{3}$	$\frac{1}{3}$	$\frac{1}{3}$
S	Strangeness	0	0	0	-1	0	0
C	Charm	0	0	$+1$	0	0	0
B'	Bottomness	0	0	0	0	0	-1
T	Topness	0	0	0	0	$+1$	0

Antiquarks have the opposite signs. The isospin of the u and d quarks is $\frac{1}{2}$, while the other quarks have isospin 0

flavours,

$$(u, d),\ (c, s),\ (t, b), \tag{1.1}$$

and the three families of corresponding antiquarks. The baryon number B is defined as $B = \frac{1}{3}$ for quarks and $B = -\frac{1}{3}$ for antiquarks. Table 1.1 lists the additive quantum numbers (the isospin is introduced in Sect. 3.1). By convention the flavour S, C, B', T of a quark has the same sign as its charge Q. Hence the flavour carried by a charged meson has the same sign as its charge, e.g. the strangeness of the $K^+(u\bar{s})$ is $+1$, the charm and bottomness of the $B_c^-(b\bar{c})$ are both -1.

The strong interaction is described by a quantum field theory, Quantum Chromodynamics (QCD), based on colour symmetry, SU(3)$_c$. Its constituents are the quarks, together with the eight gluons as gauge bosons. Colour is confined, hence free quarks and free gluons are not observed. The hadrons are states bound by the gluon fields. Since gluons are electrically neutral and do not carry any intrinsic quantum numbers (apart from colour), the quantum numbers of hadrons are given by the quantum numbers of their constituent quarks and antiquarks.

The quark constituent masses are

$$m(u) \sim 350\,\text{MeV}, m(d) \sim 350\,\text{MeV}, m(s) \sim 500\,\text{MeV},$$
$$m(c) \sim 1500\,\text{MeV}, m(b) \sim 4700\,\text{MeV and } m(t) \sim 173\,\text{GeV},$$

the t quark being as heavy as a tungsten nucleus. The Higgs boson endows the quarks with the "bare" masses:

$$m(u) \sim 2\,\text{MeV}, m(d) \sim 5\,\text{MeV}, m(s) \sim 100\,\text{MeV},$$
$$m(c) \sim 1300\,\text{MeV}, m(b) \sim 4500\,\text{MeV and } m(t) \sim 173\,\text{GeV}.$$

Most of the mass of the nucleon is not due to the Higgs boson but to gluons. In fact, without Higgs boson we and the matter in our environment would be less than 1% lighter! In the framework of chiral symmetry (an ingredient of QCD) the quark masses vanish and there are eight massless Goldstone bosons. Those bosons are identified as the lightest pseudoscalar mesons (the pion and its octet partners, Fig. 7.3 in Chap. 7) which, however, are not massless. The chiral symmetry is broken by the finite quark masses.

In the quark model the quarks carry the electric charge and the spin of the hadron, and are therefore called "valence" quarks. In addition, gluons produce virtual $q\overline{q}$ pairs, the "sea" quarks which, together with gluons, contribute to the hadron mass albeit to a lesser extent. Recent calculations and new measurements in pp collisions at high energies suggest that a large fraction of the proton spin may in fact be due to the angular momentum of the quarks and antiquarks, and/or to the gluons whose contribution rises very rapidly at low momentum fraction. Nonetheless, we shall see that numerous predictions can be made from the quark model, which is sometimes called "naive" since it emphasizes the role of the valence quarks in the hadronic wavefunction.

We now know about 180 baryons and 190 mesons (some of which are not entirely established experimentally), not counting the electric charge multiplicities. They are listed in Figs. 1.1 and 1.2, respectively. As the tables show, many mesons are actually heavier than baryons.

Since the beginning of the century a new spectroscopy has emerged beyond that of the traditional qqq and $q\overline{q}$ hadrons: evidence is mounting for the existence of baryons made of five quarks (pentaquarks, $P_c = qqqq\overline{q}$) or six quarks (hexaquarks or dibaryon) and of mesons made of two quarks and two antiquarks (tetraquarks, $qq\overline{q}\overline{q}$) or deprived of quarks (glueballs, composed only of gluons). The existence of such "exotics" is predicted by QCD, while structures such as qq or $qq\overline{q}$ are not allowed (see Chaps. 10 on colour and 16 on exotics).

The baryon number being additive, baryons (antibaryons) have $B = 1\,(-1)$ and mesons $B = 0$. The quantum numbers listed in Table 1.1 are not independent, but related through the generalized Gell-Mann-Nishijima formula, which reads

$$\boxed{Q = i_3 + \tfrac{B+S+B'+C+T}{2}}\,. \tag{1.2}$$

This relation holds for quarks and also for hadrons, for example

$$u : Q = \frac{2}{3} = \frac{1}{2} + \frac{\frac{1}{3}}{2}, \quad \Omega^-(sss) : -1 = 0 + \frac{1-3}{2}, \quad B_s^0(s\bar{b}) : 0 = 0 + \frac{-1+1}{2}. \tag{1.3}$$

The hypercharge y, defined as

$$\boxed{y \equiv B + S - \tfrac{1}{3}(C - B' + T)}\,, \tag{1.4}$$

LIGHT UNFLAVORED ($S = C = B' = 0$)

	$I^G(J^{PC})$		$I^G(J^{PC})$
• $\pi^\pm$	$1^-(0^-)$	• $\phi(1680)$	$0^-(1^{--})$
• π^0	$1^-(0^{-+})$	• $\rho_3(1690)$	$1^+(3^{--})$
• η	$0^+(0^{-+})$	• $\rho(1700)$	$1^+(1^{--})$
• $f_0(500)$	$0^+(0^{++})$	$a_2(1700)$	$1^-(2^{++})$
• $\rho(770)$	$1^+(1^{--})$	• $f_0(1710)$	$0^+(0^{++})$
• $\omega(782)$	$0^-(1^{--})$	$\eta(1760)$	$0^+(0^{-+})$
• $\eta'(958)$	$0^+(0^{-+})$	• $\pi(1800)$	$1^-(0^{-+})$
• $f_0(980)$	$0^+(0^{++})$	$f_2(1810)$	$0^+(2^{++})$
• $a_0(980)$	$1^-(0^{++})$	$X(1835)$	$?^?(0^{-+})$
• $\phi(1020)$	$0^-(1^{--})$	$X(1840)$	$?^?(?^{??})$
• $h_1(1170)$	$0^-(1^{+-})$	• $\phi_3(1850)$	$0^-(3^{--})$
• $b_1(1235)$	$1^+(1^{+-})$	$\eta_2(1870)$	$0^+(2^{-+})$
• $a_1(1260)$	$1^-(1^{++})$	• $\pi_2(1880)$	$1^-(2^{-+})$
• $f_2(1270)$	$0^+(2^{++})$	$\rho(1900)$	$1^+(1^{--})$
• $f_1(1285)$	$0^+(1^{++})$	$f_2(1910)$	$0^+(2^{++})$
• $\eta(1295)$	$0^+(0^{-+})$	$a_0(1950)$	$1^-(0^{++})$
• $\pi(1300)$	$1^-(0^{-+})$	• $f_2(1950)$	$0^+(2^{++})$
• $a_2(1320)$	$1^-(2^{++})$	$\rho_3(1990)$	$1^+(3^{--})$
• $f_0(1370)$	$0^+(0^{++})$	• $f_2(2010)$	$0^+(2^{++})$
$h_1(1380)$	$?^-(1^{+-})$	$f_0(2020)$	$0^+(0^{++})$
• $\pi_1(1400)$	$1^-(1^{-+})$	• $a_4(2040)$	$1^-(4^{++})$
• $\eta(1405)$	$0^+(0^{-+})$	• $f_4(2050)$	$0^+(4^{++})$
$a_1(1420)$	$1^-(1^{++})$	$\pi_2(2100)$	$1^-(2^{-+})$
• $f_1(1420)$	$0^+(1^{++})$	$f_0(2100)$	$0^+(0^{++})$
• $\omega(1420)$	$0^-(1^{--})$	$f_2(2150)$	$0^+(2^{++})$
$f_2(1430)$	$0^+(2^{++})$	$\rho(2150)$	$1^+(1^{--})$
• $a_0(1450)$	$1^-(0^{++})$	• $\phi(2170)$	$0^-(1^{--})$
• $\rho(1450)$	$1^+(1^{--})$	$f_0(2200)$	$0^+(0^{++})$
• $\eta(1475)$	$0^+(0^{-+})$	$f_J(2220)$	$0^+(2^{++}$ or $4^{++})$
• $f_0(1500)$	$0^+(0^{++})$		
$f_1(1510)$	$0^+(1^{++})$	$\eta(2225)$	$0^+(0^{-+})$
• $f'_2(1525)$	$0^+(2^{++})$	$\rho_3(2250)$	$1^+(3^{--})$
$f_2(1565)$	$0^+(2^{++})$	• $f_2(2300)$	$0^+(2^{++})$
$\rho(1570)$	$1^+(1^{--})$	$f_4(2300)$	$0^+(4^{++})$
$h_1(1595)$	$0^-(1^{+-})$	$f_0(2330)$	$0^+(0^{++})$
• $\pi_1(1600)$	$1^-(1^{-+})$	• $f_2(2340)$	$0^+(2^{++})$
$a_1(1640)$	$1^-(1^{++})$	$\rho_5(2350)$	$1^+(5^{--})$
$f_2(1640)$	$0^+(2^{++})$	$a_6(2450)$	$1^-(6^{++})$
• $\eta_2(1645)$	$0^+(2^{-+})$	$f_6(2510)$	$0^+(6^{++})$
• $\omega(1650)$	$0^-(1^{--})$		
• $\omega_3(1670)$	$0^-(3^{--})$		
• $\pi_2(1670)$	$1^-(2^{-+})$		

STRANGE ($S = \pm1$, $C = B' = 0$)

• $K^\pm$	$1/2(0^-)$
• K^0	$1/2(0^-)$
• K^0_S	$1/2(0^-)$
• K^0_L	$1/2(0^-)$
• $K^*_0(700)$	$1/2(0^+)$
• $K^*(892)$	$1/2(1^-)$
• $K_1(1270)$	$1/2(1^+)$
• $K_1(1400)$	$1/2(1^+)$
• $K^*(1410)$	$1/2(1^-)$
• $K^*_0(1430)$	$1/2(0^+)$
• $K^*_2(1430)$	$1/2(2^+)$
$K(1460)$	$1/2(0^-)$
$K_2(1580)$	$1/2(2^-)$
$K(1630)$	$1/2(?^?)$
$K_1(1650)$	$1/2(1^+)$
• $K^*(1680)$	$1/2(1^-)$
• $K_2(1770)$	$1/2(2^-)$
• $K^*_3(1780)$	$1/2(3^-)$
• $K_2(1820)$	$1/2(2^-)$
$K(1830)$	$1/2(0^-)$
$K^*_0(1950)$	$1/2(0^+)$
$K^*_2(1980)$	$1/2(2^+)$
• $K^*_4(2045)$	$1/2(4^+)$
$K_2(2250)$	$1/2(2^-)$
$K_3(2320)$	$1/2(3^+)$
$K^*_5(2380)$	$1/2(5^-)$
$K_4(2500)$	$1/2(4^-)$
$K(3100)$	$?^?(?^{??})$

CHARMED ($C = \pm1$)

• $D^\pm$	$1/2(0^-)$
• D^0	$1/2(0^-)$
• $D^*(2007)^0$	$1/2(1^-)$
• $D^*(2010)^\pm$	$1/2(1^-)$
• $D^*_0(2400)^0$	$1/2(0^+)$
$D^*_0(2400)^\pm$	$1/2(0^+)$
• $D_1(2420)^0$	$1/2(1^+)$
$D_1(2420)^\pm$	$1/2(?^?)$
$D_1(2430)^0$	$1/2(1^+)$
• $D^*_2(2460)^0$	$1/2(2^+)$
• $D^*_2(2460)^\pm$	$1/2(2^+)$
$D(2550)^0$	$1/2(?^?)$
$D^*_J(2600)$	$1/2(?^?)$
$D^*(2640)^\pm$	$1/2(?^?)$
$D(2740)^0$	$1/2(?^?)$
$D^*_3(2750)$	$1/2(3^-)$
$D(3000)^0$	$1/2(?^?)$

CHARMED, STRANGE ($C = S = \pm1$)

• $D^\pm_s$	$0(0^-)$
• $D^{*\pm}_s$	$0(?^?)$
• $D^*_{s0}(2317)^\pm$	$0(0^+)$
• $D_{s1}(2460)^\pm$	$0(1^+)$
• $D_{s1}(2536)^\pm$	$0(1^+)$
• $D^*_{s2}(2573)$	$0(2^+)$
• $D^*_{s1}(2700)^\pm$	$0(1^-)$
$D^*_{s1}(2860)^\pm$	$0(1^-)$
$D^*_{s3}(2860)^\pm$	$0(3^-)$
$D_{sJ}(3040)^\pm$	$0(?^?)$

BOTTOM ($B' = \pm1$)

• $B^\pm$	$1/2(0^-)$
• B^0	$1/2(0^-)$
• $B^\pm/B^0$ ADMIXTURE	
• $B^\pm/B^0/B^0_s/b$-baryon ADMIXTURE	
V_{cb} and V_{ub} CKM Matrix Elements	
• B^*	$1/2(1^-)$
• $B_1(5721)^+$	$1/2(1^+)$
• $B_1(5721)^0$	$1/2(1^+)$
$B^*_J(5732)$	$?(?^?)$
• $B^*_2(5747)^+$	$1/2(2^+)$
• $B^*_2(5747)^0$	$1/2(2^+)$
$B_J(5840)^+$	$1/2(?^?)$
$B_J(5840)^0$	$1/2(?^?)$
• $B_J(5970)^+$	$1/2(?^?)$
• $B_J(5970)^0$	$1/2(?^?)$

BOTTOM, STRANGE ($B' = \pm1$, $S = \mp1$)

• B^0_s	$0(0^-)$
• B^*_s	$0(1^-)$
$X(5568)^\pm$	$?(?^?)$
• $B_{s1}(5830)^0$	$0(1^+)$
• $B^*_{s2}(5840)^0$	$0(2^+)$
$B^*_{sJ}(5850)$	$?(?^?)$

BOTTOM, CHARMED ($B' = C = \pm1$)

• B^+_c	$0(0^-)$
$B_c(2S)^\pm$	$0(0^-)$

$c\bar{c}$

• $\eta_c(1S)$	$0^+(0^{-+})$
• $J/\psi(1S)$	$0^-(1^{--})$
• $\chi_{c0}(1P)$	$0^+(0^{++})$
• $\chi_{c1}(1P)$	$0^+(1^{++})$
• $h_c(1P)$	$?^?(1^{+-})$
• $\chi_{c2}(1P)$	$0^+(2^{++})$
• $\eta_c(2S)$	$0^+(0^{-+})$
• $\psi(2S)$	$0^-(1^{--})$
• $\psi(3770)$	$0^-(1^{--})$
• $\psi_2(3823)$	$0^-(2^{--})$
$\chi_{c0}(3860)$	$0^+(0^{++})$
• $\chi_{c1}(3872)$	$0^+(1^{++})$
• $Z_c(3900)$	$1^+(1^{+-})$
• $X(3915)$	$0^+(0/2^{++})$
• $\chi_{c2}(3930)$	$0^+(2^{++})$
$X(3940)$	$?^?(?^{??})$
• $X(4020)$	$1^+(?^{?-})$
• $\psi(4040)$	$0^-(1^{--})$
$X(4050)^\pm$	$1^-(?^{?+})$
$X(4055)^\pm$	$1^+(?^{?-})$
• $\chi_{c1}(4140)$	$0^+(1^{++})$
• $\psi(4160)$	$0^-(1^{--})$
$X(4160)$	$?^?(?^{??})$
$Z_c(4200)$	$1^+(1^{+-})$
$\psi(4230)$	$0^-(1^{--})$
$R_{c0}(4240)$	$1^+(0^{--})$
$X(4250)^\pm$	$1^-(?^{?+})$
• $\psi(4260)$	$0^-(1^{--})$
• $\chi_{c1}(4274)$	$0^+(1^{++})$
$X(4350)$	$0^+(?^{?+})$
• $\psi(4360)$	$0^-(1^{--})$
$\psi(4390)$	$0^-(1^{--})$
• $\psi(4415)$	$0^-(1^{--})$
• $Z_c(4430)$	$1^+(1^{+-})$
$\chi_{c0}(4500)$	$0^+(0^{++})$
• $\psi(4660)$	$0^-(1^{--})$
$\chi_{c0}(4700)$	$0^+(0^{++})$

$b\bar{b}$

• $\eta_b(1S)$	$0^+(0^{-+})$
• $\Upsilon(1S)$	$0^-(1^{--})$
• $\chi_{b0}(1P)$	$0^+(0^{++})$
• $\chi_{b1}(1P)$	$0^+(1^{++})$
• $h_b(1P)$	$?^?(1^{+-})$
• $\chi_{b2}(1P)$	$0^+(2^{++})$
$\eta_b(2S)$	$0^+(0^{-+})$
• $\Upsilon(2S)$	$0^-(1^{--})$
• $\Upsilon_2(1D)$	$0^-(2^{--})$
• $\chi_{b0}(2P)$	$0^+(0^{++})$
• $\chi_{b1}(2P)$	$0^+(1^{++})$
$h_b(2P)$	$?^?(1^{+-})$
• $\chi_{b2}(2P)$	$0^+(2^{++})$
• $\Upsilon(3S)$	$0^-(1^{--})$
• $\chi_{b1}(3P)$	$0^+(1^{++})$
• $\Upsilon(4S)$	$0^-(1^{--})$
• $Z_b(10610)$	$1^+(1^{+-})$
$Z_b(10650)$	$1^+(1^{+-})$
• $\Upsilon(10860)$	$0^-(1^{--})$
• $\Upsilon(11020)$	$0^-(1^{--})$

Fig. 1.1 Table of known mesons [2]. Particles without a dot are not fully established yet. The particle name (Chap. 4) is related to its quantum numbers $I^G(J^{PC})$ which will be defined in the next chapters. The masses in brackets (in MeV) are usually omitted for the ground state mesons

	J^P	
p	$1/2^+$	****
n	$1/2^+$	****
$N(1440)$	$1/2^+$	****
$N(1520)$	$3/2^-$	****
$N(1535)$	$1/2^-$	****
$N(1650)$	$1/2^-$	****
$N(1675)$	$5/2^-$	****
$N(1680)$	$5/2^+$	****
$N(1700)$	$3/2^-$	***
$N(1710)$	$1/2^+$	****
$N(1720)$	$3/2^+$	****
$N(1860)$	$5/2^+$	**
$N(1875)$	$3/2^-$	***
$N(1880)$	$1/2^+$	***
$N(1895)$	$1/2^-$	****
$N(1900)$	$3/2^+$	****
$N(1990)$	$7/2^+$	**
$N(2000)$	$5/2^+$	**
$N(2040)$	$3/2^+$	*
$N(2060)$	$5/2^-$	***
$N(2100)$	$1/2^+$	***
$N(2120)$	$3/2^-$	***
$N(2190)$	$7/2^-$	****
$N(2220)$	$9/2^+$	****
$N(2250)$	$9/2^-$	****
$N(2300)$	$1/2^+$	**
$N(2570)$	$5/2^-$	**
$N(2600)$	$11/2^-$	***
$N(2700)$	$13/2^+$	**

	J^P	
$\Delta(1232)$	$3/2^+$	****
$\Delta(1600)$	$3/2^+$	****
$\Delta(1620)$	$1/2^-$	****
$\Delta(1700)$	$3/2^-$	****
$\Delta(1750)$	$1/2^+$	*
$\Delta(1900)$	$1/2^-$	***
$\Delta(1905)$	$5/2^+$	****
$\Delta(1910)$	$1/2^+$	****
$\Delta(1920)$	$3/2^+$	***
$\Delta(1930)$	$5/2^-$	***
$\Delta(1940)$	$3/2^-$	**
$\Delta(1950)$	$7/2^+$	****
$\Delta(2000)$	$5/2^+$	**
$\Delta(2150)$	$1/2^-$	*
$\Delta(2200)$	$7/2^-$	***
$\Delta(2300)$	$9/2^+$	**
$\Delta(2350)$	$5/2^-$	*
$\Delta(2390)$	$7/2^+$	*
$\Delta(2400)$	$9/2^-$	**
$\Delta(2420)$	$11/2^+$	****
$\Delta(2750)$	$13/2^-$	**
$\Delta(2950)$	$15/2^+$	**
Λ	$1/2^+$	****
$\Lambda(1405)$	$1/2^-$	****
$\Lambda(1520)$	$3/2^-$	****
$\Lambda(1600)$	$1/2^+$	***
$\Lambda(1670)$	$1/2^-$	****
$\Lambda(1690)$	$3/2^-$	****
$\Lambda(1710)$	$1/2^+$	*
$\Lambda(1800)$	$1/2^-$	***
$\Lambda(1810)$	$1/2^+$	***
$\Lambda(1820)$	$5/2^+$	****
$\Lambda(1830)$	$5/2^-$	****
$\Lambda(1890)$	$3/2^+$	****
$\Lambda(2000)$		*
$\Lambda(2020)$	$7/2^+$	*
$\Lambda(2050)$	$3/2^-$	*
$\Lambda(2100)$	$7/2^-$	****
$\Lambda(2110)$	$5/2^+$	***
$\Lambda(2325)$	$3/2^-$	*
$\Lambda(2350)$	$9/2^+$	***
$\Lambda(2585)$		**

	J^P	
Σ^+	$1/2^+$	****
Σ^0	$1/2^+$	****
Σ^-	$1/2^+$	****
$\Sigma(1385)$	$3/2^+$	****
$\Sigma(1480)$		*
$\Sigma(1560)$		**
$\Sigma(1580)$	$3/2^-$	*
$\Sigma(1620)$	$1/2^-$	*
$\Sigma(1660)$	$1/2^+$	***
$\Sigma(1670)$	$3/2^-$	****
$\Sigma(1690)$		**
$\Sigma(1730)$	$3/2^+$	*
$\Sigma(1750)$	$1/2^-$	***
$\Sigma(1770)$	$1/2^+$	*
$\Sigma(1775)$	$5/2^-$	****
$\Sigma(1840)$	$3/2^+$	*
$\Sigma(1880)$	$1/2^+$	**
$\Sigma(1900)$	$1/2^-$	*
$\Sigma(1915)$	$5/2^+$	****
$\Sigma(1940)$	$3/2^+$	*
$\Sigma(1940)$	$3/2^-$	***
$\Sigma(2000)$	$1/2^-$	*
$\Sigma(2030)$	$7/2^+$	****
$\Sigma(2070)$	$5/2^+$	*
$\Sigma(2080)$	$3/2^+$	**
$\Sigma(2100)$	$7/2^-$	*
$\Sigma(2250)$		***
$\Sigma(2455)$		**
$\Sigma(2620)$		**
$\Sigma(3000)$		*
$\Sigma(3170)$		*

	J^P	
Ξ^0	$1/2^+$	****
Ξ^-	$1/2^+$	****
$\Xi(1530)$	$3/2^+$	****
$\Xi(1620)$		*
$\Xi(1690)$		***
$\Xi(1820)$	$3/2^-$	***
$\Xi(1950)$		***
$\Xi(2030)$	$\geq \frac{5}{2}?$	***
$\Xi(2120)$		*
$\Xi(2250)$		**
$\Xi(2370)$		**
$\Xi(2500)$		*
Ω^-	$3/2^+$	****
$\Omega(2250)^-$		***
$\Omega(2380)^-$		**
$\Omega(2470)^-$		**

	J^P	
Λ_c^+	$1/2^+$	****
$\Lambda_c(2595)^+$	$1/2^-$	***
$\Lambda_c(2625)^+$	$3/2^-$	***
$\Lambda_c(2765)^+$		*
$\Lambda_c(2860)^+$	$3/2^+$	***
$\Lambda_c(2880)^+$	$5/2^+$	***
$\Lambda_c(2940)^+$	$3/2^-$	***
$\Sigma_c(2455)$	$1/2^+$	****
$\Sigma_c(2520)$	$3/2^+$	***
$\Sigma_c(2800)$		***
Ξ_c^+	$1/2^+$	***
Ξ_c^0	$1/2^+$	***
$\Xi_c'^+$	$1/2^+$	***
$\Xi_c'^0$	$1/2^+$	***
$\Xi_c(2645)$	$3/2^+$	***
$\Xi_c(2790)$	$1/2$	***
$\Xi_c(2815)$	$3/2^-$	***
$\Xi_c(2930)$		*
$\Xi_c(2970)$		***
$\Xi_c(3055)$		***
$\Xi_c(3080)$		***
$\Xi_c(3123)$		*
Ω_c^0	$1/2^+$	***
$\Omega_c(2770)^0$	$3/2^+$	***
$\Omega_c(3000)^0$		***
$\Omega_c(3050)^0$		***
$\Omega_c(3065)^0$		***
$\Omega_c(3090)^0$		***
$\Omega_c(3120)^0$		***
Ξ_{cc}^+		*
Ξ_{cc}^{++}		***
Λ_b^0	$1/2^+$	***
$\Lambda_b(5912)^0$	$1/2^-$	***
$\Lambda_b(5920)^0$	$3/2^-$	***
Σ_b	$1/2^+$	***
Σ_b^*	$3/2^+$	***
Ξ_b^0, Ξ_b^-	$1/2^+$	***
$\Xi_b'(5935)^-$	$1/2^+$	***
$\Xi_b(5945)^0$	$3/2^+$	***
$\Xi_b(5955)^-$	$3/2^+$	***
Ω_b^-	$1/2^+$	***
$P_c(4380)^+$		*
$P_c(4450)^+$		*

Fig. 1.2 Table of baryons (adapted from [2]). Masses are given in brackets in MeV for baryons that decay through the strong interaction. The half-integer number is the baryon spin J, P its parity (when known). Baryons with 1 and 2 stars are not well established. The baryon name is related to the quark content and isospin (see Chap. 13 for the nomenclature). A hyperon is a baryon containing at least one s quark, such as the Λ (uds)

is equal to $\frac{1}{3}$ for the u and d quarks, $-\frac{2}{3}$ for the s and $y = 0$ for the heavier quarks. Thus for the u, d and s quarks the hypercharge is equal to $B + S$. The Gell-Mann-Nishijima formula

$$Q = i_3 + \frac{y}{2} \tag{1.5}$$

(in its original form) holds only for the three lightest quarks.

Throughout these lectures we shall use natural units by setting $\hbar = c = 1$ (e.g. the projection of the proton spin onto the quantization axis is $m = \pm\frac{1}{2}$). Hence

$$\hbar c = 197.3 \text{ MeV} \cdot \text{fm} = 1, \tag{1.6}$$

where 1 fm = 10^{-15} m, therefore

$$1 \text{ MeV} \equiv \frac{1}{197.3} \text{ fm}^{-1}. \tag{1.7}$$

One obtains similarly with $c = 1$

$$1 \text{ s} = 2.9979 \times 10^{23} \text{ fm} \tag{1.8}$$

and

$$1 \text{ MeV} = 1.5194 \times 10^{21} \text{ s}^{-1} . \tag{1.9}$$

Final results are easily expressed into m, s or MeV by using the aforementioned conversions. Energies and masses will be given in MeV, but momenta in MeV/c to avoid confusion with energy.

Let us complete this introduction with a summary of the conservation laws that hold for the strong, electromagnetic and weak interactions between hadrons and in hadron decays. Conservation laws, symmetries and gauge invariance are closely related (see e.g. [3]).

- The electric charge Q, the baryon number B and the total angular momentum j are conserved in all interactions. CPT is also believed to be a universal symmetry.
- Colour is conserved in all quark and gluon interactions (leptons do not carry colour).
- Flavour is conserved in strong and electromagnetic interactions between hadrons and so are parity P and charge conjugation C. These symmetries, as well as their CP combination and time reversal invariance T, are violated in weak interactions.
- The isospin and G parity (Sects. 3.1 and 3.4) are conserved only in strong interactions, while the electromagnetic interaction violates the isospin and G, but conserves the projection i_3 of the isospin. Neither isospin, i_3 nor G are conserved in weak interactions.
- The lepton number (which is not relevant in these lectures) is conserved, except in neutrino oscillations.

References

1. Mayer-Kuckuk, T.: Kernphysik, p. 183. Teubner, Stuttgart (1984)
2. Tanabashi, M., et al. (Particle Data Group): Phys. Rev. D 98, 030001 (2018)
3. Amsler, C.: Nuclear and Particle Physics. IOP Publishing, Bristol (2015)

Chapter 2
Mesons

The spin vector $\vec{J}$ of a hadron is constructed by adding the spins of the constituent quarks and their angular momenta $\vec{\ell}$. For $q\overline{q}$ mesons the total quark spin is $\vec{s} = \vec{s}_q + \vec{s}_{\overline{q}}$ and $\vec{J} = \vec{\ell} + \vec{s}$. The corresponding quantum numbers are $s = 0$ or 1 and $\ell = 0, 1, 2$, etc. Mesons with $\ell > 0$ are orbital excitations of the ground state $\ell = 0$ (Fig. 2.1). The meson spin j is an integer number with

$$|\ell - s| \leq j \leq \ell + s. \tag{2.1}$$

Radial excitations (vibrations) are denoted by the quantum number $n \geq 1$. The quantum numbers of a $q\overline{q}$ meson are specified with the notation

$$n^{2s+1}\ell_j \text{ or } i^G(j^{PC}), \tag{2.2}$$

where P is the parity, C the charge conjugation (or C parity), i the isospin, and G the G parity. The two notations are not quite equivalent since in the former the isospin i is omitted and in the latter n is not specified. The hadron spin and isospin are often written in capital letters, hence the notation $I^G(J^{PC})$ [1]. In this case I and J are obviously quantum numbers and not moduli of the vector operators $\vec{I}$ and $\vec{J}$.

For example, the ground states that can be built with the three lightest flavours u, d, s are the nine pseudoscalar mesons with quantum numbers 1^1S_0 (or $J^{PC} = 0^{-+}$)

$$\pi^-, \pi^0, \pi^+, K^-, \overline{K}^0, K^0, K^+, \eta, \eta', \tag{2.3}$$

and the nine vector mesons with quantum numbers 1^3S_1 (or $J^{PC} = 1^{--}$)

$$\rho^-, \rho^0, \rho^+, K^{*-}, \overline{K}^{*0}, K^{*0}, K^{*+}, \phi, \omega. \tag{2.4}$$

© The Author(s) 2018

C. Amsler, *The Quark Structure of Hadrons*, Lecture Notes in Physics 949,
https://doi.org/10.1007/978-3-319-98527-5_2

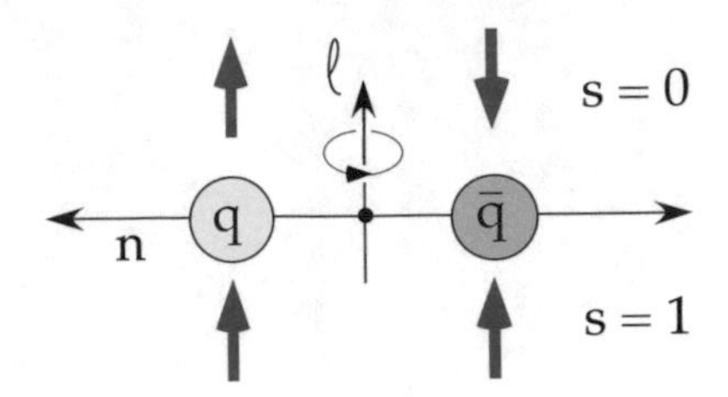

Fig. 2.1 A $q\overline{q}$ meson is in a spin-singlet state ($s = 0$, antiparallel quark spins) or in a spin-triplet state ($s = 1$, parallel quark spins). Excited states are obtained by switching angular momentum $\ell > 0$ within the pair (orbital excitation) or by inducing vibrations $n > 1$ (radial excitations)

2.1 Internal Parity

In atomic processes the dominant electromagnetic transitions are of the E1 type, in which the angular momentum changes by one unit between initial and final state levels. Hence the parity of the photon must be negative since parity is conserved in electromagnetic interactions.

The electric charge is conserved in any reaction between initial and final states. Similarly, to ensure parity conservation in strong interactions, a quantum number is assigned to every hadron, its internal (or intrinsic) parity. The parity of a quark can be chosen arbitrarily, +1 or −1. Furthermore, the relative parity between quarks of different flavours cannot be established experimentally, because quark flavours do not change in strong nor in electromagnetic interactions. However, the relative parity between a fermion and its antifermion is predicted by the Dirac equation to be negative (an experimental proof will be described below). The convention is to assign a positive parity to the quarks and a negative parity to the antiquarks.

The parity is a multiplicative quantum number. For example, for a system made of two sub-systems with parities P_1 and P_2 and relative angular momentum ℓ, the parity is given by

$$P = (-1)^{\ell} P_a P_b. \tag{2.5}$$

Hence one gets for the ground state baryons, for which all angular momenta vanish:

$$P(p) = +1, \quad P(n) = +1, \quad P(\Lambda) = +1, \tag{2.6}$$

while for antibaryons or antibaryons

$$P(\overline{p}) = -1, \quad P(\overline{n}) = -1, \quad P(\overline{\Lambda}) = -1. \tag{2.7}$$

According to (2.5) the internal parity of a $q\overline{q}$ meson is then given by

$$P(q\overline{q}) = -(-1)^{\ell}, \tag{2.8}$$

where ℓ is the angular momentum carried by the $q\bar{q}$ pair. We will show how the internal parity of the pion was determined experimentally to be negative. From (2.8) then follows that the angular momentum ℓ of the $q\bar{q}$ pair is even. The pion is spinless and therefore the $q\bar{q}$ pair is in the singlet state ($s = 0$ and $\ell = 0$, 1^1S_0 state).

The experiment of Wu and Shaknov [2] established experimentally that the relative parity between fermion and antifermion is negative. Consider the electromagnetically bound pair of a positron and an electron. The hyperfine interaction splits the ground state into a spin singlet ($^1S_0$, parapositronium) which decays rapidly with a mean lifetime of 124 ps into 2γ, and a spin triplet ($^3S_1$, orthopositronium) decaying into 3γ with a lifetime of 142 ns. We remind the reader an important rule that will be used during these lectures, namely that a spin-1 system does not decay into two massless spin-1 objects (e.g. photons or gluons), by virtue of the Landau-Yang theorem [3] (a proof is given in Appendix A). Therefore orthopositronium does not decay into 2γ. According to (2.8) for a fermion-antifermion pair with $\ell = 0$, one expects parapositronium (as well as orthopositronium) to have negative internal parity.

Figure 2.2 shows a sketch of the apparatus [2]. Positronium is produced in a ^{64}Cu positron source. The two 511 keV back-to-back photons from parapositronium annihilation are collimated, scattered by two aluminium blocks and detected by the scintillation counters S_1 and S_2. The cross section for Compton scattering depends on the azimuthal angle with respect to the photon polarization: the photons are re-emitted preferably in the direction orthogonal to the incident polarization (oscillating dipole $\vec{E}$-field). By measuring the coincidence rate $S_1 S_2$ as a function of angle ϕ between the two detectors one obtains a distribution of the relative orientations of the electric fields of the two photons. The result [2]

$$\frac{S_1 S_2(\phi = \pi/2)}{S_1 S_2(\phi = 0)} = 2.04 \pm 0.08 \tag{2.9}$$

shows that the polarizations are preferably orthogonal. In Chap. 18 we demonstrate that a negative parity spin-0 system decays into two spin-1 particles with their polarizations preferably orthogonal. One can show that, in contrast, the polarizations of the two spin-1 particles would be preferably parallel for an initial spin-0 state of positive parity (see Sect. 2.4 on the internal parity of the π^0).

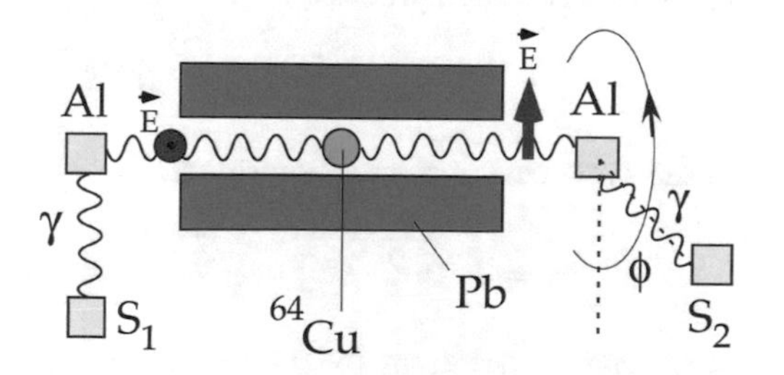

Fig. 2.2 Relative parity between fermion and antifermion determined in parapositronium decay [2]

2.2 The Pion

The existence of a massive particle mediating the short range nuclear strong force was postulated in 1935 by Hideki Yukawa. The range R of the interaction is related to the Compton wavelength of the hypothetical particle, the pion,

$$R = \frac{1}{m_\pi} \sim 1\text{ fm} \Rightarrow m_\pi \simeq 200\,\text{MeV}, \tag{2.10}$$

where we have used the unit conversion (1.7).

In 1936 a new particle was discovered in the cosmic radiation at the Earth's surface by Neddermeyer and Anderson, which was initially thought to be Yukawa's particle [4]. However, a negatively charged pion interacting with nuclear matter would be captured in the atomic orbits before being absorbed by the nucleus, leading to its disintegration (while positive pions would decay). Indeed, the negatively charged cosmics observed at ground level induced nuclear disintegration when being stopped in iron. In carbon, however, they decayed before being captured, at variance with the behaviour expected for the Yukawa particle [5]. On the other hand, capture by light nuclei and followed by nuclear splitting was observed at very high altitudes by submitting photographic emulsions (which consist mainly of carbon, oxygen and nitrogen atoms) to cosmic rays (Fig. 2.3, left) [6]. These findings led to the conclusion that the particle originally detected at ground level and the one observed at high altitudes were different entities, the muon and the pion, respectively, the latter qualifying to be Yukawa's particle. The charged pion with a mass $m = 139.6\,\text{MeV}$, was discovered in 1947 by Lattes, Occhialini and Powell [7, 8], also in cosmic rays (Fig. 2.3, right).

2.2.1 *The Spin of the Charged Pion*

The spin of the positively charged pion was established by comparing the reaction

$$A : pp \to d\pi^+ \tag{2.11}$$

to its time-reversed process

$$B : \pi^+ d \to pp \tag{2.12}$$

at the same center-of-mass energy E:

$$E = 2\sqrt{m_p^2 + k_p^2} = \sqrt{m_d^2 + k_d^2} + \sqrt{m_\pi^2 + k_\pi^2 (= k_d^2)}, \tag{2.13}$$

with obvious notations for the particle masses and momenta in the center-of-mass system. For a two-body reaction $1 + 2 \to 3 + 4$ the differential cross section in the

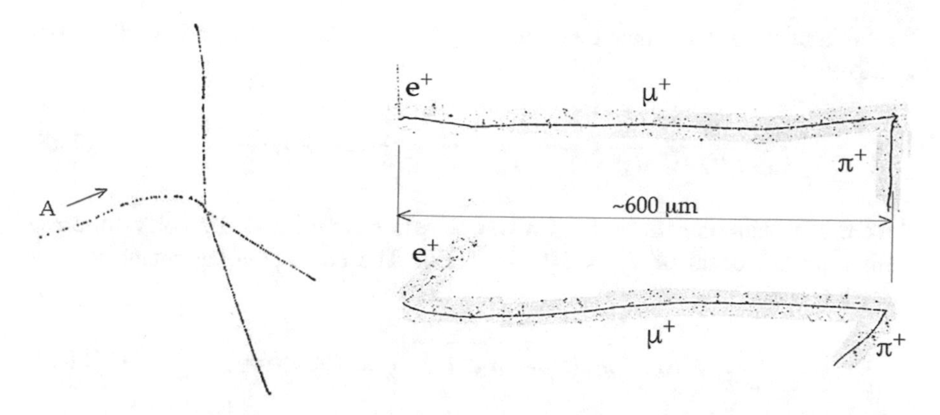

Fig. 2.3 Left: particle A interacts with a light nucleus in a photographic emulsion [6]. Note the increasing energy loss near the interaction vertex which shows that the particle entered from the left before being stopped. The emerging tracks are due to protons and nuclear fragments. The multiple scattering of the incident track excluded the projectile to be a proton. The mass was estimated between 120 and 200 electron masses. Right: discovery of the pion with photographic emulsions at the Pic du Midi. The short tracks on the right stem from stopping positive pions which decay into muons (negative pions would be absorbed by the nuclei). The track lengths of the muons are both $\simeq 600\,\mu$m, hence the muon is emitted in a two-body decay ($\pi^+ \to \mu^+ \nu_\mu$). Note the ionizing power of the stopping π^+ or μ^+, which is larger than that of the faster and lighter positron (adapted from [8])

center-of-mass reference frame is given by

$$\frac{d\sigma}{d\Omega} = \frac{1}{K \times 64\pi^2 E^2} \frac{k_3}{k_1} |\mathcal{M}|^2 , \tag{2.14}$$

where K is the initial spin multiplicity factor $K = (2s_1 + 1)(2s_2 + 1)$ and $\mathcal{M}$ is the transition amplitude (for a derivation see [9] p. 12–2). For reaction A the multiplicity factor is $K = (2s_p + 1)^2 = 4$ and the differential cross section is given by

$$\frac{d\sigma_A}{d\Omega} = \frac{1}{4 \times 64\pi^2 E^2} \frac{k_d}{k_p} |\mathcal{M}_A|^2. \tag{2.15}$$

For reaction B with $K = (2s_d+1)(2s_\pi+1) = 3(2s_\pi+1)$ and $s_d = 1$, the differential cross section is given by

$$\frac{d\sigma_B}{d\Omega} = \frac{1}{3(2s_\pi + 1) \times 64\pi^2 E^2} \frac{k_p}{k_d} |\mathcal{M}_B|^2. \tag{2.16}$$

The transition amplitudes are related by time reversal invariance, namely

$$|\mathcal{M}_A|^2 = |\mathcal{M}_B|^2. \tag{2.17}$$

The spin of the pion is then determined by building the ratio of differential cross sections

$$\frac{d\sigma_A/d\Omega}{d\sigma_B/d\Omega} = \frac{k_d}{4k_p} \times \frac{3k_d(2s_\pi + 1)}{k_p} = \frac{3}{4}(2s_\pi + 1)\frac{k_d^2}{k_p^2}. \tag{2.18}$$

An early measurement of reaction A was performed at the 184" Berkeley cyclotron with a proton beam of $T_p = 340\,\mathrm{MeV}$ [10]. The corresponding center-of-mass energy is given by

$$E = \sqrt{2m_p(T_p + m_p) + 2m_p^2} = 2039\,\mathrm{MeV}. \tag{2.19}$$

The pion and deuteron momenta in the center-of-mass can be calculated from (2.13),

$$k_\pi = k_d = \frac{\sqrt{[E^2 - (m_\pi + m_d)^2][E^2 - (m_\pi - m_d)^2]}}{2E} = 81\,\mathrm{MeV/c}, \tag{2.20}$$

while the proton momenta are

$$k_p = \frac{1}{2}\sqrt{E^2 - 4m_p^2} = 399\,\mathrm{MeV/c}. \tag{2.21}$$

For reaction B the required laboratory kinetic energy of the pion is

$$T_\pi = \frac{E^2 - m_d^2 - m_\pi^2}{2m_d} - m_\pi = 26\,\mathrm{MeV}. \tag{2.22}$$

Figure 2.4 shows the differential cross section for reaction B measured at the Nevis cyclotron of Columbia University [11], together with the prediction from reaction A. The relation (2.18) was used for $s_\pi = 0$ and alternatively for $s_\pi = 1$. These early data clearly favour $s_\pi = 0$.

We now integrate (2.15) and (2.16) over the full solid angle to obtain the total cross sections. For reaction B the integration over the scattering angle runs only from 0 to $\frac{\pi}{2}$ since protons emitted in the center-of-mass into the forward hemisphere cannot be distinguished from those emitted into the backward hemisphere. This leads to a multiplicative factor of $\frac{1}{2}$. The ratio of total cross sections is then given by

$$\frac{\sigma_A}{\sigma_B} = \frac{3}{2}(2s_\pi + 1)\frac{k_d^2}{k_p^2}. \tag{2.23}$$

Figure 2.5 shows a compilation of the measured total cross sections as a function of incident laboratory energy. The measured cross section σ_B is converted into σ_A by using (2.23), assuming that $s_\pi = 0$. The excellent agreement follows from time

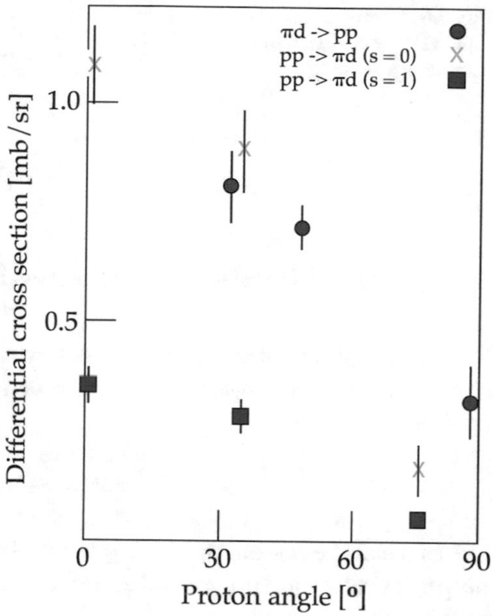

Fig. 2.4 Differential cross section in the center-of-mass system for the reaction $\pi^+ d \rightarrow pp$ for 28 MeV pions as a function of proton angle (red circles). The predictions from the reaction $pp \rightarrow \pi^+ d$ are shown for a spin-0 pion (green crosses) and for a spin-1 pion [11]

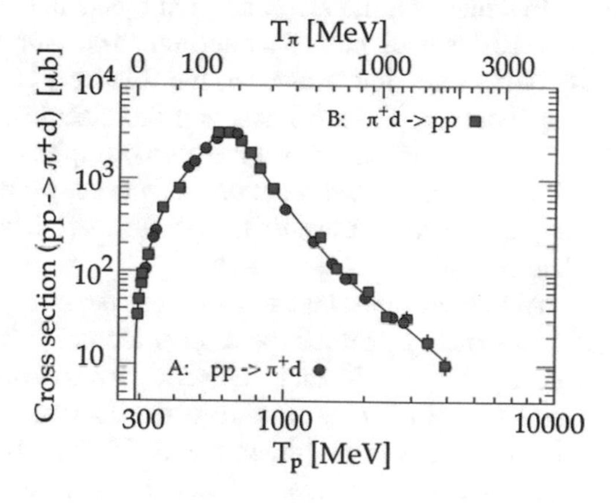

Fig. 2.5 Measured total cross sections $\sigma_A(pp \rightarrow \pi^+ d)$ vs. proton kinetic energy T_p in the laboratory (red dots) and $\sigma_B(\pi^+ d \rightarrow pp)$ vs. pion kinetic energy T_π (blue squares). The cross section σ_B has been scaled to σ_A by assuming that the pion has spin $s_\pi = 0$ [12]

reversal invariance in strong interactions, and shows without doubt that the spin of the pion is equal to zero.

Incidentally, Fig. 2.5 shows that the cross section for $pp \rightarrow d\pi^+$ reaches a maximum for proton energies of about 600 MeV. Hence, in order to maximize pion production, accelerators (such as the isochronous cyclotron of the Paul Scherrer Institute or the former CERN synchrocyclotron) have been built to accelerate protons to 600 MeV. According to (2.19) the corresponding center-of-mass energy is $E = 2160$ MeV, which is roughly equal to the mass of the $\Delta(1232)$ + neutron (Fig. 2.6).

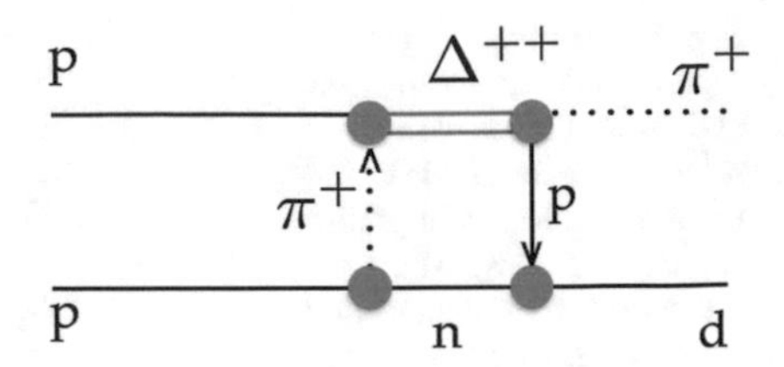

Fig. 2.6 Contribution of the $\Delta(1232)^{++}$ resonance to the reaction $pp \to \pi^+ d$

2.3 Internal Parity of the Charged Pion

The parity of the negatively charged pion was determined by observing the absorption of pions stopped in liquid deuterium, followed by the emission of two neutrons:

$$\pi^- d \to nn. \tag{2.24}$$

Let us recall the reasoning which gives us the opportunity to review the symmetry properties between fermions and to become familiar with the use of Clebsch-Gordan coefficients.

In deuterium the stopping pion ejects the shell electron and is captured in the n = 16 level of pionic deuterium. The pion then cascades to the lower levels by X-ray emission. However, the relatively small pionic deuterium interacts with neighbouring ^{2}H molecules, and the electric fields induce Stark mixing between $\pi^- d$ atomic orbitals. Once the pion occupies one of the higher s levels, the overlap with the nuclear wavefunction leads to the prompt absorption by the deuteron, thus suppressing X-ray transitions to the lowest n levels. This is the dominant effect when absorbing pions in liquid targets.

At low densities, that is in gaseous targets, Stark mixing becomes less important. For instance, in gaseous hydrogen the antiprotonic hydrogen atom (protonium) is formed at $n = 30$ and de-excites by emitting X-rays (Fig. 2.7). Transitions to the $2p$ level (L X-rays) and to the $1s$ level (K X-rays) have been observed by stopping pions in hydrogen gas at NTP [13]. Proton-antiproton annihilation and meson spectroscopy from s states has been extensively studied in liquid hydrogen by bubble chamber experiments and by the Crystal Barrel experiment (for a review see [14]), while annihilation from p states was investigated by the ASTERIX experiment using gaseous targets [15]. Both experiments were performed at CERN's Low Energy Antiproton Ring (LEAR).

We have already shown that the spin of the pion is $J_\pi = 0$. The spin $J_d = 1$ of the deuteron has been determined from the ground state hyperfine splitting of the deuterium atom [16]. The deuteron is a bound system of a proton and a neutron with even angular momentum (dominantly S wave, with $\simeq$2.5% D wave admixture). Therefore the internal parity of the deuteron is $P(d) = +1$. Thus the initial state $|i\rangle$

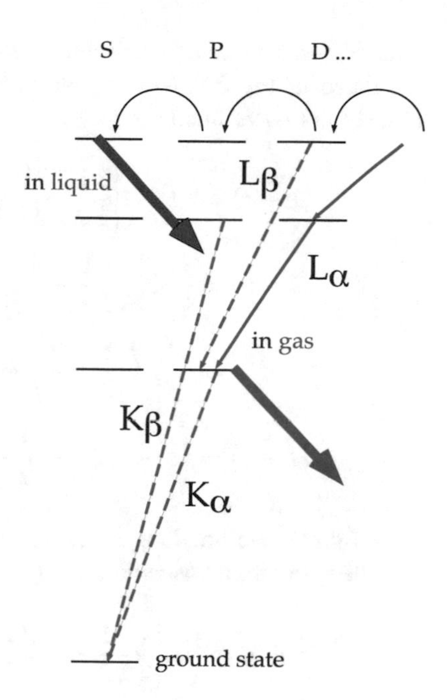

Fig. 2.7 In liquid hydrogen (or deuterium) the antiprotonic atom annihilates mostly from s states (red arrow), while in gas the suppression of the X-ray cascade is less pronounced, so that the atom can also annihilate from p states (blue arrow)

has the total spin $J_i = 1$ and the internal parity is

$$P_i = (-1)^\ell P_\pi P_d = P_\pi, \tag{2.25}$$

since in liquid the relative angular momentum between the pion and the deuteron is $\ell = 0$ (annihilation from the s levels).

Let us also determine the parity of the final state $|f\rangle$. The wavefunction of the neutron pair is given by

$$\psi(n, n) = |sm\rangle\phi_L(n, n), \tag{2.26}$$

where $|sm\rangle$ is the spin wavefunction (spin $s = 0$ or 1, projection m) and $\phi_L(n, n)$ is the orbital wavefunction with the angular momentum L between the neutrons. The wavefunction $|sm\rangle$ can be written as the linear superposition of the wavefunctions of the two subsystems with spins s_1 and s_2:

$$|sm\rangle = \sum_{m_1, m_2} \langle sm|s_1 s_2 m_1 m_2\rangle |m_1 m_2\rangle. \tag{2.27}$$

The quantum numbers s and m vary between $|s_1 - s_2|$ and $s_1 + s_2$ and between $-s$ and $+s$, respectively. The sum extends over all m_1 and m_2 such that

$$m = m_1 + m_2. \tag{2.28}$$

The Clebsch-Gordan coefficients with the generic names $\langle jm|j_1 j_2 m_1 m_2\rangle$ are tabulated in Fig. 2.8. With $j_1 \equiv s_1 = \frac{1}{2}$ and $j_2 \equiv s_2 = \frac{1}{2}$ one gets the three spin-triplet wavefunctions

$$\begin{aligned}
|1+1\rangle &= \left\langle 1+1 \middle| \frac{1}{2}\frac{1}{2}\frac{1}{2}\frac{1}{2} \right\rangle \left| \frac{1}{2}\frac{1}{2} \right\rangle = \left| \frac{1}{2}\frac{1}{2} \right\rangle, \\
|1 \quad 0\rangle &= \left\langle 1\,0 \middle| \frac{1}{2}\frac{1}{2}\frac{1}{2} - \frac{1}{2} \right\rangle \left| \frac{1}{2} - \frac{1}{2} \right\rangle + \left\langle 10 \middle| \frac{1}{2}\frac{1}{2} - \frac{1}{2}\frac{1}{2} \right\rangle \left| - \frac{1}{2}\frac{1}{2} \right\rangle \\
&= \frac{1}{\sqrt{2}} \left(\left| \frac{1}{2} - \frac{1}{2} \right\rangle + \left| - \frac{1}{2}\frac{1}{2} \right\rangle \right), \\
|1-1\rangle &= \left\langle 1-1 \middle| \frac{1}{2}\frac{1}{2} - \frac{1}{2} - \frac{1}{2} \right\rangle \left| - \frac{1}{2} - \frac{1}{2} \right\rangle = \left| - \frac{1}{2} - \frac{1}{2} \right\rangle.
\end{aligned} \tag{2.29}$$

The triplet wavefunctions are symmetric under the permutation of the two neutrons. The spin-singlet wavefunction ($j \equiv s = 0$)

$$\begin{aligned}
|00\rangle &= \left\langle 00 \middle| \frac{1}{2}\frac{1}{2}\frac{1}{2} - \frac{1}{2} \right\rangle \left| \frac{1}{2} - \frac{1}{2} \right\rangle + \left\langle 00 \middle| \frac{1}{2}\frac{1}{2} - \frac{1}{2}\frac{1}{2} \right\rangle \left| - \frac{1}{2}\frac{1}{2} \right\rangle \\
&= \frac{1}{\sqrt{2}} \left(\left| \frac{1}{2} - \frac{1}{2} \right\rangle - \left| - \frac{1}{2}\frac{1}{2} \right\rangle \right),
\end{aligned} \tag{2.30}$$

is antisymmetric under the permutation of the two neutrons. For the orbital wavefunction $\phi_L(n, n)$ in (2.26) the permutation is equivalent to a parity transformation, that is a space inversion:

$$P\phi(n, n) = (-1)^L \phi(n, n). \tag{2.31}$$

Hence ϕ is symmetric for even L, and antisymmetric for odd L. With identical fermions the total wavefunction (2.26) must be antisymmetric. Therefore, if L is even (odd) then $s = 0$ ($s = 1$). The spin singlet state can be excluded by the conservation of total angular momentum conservation, since in the final state $J_f = J_i = 1$. Hence L is odd ($L = 1$). The parity of the final state is then $P_f = -1$. Finally, parity conservation requires that $P_i = P_f$. One concludes with (2.25) that the internal parity of the pion is negative and therefore we shall write for the pion that

$$\boxed{J_\pi^P = 0^-}\,. \tag{2.32}$$

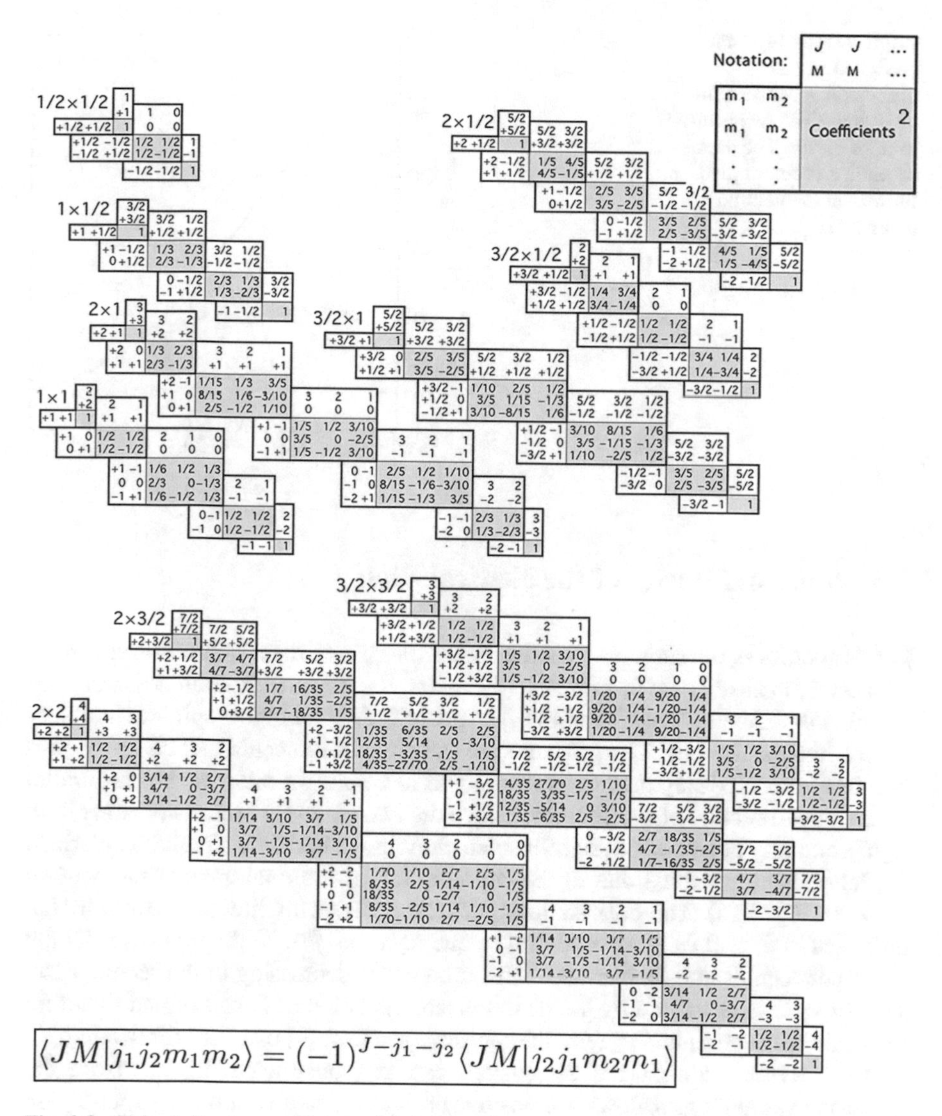

Fig. 2.8 Clebsch-Gordan coefficients for the couplings $j_1 \times j_2$ (note the boxed relation between coefficients). A square-root sign is understood over the coefficient, e.g. $-\frac{1}{2}$ stands for $-\frac{1}{\sqrt{2}}$ [1]

The reaction $\pi^- d \to nn$ was observed by stopping pions in liquid deuterium and recording the two neutrons in coincidence [17]. Figure 2.9 shows the measured rate as a function of the angle between the neutron directions. As expected, the two neutrons are preferably emitted back-to-back. The rate for this reaction is about twice that for radiative capture $\pi^- d \to nn\gamma$.

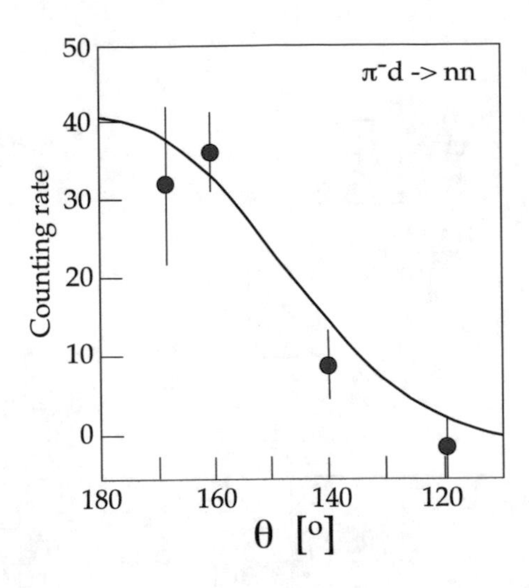

Fig. 2.9 Measured and predicted angular distributions between the direction of the two neutrons in $\pi^- d \to nn$. The angular resolution is determined by the size of the neutron counters [17]

2.4 Spin and Parity of the Neutral Pion

Let us assume as the simplest hypothesis that the spin of the π^0, the neutral partner of the $\pi^\pm$, is also $J_\pi = 0$. The main decay $\pi^0 \to 2\gamma$ already excludes spin 1 by virtue of the Landau-Yang theorem [3] (Appendix A), and higher spins are unlikely for the lightest mesons. The projection of the angular momenta carried by the two photons from π^0 decay on the z-axis is equal to zero if the latter is chosen parallel to the flight direction of the photons. Hence the two (spin-1) photons are either both right-circularly polarized (R, right-handed photons) or both left-circularly polarized (L, left-handed photons) due to the conservation of the total angular momentum $J_\pi = 0$ (Fig. 2.10). The polarizations (electric fields) with directions given by the unit vectors $\vec{e}_1$ and $\vec{e}_2$ rotate with the frequency $\omega = \frac{m_\pi}{2}$. In the first case (R) the angle between $\vec{e}_1$ and $\vec{e}_2$ increases with time while decreasing in the second case. The 2γ state is described by the quantum superposition of R and L ,and therefore the angle between $\vec{e}_1$ and $\vec{e}_2$ remains constant (see Ref. [9] p. 11–18 for details).

The wavefunction of the 2γ system can be constructed directly from the kinematic variables as follows: for positive parity the wavefunction of a spin-0 pion is a scalar, that is the sign does not change under space inversion, and is symmetric under permutations ($\vec{k}_1 \leftrightarrow \vec{k}_2$ and $\vec{e}_1 \leftrightarrow \vec{e}_2$):

$$\psi_+ \propto (\vec{k}_1 \cdot \vec{k}_2)(\vec{e}_1.\vec{e}_2) \propto \cos\phi, \tag{2.33}$$

where $\vec{k}_1 = -\vec{k}_2$ denote the photon momenta and ϕ is the angle between $\vec{e}_1$ and $\vec{e}_2$. For negative parity the wavefunction is pseudoscalar, that is the sign changes under space inversion, and is symmetric under permutations:

$$\psi_- \propto (\vec{k}_2 - \vec{k}_1)(\vec{e}_1 \times \vec{e}_2) \propto |\vec{e}_1 \times \vec{e}_2| \propto \sin\phi. \tag{2.34}$$

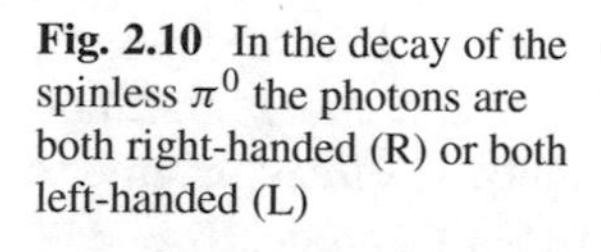
Fig. 2.10 In the decay of the spinless π^0 the photons are both right-handed (R) or both left-handed (L)

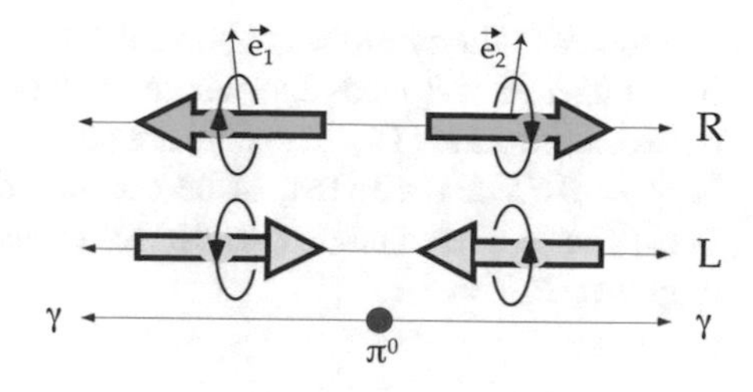

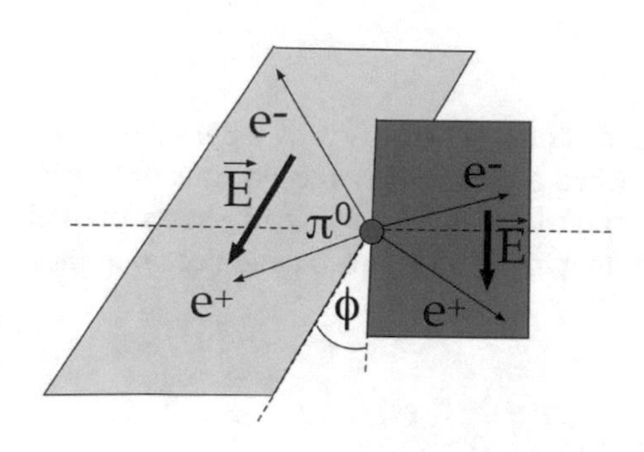

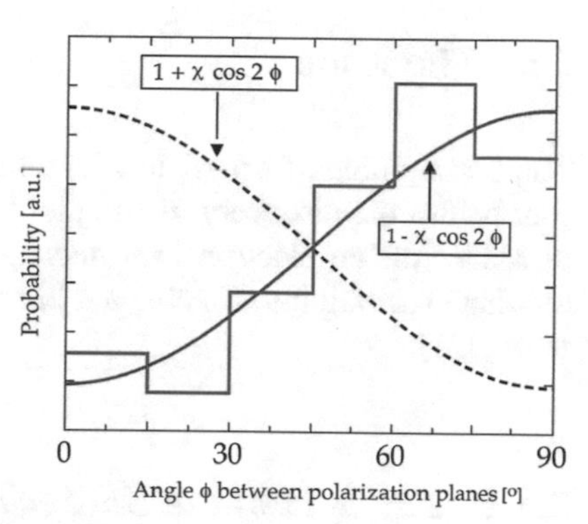

Fig. 2.11 Left: in the decay $\pi^0 \to (e^+e^-)(e^+e^-)$ the leptons are accelerated along the electric fields $\vec{E}$ of the virtual photons. One measures the angular distribution between the planes spanned by the e^+e^- pairs. Right: the histogram shows the measured ϕ distribution and the solid red curve the fit for $J^P_{\pi^0} = 0^-$ (64 decays) with $\chi \simeq 0.75$. The dashed curve would correspond to $J^P_{\pi^0} = 0^+$ [18]

The angular distribution between $\vec{e}_1$ and $\vec{e}_2$ is then given by

$$|\psi_+|^2 \propto \cos^2\phi = \frac{1}{2}(1 + \cos 2\phi) \text{ for positive parity,} \tag{2.35}$$

$$|\psi_-|^2 \propto \sin^2\phi = \frac{1}{2}(1 - \cos 2\phi) \text{ for negative parity.} \tag{2.36}$$

Hence for positive parity the electric fields of the photons prefer to be parallel ($\phi = 0$), while for negative parity they are preferably orthogonal ($\phi = 90°$).

This has been tested by using the decay $\pi^0 \to (e^+e^-)(e^+e^-)$ into two Dalitz pairs [18] which occurs with the small branching ratio (decay probability) of 3.3×10^{-5}. The neutral pions were produced in a hydrogen bubble chamber by the charge exchange reaction $\pi^- p \to \pi^0 n$. The Dalitz pairs can be thought as being generated by the internal conversion of the two photons with an expected branching ratio proportional to $(\sqrt{\alpha}\sqrt{\alpha})^2 \approx 5 \times 10^{-5}$. In the π^0 rest frame the Dalitz pairs are emitted in opposite directions under small opening angles (a few degrees). The leptons are accelerated by the electric fields of the photons and the pairs span two planes lying preferably parallel to the polarization of the photon. One then measures the angle ϕ between the e^+e^- planes, as shown in Fig. 2.11 (left).

However, the photons are virtual and the angular dependence becomes proportional to $1 \pm \chi P_{\pi^0} \cos 2\phi$, where χ depends on the angles and energy partition between the pairs [19]. The data are shown in Fig. 2.11 (right). The fitted parameter is $\chi = 0.75 \pm 0.42$ [18], to be compared with the expected theoretical value of 0.48. The preferred angle ϕ is 90° and thus the internal parity of the neutral pion is negative, $P_{\pi^0} = -1$.

2.5 The Kaon

The first sighting of a particle with a mass of about 500 MeV was reported in 1944 (i.e. before the discovery of the pion), when a charged cosmic ray was observed to scatter off an electron in a nuclear emulsion [20]. Figure 2.12 shows one of the first observations of a charged kaon stopping in an emulsion and decaying into $\mu^+\nu_\mu$ [21].

2.5.1 The Spin of the Charged Kaon

The K^+ decays into $\mu^+\nu_\mu$ with a branching ratio of 63.6%. In the rest frame of the kaon the muon is 100% polarized, see e.g. [22]. This follows as a consequence of the neutrino left-handedness and angular momentum conservation if the kaon is spinless, and is analogous to the decay $\pi^+ \to \mu^+\nu_\mu$. We shall nonetheless examine the energy distribution of the pions in the hadronic decay

$$K^\pm \to \pi^\pm\pi^+\pi^- \tag{2.37}$$

(branching ratio of 5.6%) and deduce that the spin of the kaon is indeed equal to zero. This also gives us the opportunity to rehearse some of the fundamental properties of Dalitz plots that will be used throughout these lectures.

Fig. 2.12 Decay $K^+ \to \mu^+\nu$ of a stopping kaon in a nuclear emulsion. A mass of 562 ± 70 MeV was estimated from multiple scattering and the energy loss derived from the grain density (adapted from [21])

Consider the three-body decay of a particle $a \rightarrow 1 + 2 + 3$. In the rest frame of a the directions of the three particles are coplanar and hence two parameters are required to describe the kinematics of the decay (3×2 momentum components, minus 3 constraints for energy and momentum conservation, minus 1 arbitrary angle for the orientation of the coordinate system). These two parameters are usually chosen as the invariant masses m_{12} and m_{13} (see Appendix B and Problem 2.1) or as the kinetic energies T_1 and T_2. The latter satisfy the relation

$$T_1 + T_2 + T_3 = M_a - m_1 - m_2 - m_3 \equiv Q, \tag{2.38}$$

the Q-value of the decay, with obvious notations. From (2.38) is it natural to introduce an equilateral triangle of unit height to represent the decay events as dots at the distances T_1/Q, T_2/Q and T_3/Q from the sides of the triangle, hence $(T_1 + T_2 + T_3)/Q = 1$. Such a Dalitz plot is shown in Fig. 2.13 for the special case of equal final state masses m. In the absence of interactions between daughter particles (e.g. without intermediate two-body resonances) the Dalitz plot is uniformly populated (for a proof see e.g. [9] p. 12–7).

The coordinate system in Fig. 2.13 is chosen so that

$$\boxed{x \equiv \frac{1}{\sqrt{3}Q}(T_1 - T_2) \text{ and } y \equiv \frac{T_3}{Q} - \frac{1}{3}}. \tag{2.39}$$

At point A particle 3 is at rest while 1 and 2 are emitted in opposite directions, hence $y_A = -\frac{1}{3}$. The triangle is centered at the origin of the coordinate system. At point C, $T_2 = 0$, $y_C = \frac{1}{6}$ and

$$x_C = \frac{1}{\sqrt{3}Q} T_1 = \frac{1}{\sqrt{3}Q} T_3 = \frac{1}{\sqrt{3}}\left(y_C + \frac{1}{3}\right) = \frac{1}{2\sqrt{3}}, \tag{2.40}$$

hence a triangle side length of $\frac{2}{\sqrt{3}}$.

On the other hand, particle 3 reaches its maximum possible energy at point B, recoiling against 1 and 2. From momentum conservation follows in the highly relativistic limit $m \rightarrow 0$ that

$$T_1 = T_2 = \frac{T_3}{2} \Rightarrow T_1 + T_2 + T_3 = 2T_3 = Q \tag{2.41}$$

and hence $y_B = \frac{1}{6}$. The Dalitz plot becomes a triangle (Fig. 2.13, HR), as for instance in orthopositronium decay into 3γ (the Dalitz plot is predicted to be nearly uniformly populated [23]). Another example of 3γ decay will be discussed in Sect. 3.1 (Fig. 3.2).

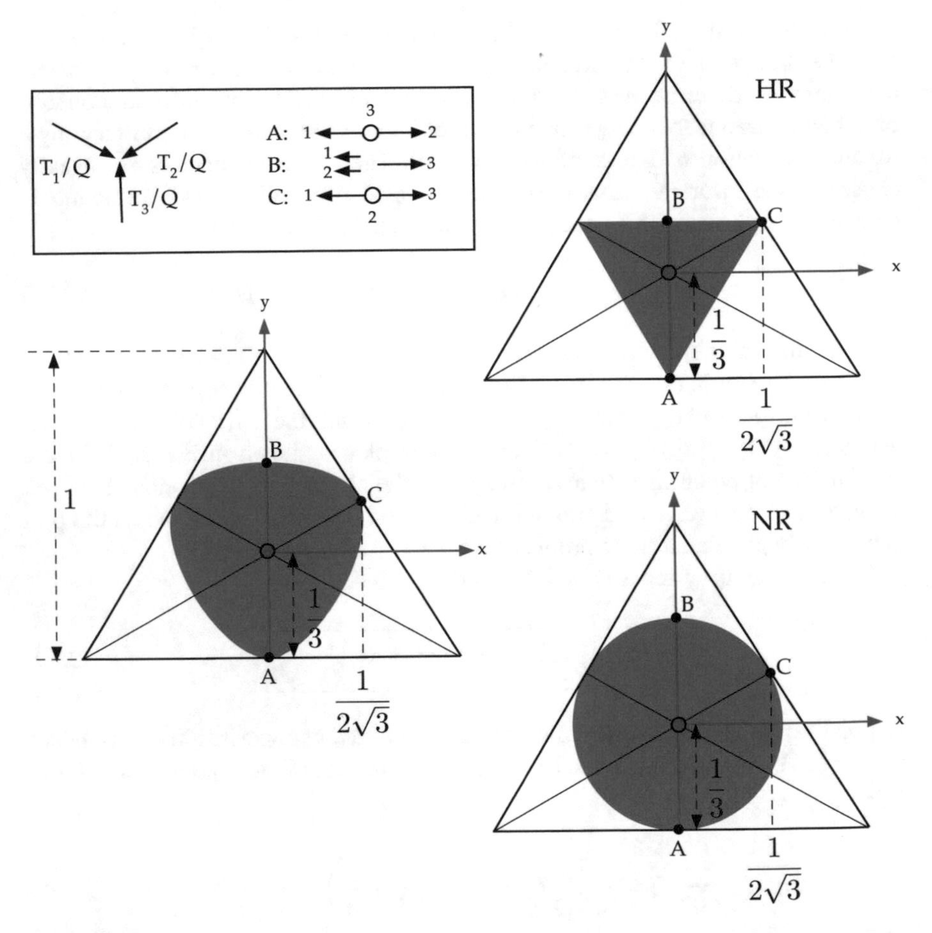

Fig. 2.13 Dalitz plot of the three-body decay for identical masses $m_1 = m_2 = m_3 = m$ (left). The physically allowed area is shown in colour. The kinetic energies divided by Q are plotted in the directions perpendicular to the triangle sides. The kinematic configurations at points A, B and C are explained in the box. The highly relativistic (HR) and non-relativistic limits (NR) are shown on the right

When the Q value is smaller than the daughter masses, the non-relativistic approximation for the kinetic energies becomes a good approximation. At point B

$$T_1 + T_2 + T_3 = \frac{k_1^2}{2m} + \frac{k_2^2}{2m} + T_3 = Q, \tag{2.42}$$

where the momenta are $k_1 = k_2 = \frac{k_3}{2}$, and therefore

$$\frac{k_3^2}{4m} + T_3 = \frac{3}{2}T_3 = Q. \tag{2.43}$$

Fig. 2.14 Dalitz plot of the decay $K^{\pm} \to \pi^{\pm}\pi^{+}\pi^{-}$. The dotted curve shows the boundary of the physically allowed area, which becomes a circle in the non-relativistic limit [24]

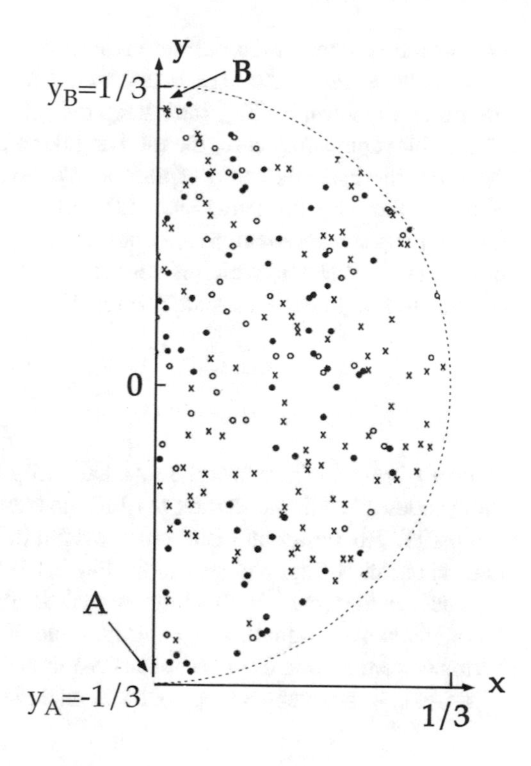

From (2.39) the point B is located at the ordinate $y_B = +\frac{1}{3}$ and hence the Dalitz plot reduces to a circle (Fig. 2.13, NR). This is nearly the case for the kaon decay (2.37) with $Q = 75\,\text{MeV} < m_\pi$. Figure 2.14 shows the Dalitz plot for 219 decays from three different experiments [24]. Let us assign T_1 and T_2 to the two like-charge pions and plot negative values of x on the positive side. We denote by ℓ the angular momentum between the two like-charge pions and by L that between the $\pi^{\mp}$ and the $\pi^{\pm}\pi^{\pm}$ dipion. At point B $\ell = 0$ since the two like-charge pions have equal and parallel momenta (see inset in Fig. 2.13). At point A the $\pi^{\mp}$ is at rest with respect to the $\pi^{\pm}\pi^{\pm}$ dipion and hence $L = 0$. No events should be observed in the vicinity of A if $L > 0$, nor near B if $\ell > 0$, in contrast to data. Therefore $\ell = L = 0$, which leads to the conclusion that the spin of the charged kaon is $J_K = 0$, the pions being spinless.

2.6 Internal Parity of the Kaon

The internal parity of the charged kaon can be determined by stopping negative pions in helium and analyzing the following reactions

$$K^{-}\,{}^{4}\text{He} \to {}^{4}\text{H}_{\Lambda}\,\pi^{0},\quad {}^{4}\text{H}_{\Lambda} \to {}^{4}\text{He}\,\pi^{-}. \tag{2.44}$$

One of the protons in the helium nucleus is replaced by a Λ hyperon to form a Λ hypernucleus. From the $^4\mathrm{H}_\Lambda$ decay branching ratio into $^4\mathrm{He}\ \pi^-$ one infers that the decay is isotropic [25], indicating that no angular momentum is carried away. The helium spin being zero, the total angular momentum in the decay also vanishes, therefore the hypernucleus is spinless. We have demonstrated that the spins of the pion and the kaon are zero. Since all particles in the first reaction have spin zero, the total angular momentum is equal to the orbital angular momentum L between initial and final states, which is conserved. The parity of the initial state ($K^-\,{}^4\mathrm{He}$) is equal to that of the final state ($^4\mathrm{H}_\Lambda\pi^0$):

$$P = (-1)^L P_K \underbrace{P_{^4\mathrm{He}}}_{+1} = (-1)^L \underbrace{P_{^4\mathrm{H}_\Lambda}}_{+1} \underbrace{P_{\pi^0}}_{-1}, \tag{2.45}$$

since $P_\Lambda = +1$. Therefore the internal parity of the kaon is negative, $P_K = -1$, independently of L. According to (2.8) the kaon is also a $1^1S_0\ q\overline{q}$ state.

In 1947 Rochester and Butler discovered in the cosmic radiation the electrically neutral partner of the charged kaon (Fig. 2.15). The K^0 (to be precise the K_S state) also decays into $\pi^0\pi^0$ with a branching ratio about half that for $\pi^+\pi^-$. Figure 2.15 (right) shows the signal from a large sample of $K_S \to \pi^0\pi^0$ decays. Bose-Einstein symmetry for identical spin 0 bosons requires the relative angular momentum of the $\pi^0\pi^0$ pair to be even. Hence spin 1 is excluded and from the nearly equal masses

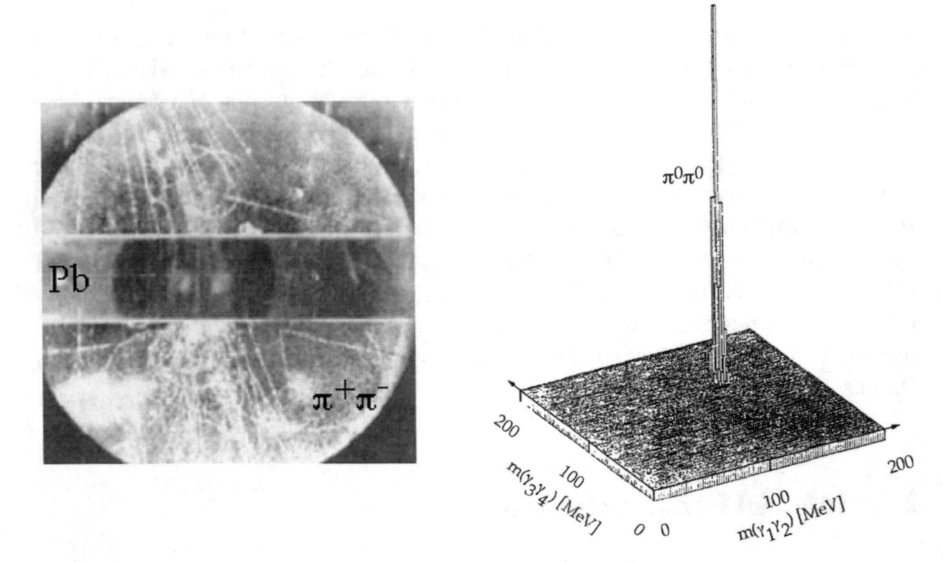

Fig. 2.15 Left: discovery of the K^0 decaying into $\pi^+\pi^-$ in a cloud chamber. The K^0 was produced by cosmic rays in the lead plate [26]; right: mass distribution of the 2γ pairs measured by a liquid argon calorimeter in a 100 GeV K_S beam [27]

Table 2.1 The ($\overline{K}^0$, K^-) and (K^+, K^0) isospin doublets

	$\overline{K}^0$	K^-	K^+	K^0
i_3	$\frac{1}{2}$	$-\frac{1}{2}$	$\frac{1}{2}$	$-\frac{1}{2}$
Quark content	$s\overline{d}$	$s\overline{u}$	$u\overline{s}$	$d\overline{s}$
Strangeness	-1		1	
Mass [MeV]	497.6	493.7	493.7	497.6

of the neutral and charged kaons it is natural to also assume spin 0 for the neutral kaon, hence

$$\boxed{J_K^P = 0^-}. \tag{2.46}$$

The K^0 is produced at accelerators, for example by the reaction $\pi^- p \to \Lambda K^0$, where the Λ hyperon (uds) contains an s quark. Flavour conservation is strong interactions then implies that the K^0 contains an $\overline{s}$ quark and its charge conjugated partner, the $\overline{K}^0$, an s quark. Table 2.1 summarizes the two kaon isospin doublets and their quark contents. The K^- is sometimes called "antikaon" because it contains a $\overline{u}$ antiquark and is the isospin partner of the $\overline{K}^0$.

References

1. Tanabashi, M., et al. (Particle Data Group): Phys. Rev. D 98, 030001 (2018)
2. Wu, C.S., Shaknov, I.: Phys. Rev. 77, 136 (1950)
3. Landau, L.D.: Dokl. Akad. Nauk. 60, 207 (1948); Yang, C.N.: Phys. Rev. 77, 242 (1950)
4. Neddermeyer, S.H., Anderson, C.D.: Phys. Rev. 51, 884 (1937)
5. Conversi, M., Pancini, E., Piccioni, O.: Phys. Rev. 71, 209 (1947)
6. Perkins, D.H.: Nature 159, 126 (1947)
7. Lattes, C.M.G., Occhialini, G.P.S., Powell, C.F.: Nature 160, 453 (1947)
8. Powell, C.F.: Rep. Prog. Phys. 13, 350 (1950)
9. Amsler, C.: Nuclear and Particle Physics. IOP Publishing, Bristol (2015)
10. Peterson, V.Z.: Phys. Rev. 79, 407 (1950)
11. Durbin, R., Loar, H., Steinberger, J.: Phys. Rev. 83, 646 (1951)
12. Lock, W.O., Measday, D.F.: Intermediate Energy Nuclear Physics. Methuen, London (1970)
13. Ahmad, S., et al.: Phys. Lett. 157B, 333 (1985)
14. Amsler, C.: Rev. Mod. Phys. 70, 1293 (1998)
15. May, B., et al.: Z. Phys. C 46, 203 (1990)
16. Nafe, J.E., Nelson, E.B., Rabi, I.I.: Phys. Rev. 71, 914 (1947)
17. Chinowsky, W., Steinberger, J.: Phys. Rev. 95, 1561 (1954)
18. Plano, R., et al.: Phys. Rev. Lett. 3, 525 (1959)
19. Kroll, N.M., Wada, W.: Phys. Rev. 98, 1355 (1955)
20. Leprince-Ringuet, L., L'Héritier, M.: C. R. Acad. Sci. Paris (13 décembre) 618 (1944)
21. O'Ceallaigh, C.: Philos. Mag. XLII, 1032 (1951)
22. Aoki, M., et al., Phys. Rev. D 50, 69 (1994)
23. Ore, A., Powell, J.L.: Phys. Rev. 75, 1696 (1949)
24. Orear, J., Harris, G., Taylor, S.: Phys. Rev. 102, 1676 (1956)
25. Dalitz, R.H., Downs, B.W.: Phys. Rev. 111, 967 (1958)
26. Rochester, G.D., Butler, C.C.: Nature 160, 855 (1947)
27. Burkhardt, H., et al.: Nucl. Instrum. Methods Phys. Res. A 268, 116 (1988)

Chapter 3
Isospin

3.1 Charge Independence

Hadrons with similar masses and same spin and parity are grouped in multiplets of the electric charge. For example, the two charged pions π^+ and π^- (139.6 MeV) and the π^0 (135.0 MeV) form a triplet with similar masses. As Table 2.1 shows, the four kaons with masses $m_{K^\pm} = 493.7$ MeV and $m_{K^0} = m_{\overline{K}^0} = 497.6$ MeV form two doublets, (K^+, K^0) and $(\overline{K}^0, K^-)$.

The nucleon also exists in two varieties, the proton (938.3 MeV) and the neutron (939.6 MeV) with almost equal masses. They form a doublet of the isospin, a concept that was introduced in 1932 by Werner Heisenberg, inspired by an analogy with spin: in the 3-dimensional vector space the projection m of the spin $\vec{s}$ onto the z-axis assumes $2s + 1$ possible values between $-s$ and $+s$, where s is the spin quantum number. In the absence of external fields the corresponding $2s + 1$ spin wavefunctions (spinors) are degenerate in energy.

Similarly, one introduces a vector $\vec{I}$ in the (abstract) isospin space with corresponding isospin quantum number i. The projection i_3 of $\vec{I}$ onto the z-axis in isospin space takes $2i + 1$ values between $-i$ and $+i$. The associated $2i + 1$ isospin wavefunctions (isospinors) describe the $2i + 1$ hadrons with different charges but degenerate masses.

In the absence of electric charge, isospin would be a perfect symmetry. However the hadrons in the multiplets, such as the proton and the neutron, have slightly different masses. We have seen that most of the nucleon mass (99%) stems from the gluonic interaction between quarks. The slightly different masses between proton and neutron is due to the difference in quark masses, the d quark being heavier than the u, because the Higgs boson couples more strongly to the former.[1]

[1]The electromagnetic repulsion acts in the opposite direction because it reduces the binding energy and hence increases the mass of the proton.

© The Author(s) 2018

C. Amsler, *The Quark Structure of Hadrons*, Lecture Notes in Physics 949,
https://doi.org/10.1007/978-3-319-98527-5_3

With the help of the Gell-Mann-Nishijima relation (1.2), which relates Q to i_3, one assigns the i_3 values to the corresponding hadrons. For example, for the pion

$$\pi(i=1): \quad \pi^+ : i_3 = +1,\ \pi^0 : i_3 = 0,\ \pi^- : i_3 = -1, \tag{3.1}$$

(see also Table 2.1 for the kaon), while for baryons e.g.

$$\begin{aligned} &N(i=\tfrac{1}{2}) : p : i_3 = +\frac{1}{2},\ n : i_3 = -\frac{1}{2}, \\ &\Delta(i=\tfrac{3}{2}) : \Delta^{++} : i_3 = +\frac{3}{2},\ \Delta^+ : i_3 = \frac{1}{2},\ \Delta^0 : i_3 = -\frac{1}{2},\ \Delta^- : i_3 = -\frac{3}{2}, \\ &\Lambda(i=0) : i_3 = 0. \end{aligned} \tag{3.2}$$

In isospin space the multiplet members are represented by $2i+1$-dimensional isospinors, e.g.

$$\pi^+ = \begin{pmatrix} 1 \\ 0 \\ 0 \end{pmatrix},\ \pi^0 = \begin{pmatrix} 0 \\ 1 \\ 0 \end{pmatrix},\ p = \begin{pmatrix} 1 \\ 0 \end{pmatrix},\ n = \begin{pmatrix} 0 \\ 1 \end{pmatrix},\ \Delta^- = \begin{pmatrix} 0 \\ 0 \\ 0 \\ 1 \end{pmatrix}. \tag{3.3}$$

Since the isospin $\vec{I}$ of the hadron is equal to the vector sum of the constituent isospins and i_3 is additive (in analogy to spin), one assigns to the quarks the isospins given in Table 1.1. The u and d quarks in the first generation form a doublet of isospin, while the other quarks are singlets. For i_3 the $\overline{u}$ and $\overline{d}$ quarks have the opposite signs:

$$i_3(u) = \frac{1}{2},\ i_3(d) = -\frac{1}{2} \quad \Rightarrow \quad i_3(\overline{u}) = -\frac{1}{2},\ i_3(\overline{d}) = +\frac{1}{2}. \tag{3.4}$$

With a quark and an antiquark from the first generation either an isoscalar ($i=0$) can be formed, such as the neutral ω, or an isovector ($i=1$) such as the (ρ^+, ρ^0, ρ^-). With three quarks of the first generation one gets an isospin doublet ($i=\frac{1}{2}$) such as the nucleon, or a quadruplet ($i=\frac{3}{2}$) such as the Δ.

The strong interaction is invariant under rotations in isospin space, hence insensitive to rotations of the isospinors (3.3) by an arbitrary angle α around the direction $\vec{\alpha}$ in isospin space. There are two types of rotations which differ by the sign of α, passive and active. The former is a rotation of the coordinate system, the latter a rotation of the physical system (Chap. 18). The active rotation is achieved by applying the matrix representation of the unitary operator $U = \mathrm{e}^{-i\vec{\alpha}\cdot\vec{I}}$. Rotational symmetry in isospin space means that $\vec{I}$ commutes with the interaction Hamiltonian H:

$$[U, H] = 0 \Rightarrow [\vec{I}, H] = 0 \Rightarrow [I_1, H] = [I_2, H] = [I_3, H] = 0. \tag{3.5}$$

Since the components I_k do not commute among themselves, only one of them is conserved during the interaction, usually chosen as I_3. Charge independence (sometimes also called "charge symmetry") states that the strong interaction does not depend on i_3 and hence does not distinguish between members of the same isospin multiplet. In particular, all masses should be equal within multiplets. The forces between two protons ($i_3 = +1$, hence $i = 1$) and two neutrons ($i_3 = -1$) are equal (obviously before adding electromagnetic repulsion), $F_{pp} = F_{nn}$, but they are not equal to the force between a proton and a neutron, since the latter is a superposition of $i = 0$ and $i = 1$ states. However, $F_{np}(i = 1) = F_{pp}$.

The isospin quantum number i is also conserved in strong interactions. For example, $\overline{p}p$ and $\overline{n}n$ states ($i_3 = 0$) are linear combinations of $i = 0$ and $i = 1$ states. However, there is no transition between $i = 0$ and $i = 1$ in the charge exchange reaction $\overline{p}p \rightarrow \overline{n}n$ (Problems 3.1 and 3.2).

Let us now switch on the electric charge. The interaction Hamiltonian becomes sensitive to any rotation that changes the value of i_3—related to the charge Q by the Gell-Mann-Nishijima relation (1.2)—that is rotations around the x- or the y-axis (α_x or $\alpha_y \neq 0$). On the other hand, H is not affected by rotations around the z-axis which leave i_3 constant, therefore

$$[I_1, H] \neq 0, [I_2, H] \neq 0, \text{ but } [I_3, H] = 0, \tag{3.6}$$

and i_3 is still conserved by the interaction.[2]

However, i is not defined in electromagnetic interactions, as the following example shows. Consider the two radiative decay channels of the ω meson

$$\omega \rightarrow \pi^0\gamma \text{ and } \omega \rightarrow \eta\gamma. \tag{3.7}$$

Let us assign to the γ the quantum number $i_3 = 0$ which is then conserved in both decays, since the ω, π^0 and η mesons all have $i_3 = 0$. However, conservation of $i = 0$ (ω) would imply that the γ has $i = 1$ when associated with a π^0 and $i = 0$ when emitted with an η. The contradiction shows that i is not, in contrast to i_3, a good quantum number in electromagnetic interactions.

These radiative decay modes have been studied by the Crystal Barrel experiment at CERN's Low Energy Antiproton ring LEAR. (Since other results from the Crystal Barrel will be discussed during these lectures, a brief description of the apparatus will be given in the next subsection.) The ω meson was observed in $\overline{p}p$ annihilation

[2]The inequalities (3.6) lift the mass degeneracy in the isospin multiplet. This is analogous to the two-level splitting of a free electron when an external magnetic field is switched on: the spin precesses around the z-axis with constant projection m.

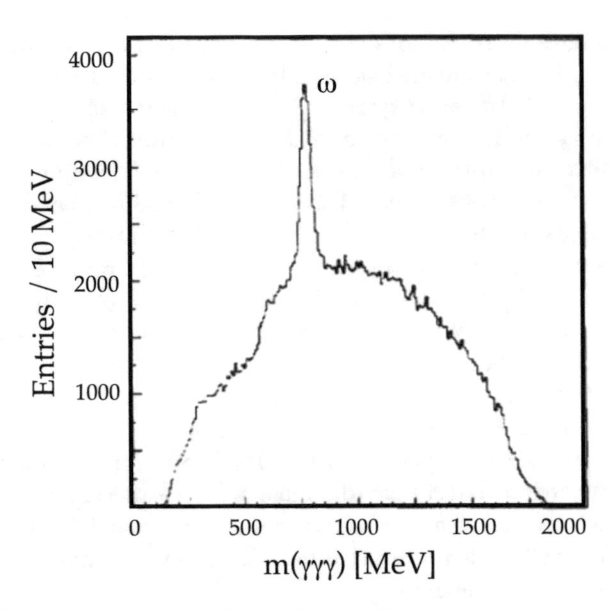

Fig. 3.1 Invariant 3γ mass in $\overline{p}p$ annihilation into 5γ [1]

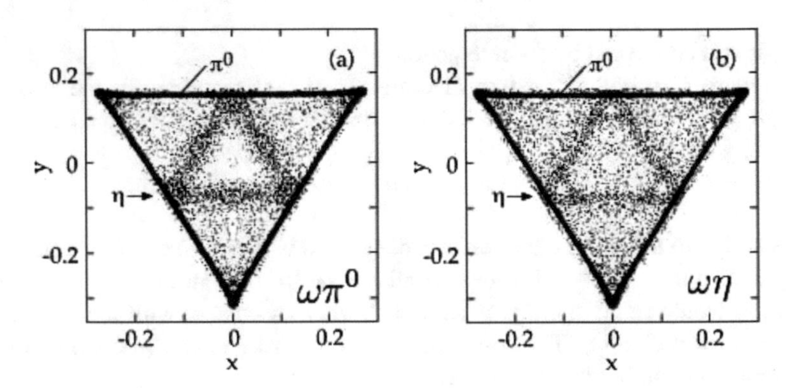

Fig. 3.2 (**a**) Dalitz plot of $\omega \to 3\gamma$ in $\overline{p}p \to \omega\pi^0$ (62,853 decays); (**b**) in $\overline{p}p \to \omega\eta$ (54,865 decays). The plots have been symmetrized: there are six possible combinations and hence six entries/event [1]

at rest into $\omega\pi^0$ and $\omega\eta$ with $\omega \to \pi^0\gamma$ or $\omega \to \eta\gamma$, leading to 5γ in the final state.[3] Figure 3.1 shows the ω signal in the 3γ invariant mass distribution and Fig. 3.2 the 3γ Dalitz plots, which are triangular in this highly relativistic case (see Fig. 2.13).

[3]The analysis of these radiative decays, in particular $\omega(\to \eta\gamma)\pi^0$, is complicated by $\rho - \omega$ mixing: the electromagnetic isospin violating decay $\omega \to 2\pi$ (Sect. 3.4), with a branching ratio f of 1.5%, interferes with the 2π decay of the ρ^0 ($f = 100\%$) through the transition $\omega \to 2\pi \to \rho^0$. This interference must be taken into account when the ρ^0 production $\times$ decay rate is much larger than the corresponding ω one (see [1] for details).

Signals from the intermediate states $\pi^0\gamma$ and $\eta\gamma$ are clearly visible in both plots. The branching ratios for $\omega \rightarrow \pi^0\gamma$ and $\omega \rightarrow \eta\gamma$ are $(8.28 \pm 0.28) \times 10^{-2}$ and $(4.6 \pm 0.4) \times 10^{-4}$, respectively [2].

The events between the bands in Fig. 3.2 are mainly due to 6γ events with a missing (undetected) γ. The single event in the center of Fig. 3.2a was used to set an upper limit to the direct decay $\omega \rightarrow 3\gamma$. By assuming phase space distribution Ref. [1] arrives at an upper limit of 1.9×10^{-4} for the non-resonant $\omega \rightarrow 3\gamma$ decay. We shall return to radiative meson decays in the context of SU(3) symmetry (Sect. 7.3).

3.2 The Crystal Barrel Experiment at LEAR

The Low Energy Antiproton Ring (LEAR) was operated at CERN between 1983 and 1996 (Fig. 3.3). The 3.5 GeV/c antiprotons from a target struck by the PS beam were stored in the antiproton accumulator (and later in the antiproton collector) before being accelerated by the PS and injected into the high energy SPS $\overline{p}p$ collider. They could also be decelerated in the PS and stored in the LEAR ring, where they were decelerated further down to 60 MeV/c (2 MeV kinetic energy) or accelerated to 1940 MeV/c (1.22 GeV), and then slowly extracted and distributed to the experiments in the South Hall. The intense and pure low energy $\overline{p}$ beam of small momentum spread ($\frac{\Delta p}{p} \leq 10^{-3}$) was achieved thanks to the invention of stochastic

Fig. 3.3 The Low Energy Antiproton Ring (LEAR) in the South Hall (image credit CERN)

cooling. The continuous antiproton flux was $10^6\ \overline{p}\ s^{-1}$ and the spill time about 1 h, after which the ring had to be replenished. Compare this flux with the one available at the time of the antiproton discovery in 1955: 1 antiproton every 15 min!

The Crystal Barrel detector took data between 1989 and until LEAR was decommissioned in 1996. The main goal of the experiment was to study $\overline{p}p$ annihilation at rest with very high statistics, in particular annihilation into final states with neutral mesons (π^0, η, η', ω, etc.) leading to multiphoton final states. These channels occur with a probability of $\simeq$50% and had not been investigated previously. As we have seen in Sect. 2.3, annihilation at rest follows the capture of the antiproton in the orbitals of the hydrogen atom. Annihilation at rest into neutral mesons strongly reduces the number of contributing initial states, mainly due to the conservation of C parity (next section) and is therefore simpler to analyze. For a review of the physics results and a comprehensive list of original publications, see [3].

Figure 3.4 shows a sketch and a photograph of the Crystal Barrel [4]. The antiprotons from LEAR fly along the axis of a 1.5 T solenoidal magnet and stop in a liquid (or gaseous) hydrogen target. The final state charge multiplicity is determined online by two cylindrical proportional wire chambers. The momenta of the charged annihilation products (mainly pions and some kaons) are measured by a jet drift chamber, which is also capable to distinguish low energy (<500 MeV/c) kaons from pions by dE/dx ionization sampling.

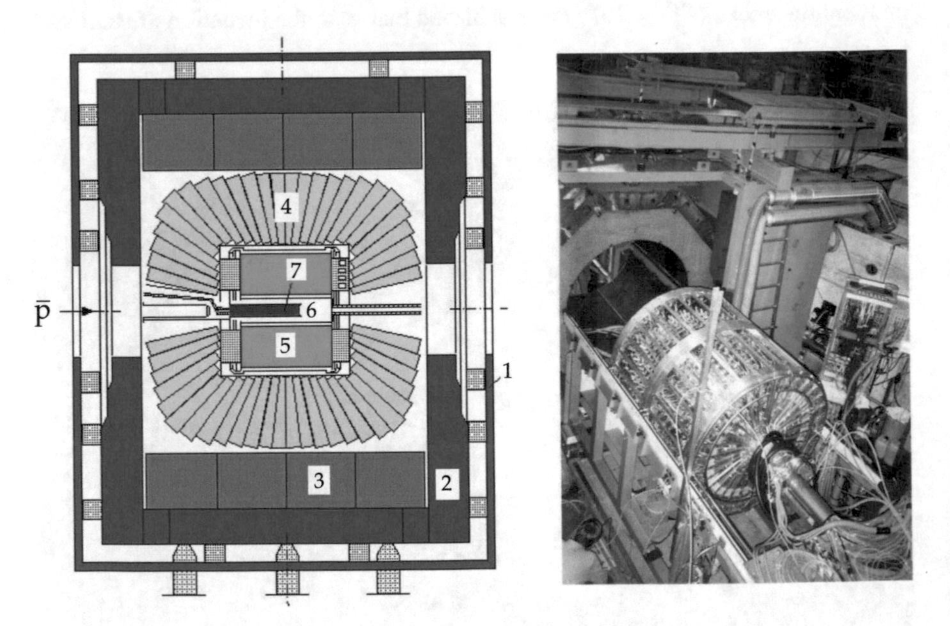

Fig. 3.4 Left: sketch of the Crystal Barrel detector [4]. 1, 2—yoke, 3—coil, 4—CsI(Tl) barrel, 5—jet drift chamber, 6—proportional wire chambers, 7—hydrogen target. Right: photograph of the CsI barrel before insertion into the magnet (image credit CERN)

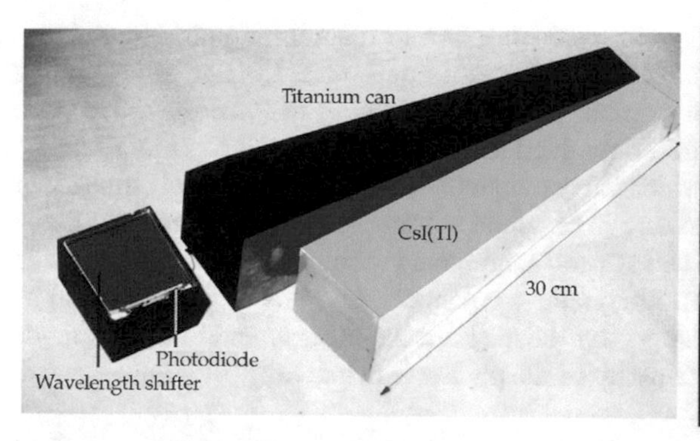

Fig. 3.5 Left: CsI(Tl) crystal wrapped in aluminium foil with titanium container and wavelength shifter. Right: photograph of the jet drift chamber showing the readout connectors for the 30 sectors and the signal cables from the multiwire proportional chambers (image credits CERN)

Photons are detected by a barrel-shaped assembly of 1380 CsI(Tl) crystals, 30 cm long (16 radiation lengths), read out by photodiodes (Fig. 3.5, left). The crystals are oriented towards the annihilation point. They are wrapped in teflon and aluminized mylar, and are enclosed in thin titanium containers. The scintillation light is converted to higher wavelengths by a wavelength shifter. The re-emitted light is detected by a photodiode glued on the edge of the wavelength shifter. Up to 10 γ's with energies as low as 4 MeV are routinely reconstructed with good efficiency, thanks to the large solid angle coverage (97% $\times 4\pi$). The angular resolution is typically $\pm 1°$, the mass resolution ± 10 MeV for γ's from $\pi^0 \rightarrow 2\gamma$ and $\pm$ 20 MeV for γ's from $\eta \rightarrow 2\gamma$ decay.

The jet drift chamber (Fig. 3.5, right) is made of 30 sectors, each with 23 sense wires, read out on both ends by flash ADCs to determine the coordinate along the z-axis through charge division. The chamber is filled with a CO_2/isobutane mixture. The position resolution in the plane transverse to the beam axis is ± 150 µm, the one along the wires ± 1 cm. The momentum resolution on pions is typically 2% at 200 MeV/c, rising to $\simeq$ 7% at 1 GeV/c.

3.3 Charge Conjugation

Charge conjugation transforms a particle into its corresponding antiparticle, whereby i_3, the electric charge Q, the baryon number B and the flavour quantum numbers (S, C, B' and T) reverse sign. For example, a π^+ becomes a π^-, a neutron (n) an antineutron ($\bar{n}$). The operation also flips the sign of the magnetic moment. Space coordinates and kinematic quantities such as momentum or angular momentum are not affected.

Electrically neutral bosons with vanishing flavour quantum numbers are eigenstates of the charge conjugation with C parity equal to ± 1, since applying the operation twice retrieves the original particle. According to (1.2) they also have $i_3 = 0$. Examples of such self-conjugated bosons are the π^0, the ω, the J/ψ ($c\bar{c}$) and the γ. Note that gluons, which carry colour and anticolour, are not eigenstates of charge conjugation: e.g. a red-antigreen gluon is transformed into an antired-green gluon. To determine the sign (± 1) of self-conjugated bosons, let us first consider the photon. The sign of the electromagnetic 4-potential ($A^\mu \to -A^\mu$) changes signs when flipping the electric charge, and hence both the electric and magnetic fields change signs. Therefore the C parity of the photon is negative

$$C|\gamma\rangle = -|\gamma\rangle. \tag{3.8}$$

The parity of the photon is also negative (Sect. 2.1) and therefore

$$J^{PC}(\gamma) = 1^{--}. \tag{3.9}$$

The C parity is a multiplicative quantum number, hence a set of n photons is an eigenstate of C parity with

$$C|n\gamma\rangle = (-1)^n|n\gamma\rangle. \tag{3.10}$$

Since C parity is conserved in electromagnetic transitions, the C parity of the π^0 meson, which decays into $\gamma\gamma$, is $C(\pi^0) = +1$, hence

$$J^{PC}(\pi^0) = 0^{-+}. \tag{3.11}$$

Likewise for the $\eta \to \gamma\gamma$, $C(\eta) = +1$. From the electromagnetic decay $\omega \to \pi^0\gamma$ or $\omega \to \eta\gamma$ (Fig. 3.2) one concludes that $C(\omega) = -1$, which is also true for $\rho^0 \to \pi^0\gamma$, $C(\rho^0) = -1$.

Conservation of C parity in electromagnetic decays is well established. For example, the C violating decay of the π^0 into 3γ ($C = -1$) has not been observed. The decay was sought at LAMPF (Los Alamos) by producing neutral pions through the charge exchange reaction $\pi^- p \to \pi^0 n$ with stopping pions, and by looking for 3γ events with a large solid angle box of NaI(Tl) crystals [5]. An upper limit of

$$\frac{f(\pi^0 \to 3\gamma)}{f(\pi^0 \to 2\gamma)} < 3.1 \times 10^{-8} \tag{3.12}$$

was obtained. A test of C conservation can also be performed by looking for J/ψ decay into $\phi\gamma$. Both J/ψ and ϕ are vector mesons with negative C parity. A search was conducted with the BESIII detector at Beijing (described in Sect. 9.1) by exciting the $\psi(2s)$ in e^+e^- collisions, and by looking at its decay into $J/\psi\,\pi^+\pi^-$

with $J/\psi \to \phi\gamma$ ($\phi \to K^+K^-$) [6]. No events were observed, setting an upper limit of 1.4×10^{-6} for the decay[4] $J/\psi \to \phi\gamma$.

Boson-antiboson ($B\bar{B}$) and fermion-antifermion ($F\bar{F}$) pairs are also eigenstates of C. Consider a $B\bar{B}$ or $F\bar{F}$ pair with relative angular momentum ℓ and total spin s and denote the corresponding wavefunctions by ϕ_ℓ and χ_s. From the properties of Clebsch-Gordan coefficients (boxed relation in Fig. 2.8), one derives the symmetry property

$$\langle s_B s_{\bar{B}} | s m_s \rangle = (-1)^{s-2s_B} \langle s_{\bar{B}} s_B | s m_s \rangle = (-1)^s \langle s_{\bar{B}} s_B | s m_s \rangle \tag{3.13}$$

for bosons (with $s_B = s_{\bar{B}}$ integer) and

$$\langle s_F s_{\bar{F}} | s m_s \rangle = (-1)^{s-2s_F} \langle s_{\bar{F}} s_F | s m_s \rangle = (-1)^{s-1} \langle s_{\bar{F}} s_F | s m_s \rangle \tag{3.14}$$

for fermions (with $s_F = s_{\bar{F}}$ half-integer). Charge conjugation permutes the two bosons and Bose-Einstein symmetry requires that $|\overline{B}B\rangle = |B\overline{B}\rangle$, therefore for bosons

$$\begin{aligned} C|B\overline{B}\rangle &= \pm \underbrace{|\overline{B}B\rangle}_{=|B\overline{B}\rangle} = \pm\phi_\ell(\overline{B}, B)\chi_s(\overline{B}, B) = \pm(-1)^\ell \phi_\ell(B, \overline{B})(-1)^s \chi_s(B, \overline{B}) \\ &= \pm(-1)^{\ell+s}|B\overline{B}\rangle, \end{aligned} \tag{3.15}$$

hence $\ell + s$ must be even. A pair of self-conjugated bosons, such as $\pi^0\pi^0$ or $\gamma\gamma$, is an eigenstate of C with eigenvalue $+1$. Then the plus sign must be chosen in (3.15).

Similarly for fermion-antifermion pairs, recalling that fermion and antifermion have opposite internal parities and taking (3.14) into account,

$$\begin{aligned} C|F\overline{F}\rangle &= \pm|\overline{F}F\rangle = \pm\phi_\ell(\overline{F}, F)\chi_s(\overline{F}, F) = \pm(-1)^{\ell+1}\phi_\ell(F, \overline{F})(-1)^{s-1}\chi_s(F, \overline{F}) \\ &= \pm(-1)^{\ell+s}|F\overline{F}\rangle. \end{aligned} \tag{3.16}$$

The ground state of parapositronium ${}^1S_0(e^+e^-)$ decays into two photons, hence $C = +1$. Since $\ell = 0$ and $s = 0$, again the plus sign in (3.16) must be chosen. This is also true for orthopositronium ${}^3S_1(e^+e^-)$ which decays into three photons ($C = -1$ therefore $\ell = 0$ and $s = 1$). Thus particle-antiparticle pairs $\Phi\overline{\Phi}$ ($F\overline{F}$ or $B\overline{B}$) are eigenstates of charge conjugation with eigenvalue

$$\boxed{C(\Phi\overline{\Phi}) = (-1)^{\ell+s}}\,. \tag{3.17}$$

[4]The C violating decay $J/\psi \to \gamma\gamma$ is forbidden by the Landau-Yang theorem. The experimental upper limit is 2.7×10^{-7} [6].

Let us apply this handy formula to several examples:

1. The neutral pion is spinless with positive C parity, hence the relative angular momentum ℓ between the quark and the antiquark is even. One expects $\ell = 0$ for the lightest known meson: the neutral pion ($J^{PC} = 0^{-+}$) is a ${}^1S_0(q\overline{q})$ state.
2. The ω meson (1^{--}) is a ${}^3S_1(q\overline{q})$ state since $s = 1$ and $C = -1$. This is also true for the ρ^0 meson. Both have $i_3 = 0$ but the former is an isoscalar while the latter is an isovector.
3. The ρ^0 decays with a branching ratio of $\simeq$100% into $\pi^+\pi^-$. The conservation of C parity in strong interactions requires the angular momentum ℓ between the two pions to be odd since $C(\pi^+\pi^-) = (-1)^\ell = -1$. In fact ℓ must be equal to 1 from total angular momentum conservation since $J_\rho = 1$. Note that ρ^0 cannot decay into $\pi^0\pi^0$, since ℓ would have to be even by virtue of Bose-Einstein symmetry. This decay not only violates C, but is also forbidden by all interactions.

3.4 *G* Parity

We have seen that neutral bosons with vanishing flavour quantum numbers are eigenstates of C and have isospin $i_3 = 0$. They can be isoscalars (such as the η or the ω) or isovectors (such as the π^0 or the ρ^0). Let us now include charged mesons with vanishing flavour quantum numbers, i.e. isovectors such as the $\pi^\pm$ or the $\rho^\pm$ which are composed of $u\overline{d}$ or $d\overline{u}$ quarks, or their orbital or radial excitations. A (passive) rotation of 180° of the coordinate system about the y-axis in isospin space is represented by the operator $e^{i\pi I_2}$, which flips the third component i_3 of the isospin and hence the charge of the meson. The G parity is the combined operation of charge conjugation and rotation. The operator

$$G = Ce^{i\pi I_2} \tag{3.18}$$

retrieves the original state, which is then an eigenstate of G. The rotation of the physical system transforms for instance a π^+ into a π^-, while C conjugation recovers the π^+ (Fig. 3.6). The charged pion is therefore an eigenstate of the G

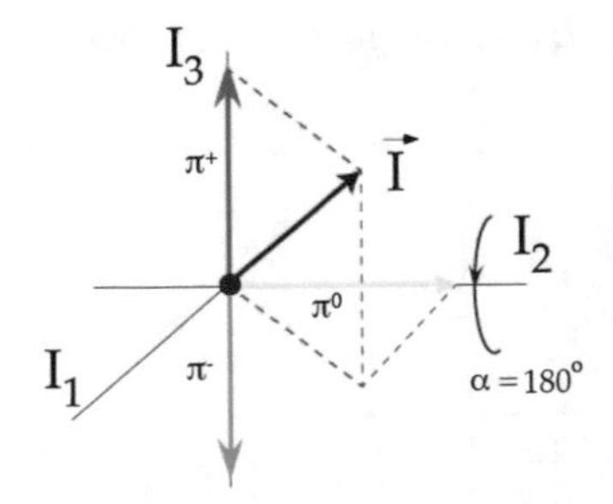

Fig. 3.6 A rotation of the coordinate system by 180° about the y-axis transforms the π^+ into a π^-

parity. For mesons the isospin i is an integer number and hence the isospinor behaves under rotations like the spherical function Y_ℓ^m under rotations around the y-axis in coordinate space. For $i_3 = 0$ the isospinor transforms as Y_i^0:

$$e^{i\pi I_2} Y_i^0(\theta) = Y_i^0(\pi - \theta) = (-1)^i Y_i^0(\theta). \tag{3.19}$$

The rotation operator $e^{i\pi I_2}$ therefore contributes the multiplicative factor $(-1)^i$. The G parity of the neutral pion is negative:

$$G|\pi^0\rangle = (-1)C|\pi^0\rangle = -|\pi^0\rangle. \tag{3.20}$$

Invoking charge independence gives also a negative G parity for the charged pion:

$$G|\pi^\pm\rangle = -|\pi^\pm\rangle. \tag{3.21}$$

The G parity of particle-antiparticle pair is with (3.17),

$$\boxed{G(\Phi\overline{\Phi}) = (-1)^{\ell+s+i}(\Phi\overline{\Phi})}\,. \tag{3.22}$$

From charge independence and C conservation follows that G parity is conserved in strong interactions. G parity is not an additional symmetry of strong interactions, but a practical concept, as the following applications show.

1. The G parities of the ρ^0 and ω mesons are

$$G|\rho\rangle = (-1)^1 C|\rho\rangle = +|\rho\rangle,$$
$$G|\omega\rangle = (-1)^0 C|\omega\rangle = -|\omega\rangle. \tag{3.23}$$

   A system made of n pions is an eigenstate of G with eigenvalue $G(n\pi) = (-1)^n$. Due to G parity conservation, the ρ decays into 2π and the ω into 3π. However, the ω does not decay into $3\pi^0$ since this violates C and isospin conservation. As we have seen, the ρ^0 does not decay into $\pi^0\pi^0$ due to J, C and isospin conservation.

2. A pion pair has positive G parity. From (3.22)

$$G|\pi^+\pi^-\rangle = (-1)^{\ell+i}|\pi^+\pi^-\rangle = +|\pi^+\pi^-\rangle. \tag{3.24}$$

   The isospin of the pair depends on the angular momentum ℓ between the pions. For ℓ even, $i = 0$ (or 2), while for ℓ odd, $i = 1$.

3. The η meson has positive G parity:

$$G|\eta\rangle = (-1)^0 C|\eta\rangle = +|\eta\rangle. \tag{3.25}$$

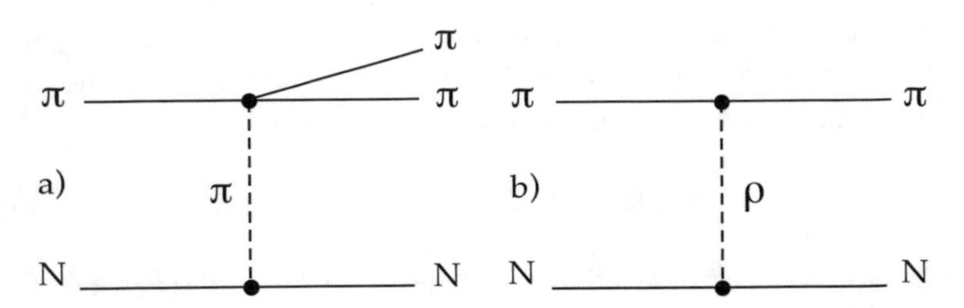

Fig. 3.7 (**a**) Pion production by OPE in πp inelastic scattering. The total number of pions emerging from the vertex must be even; (**b**) elastic πN scattering through ρ exchange

The η cannot decay into two pions since its negative parity would require ℓ to be odd, in conflict with $J_\eta = 0$. The η decays into $\pi^+\pi^-\pi^0$ and $3\pi^0$ by violating G conservation (branching ratio of 55.6%). The electromagnetic decay into 2γ occurs with a comparable branching ratio of 39.4%. Thus, $\eta \to 3\pi$ is an electromagnetic process which does not conserve isospin nor G.

4. G parity is conserved at a strong interaction vertex. For instance, in πN scattering with one pion exchange (OPE) the number of emitted pions must be even (Fig. 3.7a). On the other hand, OPE does not contribute to πN elastic scattering but proceeds e.g. through ρ exchange with positive G parity (Fig. 3.7b).

Note that the G parity is not defined for the K, D and B mesons which carry open flavours. For example, the K^+ becomes a K^0 under rotation (Table 2.1) and then a $\overline{K}^0$ under C conjugation. We shall return to the G transformation for kaons in Sect. 6.2.

References

1. Abele, A., et al.: Phys. Lett. B 411, 361 (1997)
2. Patrignani, C., et al. (Particle Data Group): Chin. Phys. C 40, 100001 (2016); 2017 update
3. Amsler, C.: Rev. Mod. Phys. 70, 1293 (1998)
4. Aker, E., et al.: Nucl. Inst. Methods Phys. Res. A 321, 69 (1992)
5. McDonough, J.E., et al.: Phys. Rev. D 38, 2121 (1988)
6. Ablikim, M., et al.: Phys. Rev. D 90, 092002 (2014)

Chapter 4
Nomenclature

Before 1986 hadron names were often inspired by fantasy or the name of the discoverer.[1] However, the increasing number of observed hadrons required the introduction of a systematic naming scheme. The idea was to assign names from which the quantum numbers J^{PC} of the hadrons could be inferred, while at the same time refraining to change the names of the well-known ones, such as the pion (π), the kaon (K) or the ρ. The new scheme was introduced in 1986 by the Particle Data Group [1]. Let us deal here with mesons and defer baryons to Chap. 13. We have seen that hadrons appear in isospin multiplets, and in the following sections we shall extend the multiplets to higher symmetries. The multiplets are labelled according to the quantum numbers J^{PC} of the neutral mesons with hidden flavour, for which the C parity is defined. Let us summarize the quantum numbers of $q\overline{q}$ mesons derived in the previous sections:

$$\text{Spin}: \quad \boxed{|\ell - s| < j < \ell + s}, \tag{4.1}$$

$$\text{Parity}: \quad \boxed{P = -(-1)^{\ell}}, \tag{4.2}$$

$$C\ \text{parity}: \quad \boxed{C = (-1)^{\ell+s}}, \tag{4.3}$$

with $s = 0$ or 1 (the C parity being defined only for neutral quark-antiquark pairs with hidden flavours $S = C = B' = 0$), and

$$G\ \text{parity}: \quad \boxed{G = (-1)^{i+\ell+s}}, \tag{4.4}$$

[1] such as A for "Andrei", B for "Buddha" (a fat resonance), F for "Felicitas", E for "Elisabeth" or perhaps "Europe"!

© The Author(s) 2018

C. Amsler, *The Quark Structure of Hadrons*, Lecture Notes in Physics 949,
https://doi.org/10.1007/978-3-319-98527-5_4

Table 4.1 Nomenclature for mesons with hidden flavours

i	J^{PC}	J^{-+}	J^{+-}	J^{--}	J^{++}
1	$d\overline{d}-u\overline{u}$, $u\overline{d}$, $d\overline{u}$	π	b	ρ	a
0[a]	$d\overline{d}+u\overline{u}$, $s\overline{s}$	η, η'	h, h'	ω, ϕ	f, f'
0	$c\overline{c}$	η_c	h_c	ψ[b]	χ_c
0	$b\overline{b}$	η_b	h_b	Υ	χ_b
1	$c\overline{c}q\overline{q}'$	Π_c	Z_c	R_c	W_c
1	$b\overline{b}q\overline{q}'$	Π_b	Z_b	R_b	W_b

$G=-1$ for $PC=+-$ and $--$, and $G=+1$ for $-+$ and $++$ mesons with $i=0$. The opposite holds for $i=1$: $G=+1$ for $+-$ and $--$ and $G=-1$ for $-+$ and $++$

[a]The superposition of $u\overline{u}$, $d\overline{d}$ and $s\overline{s}$ pairs in isoscalar mesons is discussed in Sect. 5.1

[b]The $J=1$ ground state is called J/ψ

with $i=0$ or 1, defined only for quark-antiquark pairs with hidden flavours. Note that hadrons containing the t quark do not bind due to the very short lifetime of this quark which decays into W^+b with a mean life of about 10^{-25} s.

Table 4.1 shows the nomenclature for mesons with hidden flavours for all possible J^{PC} and i. A subscript J is added for the spin except for pseudoscalar (0^{-+}) and vector (1^{--}) mesons. The mass is given in parentheses in MeV, except for the well-known ground state pseudoscalars and vectors, the π, η, η' and ρ, ω, ϕ, respectively. For the $c\overline{c}$ and $b\overline{b}$ states the mass is sometimes replaced by the label $n\ell$, see (2.2) (e.g. $1S$, $2S$, $1P$, etc.). For example, the $\rho_3^+(1690)$ is an established $\ell=2$ orbital excitation $u\overline{d}$ meson with quantum numbers 3^{--} and $i=1$.

Pseudoscalar mesons have the quantum numbers 0^{-+}, scalar mesons 0^{++}, vector mesons 1^{--}, axial-vector mesons 1^{+-} and tensor mesons 2^{++}. Note that the combinations 0^{--}, 0^{+-}, 1^{-+}, 2^{+-}, 3^{-+}... are forbidden for $q\overline{q}$ states (Problem 4.1). Candidates for such "exotic" non-$q\overline{q}$ mesons have been observed (Sect. 11.3). An example is the $\pi_1(1600)$ with quantum numbers $1^{-+}(i=1)$. Table 4.1 also includes the nomenclature introduced in 2017 for mesons which contains $c\overline{c}$ or $b\overline{b}$ pairs but are electrically charged and hence must have at least an additional $q\overline{q}'$ charged pair (tetraquarks). Candidates have been observed recently (Sect. 16.2).

States with yet unknown quantum numbers are labelled X. In the heavy quark sector several states have properties not easily compatible with the naive quark model (see Sect. 16.2). These states have been labelled X, Y or Z. The Particle Data Group now calls the states with known quantum numbers according to the list in Table 4.1, but often also appends the original name X, Y or Z.

Mesons with S, C, $B' \neq 0$ are labelled as follows:

1. The name refers to the heavier quark: $\overline{K}$ for s (e.g. $K^-[s\overline{u}]$ or $\overline{K}^0[s\overline{d}]$), D for c (e.g. $D^0[c\overline{u}]$) and $\overline{B}$ for b (e.g. $B^-[b\overline{u}]$ or $\overline{B}^0[b\overline{d}]$).
2. The lighter quark, if not u nor d, is indicated by a subscript, for instance $D_s^+[c\overline{s}]$, $B_c^-[b\overline{c}]$.
3. A superscript * is added for natural parity states, that is those with $P=(-1)^J$.

Table 4.2 Old names for some of the mesons before 1986

Old name	New name	Old name	New name	Old name	New name
$S(975)$	$f_0(980)$	$\delta(980)$	$a_0(980)$	$H(1190)$	$h_1(1170)$
$B(1235)$	$b_1(1235)$	$A_1(1270)$	$a_1(1260)$	$f(1270)$	$f_2(1270)$
$D(1285)$	$f_1(1285)$	$A_2(1320)$	$a_2(1320)$	$\epsilon(1300)$	$f_0(1370)$
$E(1420)$	$f_1(1420)$[a]	$\iota(1440)$	$\eta(1405)$	$f'(1525)$	$f_2'(1525)$
$\omega(1670)$	$\omega_3(1670)$	$F^{\pm}$	$D_s^{\pm}$	$F^*(2140)$	$D_s^{*\pm}$
$Q_1(1280)$	$K_1(1270)$	$Q_2(1400)$	$K_1(1400)$	$K^*(1430)$	$K_2^*(1430)$

[a]or $\eta(1405)$

4. The spin J appears as a subscript except for pseudoscalar and vector mesons.

For example, the B_c^{*+} would be the 1^- $[c\bar{b}]$ meson, which has not been observed yet (2017), in contrast to its 0^- partner, the B_c^+. For further examples, see Problem 4.2.

Table 4.2 recalls some of the old names quoted in the literature before 1986.

Reference

1. Aguilar-Benitez, M., et al. (Particle Data Group): Phys. Lett. 170B, 1 (1986)

Chapter 5
Quark-Antiquark Nonets

Let us first deal with mesons made of the three lightest quarks u, d, s and $\overline{u}$, $\overline{d}$, $\overline{s}$. As illustrated in Fig. 2.1, with three flavours and three antiflavours one can construct a nonet of mesons for each orbital mode, each vibrational mode and for each parallel and antiparallel spin.

Figure 5.1 shows the spectrum of the lower meson excitations which resembles the one of the hydrogen atom. Note that for the hydrogen atom the lowest vibrational states are labelled $1s$, $2p$, $3d$..., while in the quark model one uses the notation $1S$, $1P$, $1D$... We have already introduced the ground states ($\ell = 0$, $n = 1$), the $s = 0$ pseudoscalars (2.3) and the $s = 1$ vectors (2.4) which are discussed in more detail below. Their vibrational excitations also build two nonets of pseudoscalar and vector mesons for each value of $n \geq 2$. The orbital excitations $\ell \geq 1$ consist of four nonets for each value of n, since $j = \ell$ for antiparallel quark spins and $j = \ell - 1$, ℓ or $\ell + 1$ for parallel spins. Each row contains three isovectors, two strange isodoublets, and two isoscalar singlets. Mesons in the dark (blue) areas are well established, those in the white areas are not fully established, or their classification is only tentative. As the meson masses increase, the states become broad and overlap, which complicates the spin-parity determinations.

The C parity is that of the neutral members. States with the same quantum numbers mix. For instance, the 2^3S_1 and 1^3D_1 mesons have the same J^{PC} and are expected to mix. The states K_{1a} and K_{1b} in the axial vector 1^{++} and 1^{+-} nonets are mathematical constructs. The observed mesons are the K_1(1270 and K_1(1400) which are orthogonal linear combinations of K_{1a} and K_{1b}. The former decays into $K\rho$ and the latter into $K^*\pi$, leading to the same final state $K\pi\pi$. A coupled channel analysis taking into account the interference between the two decays in $K\pi\pi$ leads to a mixing angle close to 45° between K_{1a} and K_{1b} (see [1] and references therein).

The classification in the scalar nonet and its radial excitations is controversial. The $f_0(1500)$ could qualify as one of the $i = 0$ state in the nonet shown in Fig. 5.1 but has also been interpreted as a glueball (Sect. 11.1). The scalar mesons $a_0(980)$, $K_0^*(700)$ (also known as κ), the $f_0(500)$ (also known as σ), and the $f_0(980)$ are not

© The Author(s) 2018

C. Amsler, *The Quark Structure of Hadrons*, Lecture Notes in Physics 949,
https://doi.org/10.1007/978-3-319-98527-5_5

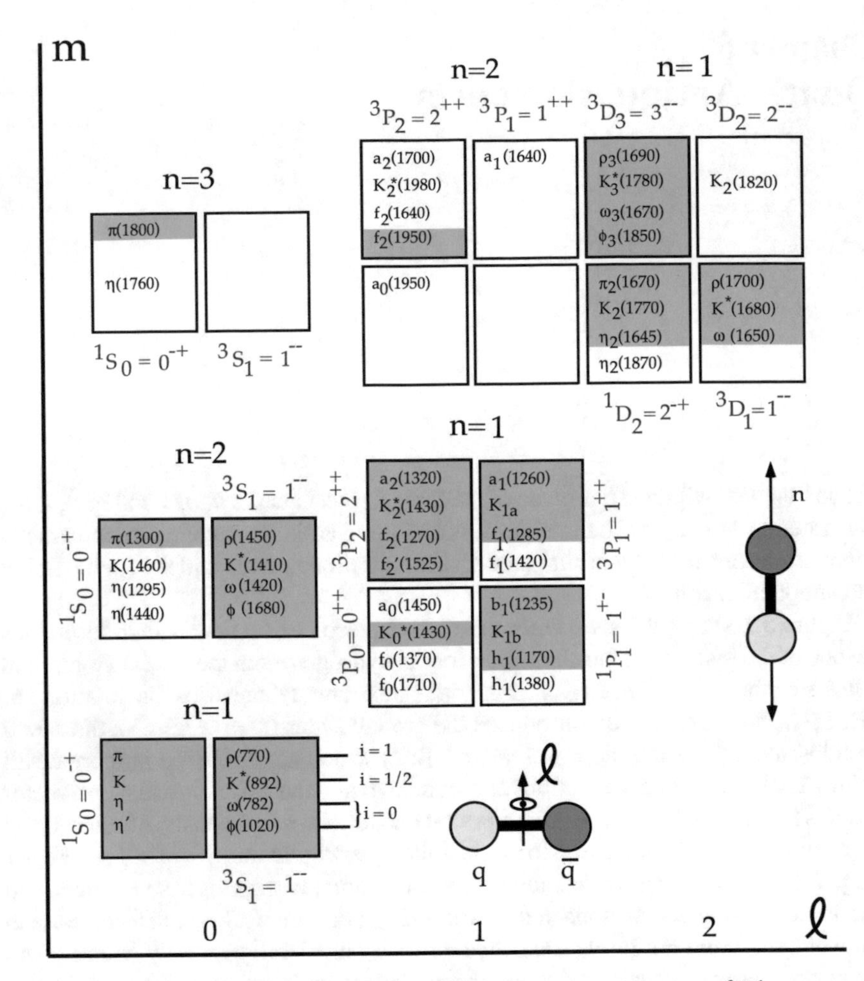

Fig. 5.1 The mesons made of the u, d, and s light quarks are organised in $n^{2s+1}\ell_j$ nonets (or J^{PC}). The well established mesons are shown in the dark (blue) areas. Strong evidence has been reported for those in the white areas, but their classification is tentative. The vertical mass scale is approximate. Many orbital and radial excitations have not been identified yet (see the text and also Fig. 1.1 for further candidates)

shown in the figure. They could build the lightest nonet, but they have been proposed to be two-meson resonances or tetraquarks (Table 11.1 and Sect. 16.1).

The pseudoscalar slot labelled $\eta(1440)$ may in fact consist of two states, one at 1405 MeV, the other at 1475 MeV, while the $\eta(1295)$ is not firmly established (for a review see [2]). On the other hand, the $q\bar{q}$ nature of the established $f_1(1420)$ has been questioned. It could be replaced in the figure by the less well established $f_1(1510)$ [2].

5.1 Nonet Mixing Angle

Let us define the notation that will be used throughout these lectures. With three quarks and three antiquarks the $q\overline{q}$ combinations are

$$|u\overline{d}\rangle, |d\overline{u}\rangle, |u\overline{u}\rangle, |d\overline{d}\rangle, |u\overline{s}\rangle, |s\overline{u}\rangle, |d\overline{s}\rangle, |s\overline{d}\rangle, |s\overline{s}\rangle \tag{5.1}$$

which then need to be symmetrized (e.g. $|u\overline{d}\rangle \pm |\overline{d}u\rangle$, see Sect. 6.2). The $q\overline{q}$ pairs are orthogonal and normalized:

$$\begin{aligned}
\langle u\overline{u}|d\overline{d}\rangle &= \langle u\overline{d}|d\overline{u}\rangle = \langle u\overline{s}|s\overline{s}\rangle \ldots = 0, \\
\langle u\overline{d}|\overline{d}u\rangle &= \langle u\overline{s}|\overline{s}u\rangle \ldots = 0, \\
\langle u\overline{u}|u\overline{u}\rangle &= \langle d\overline{d}|d\overline{d}\rangle = \langle s\overline{s}|s\overline{s}\rangle = \langle u\overline{d}|u\overline{d}\rangle \ldots = 1.
\end{aligned} \tag{5.2}$$

We will write a superposition as one single ket or bra, e.g. $|d\overline{d}\rangle - |u\overline{u}\rangle \equiv |d\overline{d} - u\overline{u}\rangle$ or $|u\overline{d}\rangle \pm |\overline{d}u\rangle \equiv |u\overline{d} \pm \overline{d}u\rangle$. Table 5.1 lists the quark assignments of the ground state mesons.

The wavefunctions are normalized and orthogonal. Missing in the table are the two isoscalar states which are expected to mix since they have identical quantum numbers ($Q = i = S = 0$). As we shall see in Chap. 7 their wavefunctions are linear superpositions of the octet and singlet wavefunctions

$$\begin{aligned}
|8\rangle &\equiv \frac{1}{\sqrt{6}}|u\overline{u} + d\overline{d} - 2s\overline{s}\rangle, \quad \text{SU(3) octet}, \\
|1\rangle &\equiv \frac{1}{\sqrt{3}}|u\overline{u} + d\overline{d} + s\overline{s}\rangle, \quad \text{SU(3) singlet}.
\end{aligned} \tag{5.3}$$

These functions are normalized and orthogonal to the ones listed in Table 5.1. The linear superposition involves an angle that can be measured, the mixing angle θ.

Table 5.1 Quark composition of the $q\overline{q}$ ground state mesons (S stands for strangeness)

0^{-+}		1^{--}		S
π^+	$\|u\overline{d}\rangle$	ρ^+	$\|u\overline{d}\rangle$	0
π^0	$\frac{1}{\sqrt{2}}\|d\overline{d} - u\overline{u}\rangle$	ρ^0	$\frac{1}{\sqrt{2}}\|d\overline{d} - u\overline{u}\rangle$	0
π^-	$-\|d\overline{u}\rangle$	ρ^-	$-\|d\overline{u}\rangle$	0
K^+	$\|u\overline{s}\rangle$	K^{*+}	$\|u\overline{s}\rangle$	+1
K^0	$\|d\overline{s}\rangle$	K^{*0}	$\|d\overline{s}\rangle$	+1
$\overline{K}^0$	$-\|s\overline{d}\rangle$	$\overline{K}^{*0}$	$-\|s\overline{d}\rangle$	−1
K^-	$-\|s\overline{u}\rangle$	K^{*-}	$-\|s\overline{u}\rangle$	−1

The π^0 and ρ^0 are linear combinations of $u\overline{u}$ and $d\overline{d}$ pairs and do not contain any $s\overline{s}$. The minus signs are discussed in Sect. 6.2

The physically observed states are

$$\psi' = |8\rangle \cos\theta - |1\rangle \sin\theta,$$
$$\psi = |8\rangle \sin\theta + |1\rangle \cos\theta. \tag{5.4}$$

For those nonets with mixing angle θ satisfying the condition of ideal mixing

$$\boxed{\tan\theta = \frac{1}{\sqrt{2}}} \Rightarrow \cos\theta = \sqrt{\frac{2}{3}}, \quad \sin\theta = \sqrt{\frac{1}{3}}, \tag{5.5}$$

or $\theta = 35.3°$, the $s\overline{s}$ component decouples from $u\overline{u}$ and $d\overline{d}$:

$$\psi' = \frac{1}{3}|u\overline{u} + d\overline{d} - 2s\overline{s}\rangle - \frac{1}{3}|u\overline{u} + d\overline{d} + s\overline{s}\rangle = -|s\overline{s}\rangle,$$
$$\psi = \frac{1}{3\sqrt{2}}|u\overline{u} + d\overline{d} - 2s\overline{s}\rangle + \frac{\sqrt{2}}{3}|u\overline{u} + d\overline{d} + s\overline{s}\rangle = \frac{1}{\sqrt{2}}|d\overline{d} + u\overline{u}\rangle. \tag{5.6}$$

Ideal mixing is also fulfilled when

$$\boxed{\tan\theta = -\sqrt{2}} \Rightarrow \cos\theta = \sqrt{\frac{1}{3}}, \quad \sin\theta = -\sqrt{\frac{2}{3}}, \tag{5.7}$$

or $\theta = -54.7°$. In this case the quark contents of ψ and ψ' are swapped:

$$\psi' = \frac{1}{3\sqrt{2}}|u\overline{u} + d\overline{d} - 2s\overline{s}\rangle + \frac{\sqrt{2}}{3}|u\overline{u} + d\overline{d} + s\overline{s}\rangle = \frac{1}{\sqrt{2}}|d\overline{d} + u\overline{u}\rangle,$$
$$\psi = -\frac{1}{3}|u\overline{u} + d\overline{d} - 2s\overline{s}\rangle + \frac{1}{3}|u\overline{u} + d\overline{d} + s\overline{s}\rangle = |s\overline{s}\rangle. \tag{5.8}$$

There is an ambiguity in the mixing (5.4): $\theta = 35.3°$ or $-54.7°$, depending on which observed state is ascribed to the ψ and which to the ψ'. Note that 180° can always be added or subtracted (which flips the signs of the wavefunctions (5.4)), but we adopt the convention that θ lies between $-90°$ and $+90°$. For vector mesons the mixing is usually written as

$$\phi = |8\rangle \cos\theta_V - |1\rangle \sin\theta_V,$$
$$\omega = |8\rangle \sin\theta_V + |1\rangle \cos\theta_V. \tag{5.9}$$

We will show in the next section that θ_V is close to ideal. Hence the ϕ becomes almost pure $s\overline{s}$ and the ω almost pure $u\overline{u} + d\overline{d}$:

$$\phi \simeq -|s\overline{s}\rangle, \quad \omega \simeq \frac{1}{\sqrt{2}}|d\overline{d} + u\overline{u}\rangle. \tag{5.10}$$

Nearly ideal mixing also occurs for the 2^{++} and 3^{--} nonets for which enough information is available to calculate θ (see Table 5.2). However, there is an exception with the pseudoscalars (an intuitive explanation will be given in the next section). The convention is to express the mixing as

$$\eta = |8\rangle \cos\theta_P - |1\rangle \sin\theta_P,$$
$$\eta' = |8\rangle \sin\theta_P + |1\rangle \cos\theta_P. \tag{5.11}$$

As we shall see, θ_P lies in the range $-10°$ to $-20°$, far from ideal mixing. Assuming that $\theta \simeq 0$ leads to the crude approximation

$$\eta \sim \frac{1}{\sqrt{6}}|u\overline{u} + d\overline{d} - 2s\overline{s}\rangle,$$
$$\eta' \sim \frac{1}{\sqrt{3}}|u\overline{u} + d\overline{d} + s\overline{s}\rangle. \tag{5.12}$$

Hence η and η' are almost pure octet and pure singlet, respectively.

5.2 Mass Formulae

The meson masses can be used to estimate the nonet mixing angle. Let us consider a 2-dimensional space spanned by the basis made of the two isoscalars (5.3), and build the mass matrix

$$M = \begin{pmatrix} m_8 & m_{81} \\ m_{18} & m_1 \end{pmatrix} \tag{5.13}$$

with

$$m_8 = \langle 8|H|8\rangle, \quad m_1 = \langle 1|H|1\rangle, \quad m_{81} = m_{18} = \langle 8|H|1\rangle. \tag{5.14}$$

The two physical states ψ_1 and ψ_2 and their corresponding mass eigenvalues λ_1 and λ_2 are obtained by diagonalizing M, that is by solving the equation $M\psi = \lambda\psi$. The two solutions of the secular equation

$$\begin{vmatrix} m_8 - \lambda & m_{81} \\ m_{18} & m_1 - \lambda \end{vmatrix} = 0 \Rightarrow (m_8 - \lambda)(m_1 - \lambda) - m_{18}^2 = 0 \tag{5.15}$$

are

$$\lambda_{1,2} = \frac{m_1 + m_8 \pm \sqrt{(m_1 + m_8)^2 + 4(m_{81}^2 - m_1 m_8)}}{2}. \tag{5.16}$$

Adding and multiplying the eigenvalues leads to the relations

$$\lambda_1 + \lambda_2 = m_1 + m_8 \quad \text{and} \quad \lambda_1\lambda_2 = m_1 m_8 - m_{81}^2, \tag{5.17}$$

which will be used below. The components of the eigenfunction ψ_1 in the $(|8\rangle, |1\rangle)$ basis are

$$\psi_1 = |8\rangle \cos\theta - |1\rangle \sin\theta = \begin{pmatrix} \cos\theta \\ -\sin\theta \end{pmatrix}. \tag{5.18}$$

The corresponding eigenvalue fulfils the equation

$$\begin{pmatrix} m_8 & m_{81} \\ m_{18} & m_1 \end{pmatrix} \begin{pmatrix} \cos\theta \\ -\sin\theta \end{pmatrix} = \lambda_1 \begin{pmatrix} \cos\theta \\ -\sin\theta \end{pmatrix}. \tag{5.19}$$

From the first row on obtains by solving for θ

$$\tan\theta = \frac{m_8 - \lambda_1}{m_{81}}. \tag{5.20}$$

The terms m_8 and m_{18} can be calculated from the constituent quark masses. By writing the octet and singlet (5.3) in components $\frac{1}{\sqrt{6}}(1, 1, -2)$ and $\frac{1}{\sqrt{3}}(1, 1, 1)$ one obtains

$$m_8 = \frac{1}{6}(1, 1, -2) \begin{pmatrix} 2m_u & 0 & 0 \\ 0 & 2m_d & 0 \\ 0 & 0 & 2m_s \end{pmatrix} \begin{pmatrix} 1 \\ 1 \\ -2 \end{pmatrix} = \frac{1}{3}(m_u + m_d + 4m_s), \tag{5.21}$$

and

$$m_{81} = m_{18} = \frac{1}{\sqrt{18}}(1, 1, -2) \begin{pmatrix} 2m_u & 0 & 0 \\ 0 & 2m_d & 0 \\ 0 & 0 & 2m_s \end{pmatrix} \begin{pmatrix} 1 \\ 1 \\ 1 \end{pmatrix} = \frac{\sqrt{2}}{3}(m_u + m_d - 2m_s). \tag{5.22}$$

We now replace the constituent quark masses by the meson masses. From Table 5.1 one finds, e.g. for the vector nonet and with the assumption that $m_u = m_d$,

$$m_8 = \frac{1}{3}(2m_u + 4[m_s + m_u] - 4m_u) = \frac{1}{3}(4m_{K^*} - m_\rho), \tag{5.23}$$

$$m_{81} = \frac{\sqrt{2}}{3}(2m_u - 2[m_s + m_u] + 2m_u) = \frac{2\sqrt{2}}{3}(m_\rho - m_{K^*}). \tag{5.24}$$

Then with $\lambda_1 \equiv m_\phi$ and after substituting into (5.20) the mass formula reads

$$\boxed{\tan\theta = \frac{4m_{K*}-m_\rho-3m_\phi}{2\sqrt{2}(m_\rho-m_{K*})}}. \tag{5.25}$$

One obtains $\theta \simeq 36.5°$ when introducing the meson masses, close to the value of 35.3° expected from ideal mixing. Formula (5.25), the so-called linear mass formula, is sometimes replaced by its quadratic version, in which the masses are replaced by their squares, although there are no compelling theoretical reasons to do so[1]:

$$\tan\theta = \frac{4m_{K*}^2 - m_\rho^2 - 3m_\phi^2}{2\sqrt{2}(m_\rho^2 - m_{K*}^2)}. \tag{5.26}$$

This leads to a somewhat larger mixing angle $\theta \simeq 42°$ (see Table 5.2).

For ideal mixing the ρ and ω mesons have the same quark content, $\frac{1}{\sqrt{2}}(d\bar{d} - u\bar{u})$ and $\frac{1}{\sqrt{2}}(d\bar{d} + u\bar{u})$, respectively, Table 5.1 and (5.10). One therefore expects that $m_\rho = m_\omega$, which is almost fulfilled experimentally (775.3 and 782.7 MeV, respectively). Furthermore, inserting $\tan\theta = \frac{1}{\sqrt{2}}$ on the left-hand side of (5.25) predicts that

$$m_\rho + m_\phi = 2\,m_{K^*}, \tag{5.27}$$

which is also in quite good agreement with the measured values [3].

The ω mass does not appear in formula (5.25). Let us therefore derive an alternative mass formula which includes all nonet members. The mass matrix M (5.13) can be written as a product $M = UDU^{-1}$ where D is the (diagonal) eigenvalue matrix and the columns of U contain the components of the eigenfunctions. Therefore

$$\begin{pmatrix} m_8 & m_{81} \\ m_{18} & m_1 \end{pmatrix} = \begin{pmatrix} \cos\theta & \sin\theta \\ -\sin\theta & \cos\theta \end{pmatrix} \begin{pmatrix} \lambda_1 & 0 \\ 0 & \lambda_2 \end{pmatrix} \begin{pmatrix} \cos\theta & -\sin\theta \\ \sin\theta & \cos\theta \end{pmatrix}. \tag{5.28}$$

The octet mass m_8 is the given by

$$m_8 = \lambda_1 \cos^2\theta + \lambda_2 \sin^2\theta = \frac{\lambda_1 + \lambda_2 \tan^2\theta}{1 + \tan^2\theta}. \tag{5.29}$$

Solving for θ yields

$$\tan^2\theta = \frac{\lambda_1 - m_8}{-\lambda_2 + m_8}. \tag{5.30}$$

[1] A reason often invoked is that boson masses enter quadratically in the Klein-Gordon wave equation. Squared masses also appear in chiral perturbation theories.

For the 1^{--} nonet masses one obtains with (5.23) and with $\lambda_1 = m_\phi$, $\lambda_2 = m_\omega$:

$$\boxed{\tan^2\theta = \frac{4m_{K^*} - m_\rho - 3m_\phi}{-4m_{K^*} + m_\rho + 3m_\omega}}, \tag{5.31}$$

a formula that involves all nonet masses. Introducing the meson masses gives $\theta \simeq 36.5°$, which is very close to ideal mixing. However, this mass relation does not provide the sign of $\tan\theta$.

The mass formulae can be applied to any nonet by substituting the corresponding mesons. For the pseudoscalars the mass formulae (5.25) and (5.26) read

$$\begin{aligned} \tan\theta &= \frac{4m_K - m_\pi - 3m_\eta}{2\sqrt{2}(m_\pi - m_K)} \Rightarrow \theta_P = -11.7°, \\ \tan^2\theta &= \frac{4m_K - m_\pi - 3m_\eta}{-4m_K + m_\pi + 3m_{\eta'}} \Rightarrow \theta_P = -24.5°, \end{aligned} \tag{5.32}$$

where we have adopted the negative sign for θ_p in the bottom relation. The mixing angle also depends slightly on the choice between neutral and charged mesons (we have introduced the masses of the π^0 and K^0). Another way to determine the pseudoscalar mixing angle will be given in Sect. 7.3, see also Problem 5.1.

Table 5.2 lists the mixing angles for the well established nonets shown in Fig. 5.1.

The preference for ideal mixing in nonets with the exception of pseudoscalars can be understood as follows. Figure 5.2 shows the oscillation between $q\bar{q}$ pairs mediated by the exchange of gluons which induces a perturbation leading to an additional term in the mass matrix (5.13). We model the perturbation by adding the constant term A to the mass matrix, hence

$$\begin{aligned} m_8 &= \frac{1}{6}(1, 1, -2)\begin{pmatrix} 2m_u + A & A & A \\ A & 2m_d + A & A \\ A & A & 2m_s + A \end{pmatrix}\begin{pmatrix} 1 \\ 1 \\ -2 \end{pmatrix} \\ &= \frac{1}{3}(m_u + m_d + 4m_s) = \frac{1}{3}(4m_{K^*} - m_\rho) \end{aligned} \tag{5.33}$$

Table 5.2 Mixing angle θ for various $q\bar{q}$ nonets, using the neutral members

		θ [°]			
		Linear	Quadratic	Linear	
J^{PC}	ψ'	(5.25)	(5.26)	(5.31)	Quadratic
1^{--}	ϕ	36.5	42.0	36.5	39.2
0^{-+}	η	−11.7	−6.3	−24.5	−11.3
2^{++}	$f_2'(1525)$	27.1	30.0	28.0	29.6
3^{--}	$\phi_3(1850)$	30.9	32.7	30.8	31.8

The sign is taken from (5.25). Masses are taken from [3]. The last column refers to (5.31) with quadratic meson masses. The column labelled ψ' specifies the isoscalar used in the numerators of the mass formulae

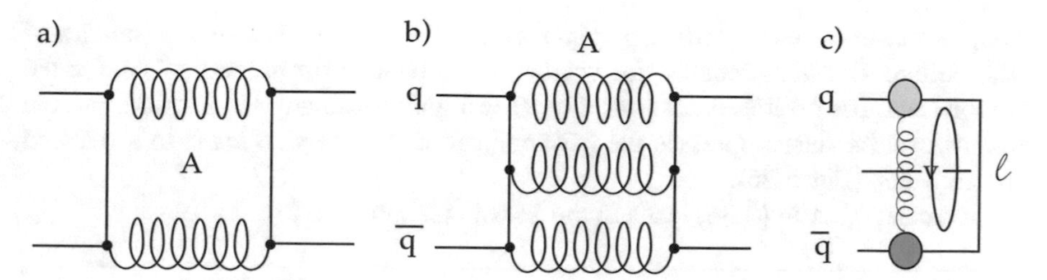

Fig. 5.2 Perturbation of the wavefunction by the intermediate gluonic state. The 2-gluon process (**a**) is forbidden for isoscalar 1^{--} mesons, but allowed for isoscalar 0^{-+} mesons. For 1^{--} mesons the oscillation proceeds via 3 or more gluons (**b**). For mesons with orbital excitations $\ell > 0$ the perturbation is suppressed by the angular momentum barrier (**c**)

for the vector nonet, and

$$\begin{aligned} m_{81} &= \frac{1}{\sqrt{18}}(1, 1, -2) \begin{pmatrix} 2m_u + A & A & A \\ A & 2m_d + A & A \\ A & A & 2m_s + A \end{pmatrix} \begin{pmatrix} 1 \\ 1 \\ 1 \end{pmatrix} \\ &= \frac{\sqrt{2}}{3}(m_u + m_d - 2m_s) = \frac{2\sqrt{2}}{3}(m_\rho - m_{K^*}). \end{aligned} \tag{5.34}$$

Hence m_8 and m_{81} are independent of A and we reproduce (5.23) and (5.24). In contrast, the perturbation enters the singlet mass

$$\begin{aligned} m_1 &= \frac{1}{3}(1, 1, 1) \begin{pmatrix} 2m_u + A & A & A \\ A & 2m_d + A & A \\ A & A & 2m_s + A \end{pmatrix} \begin{pmatrix} 1 \\ 1 \\ 1 \end{pmatrix} \\ &= \frac{(m_u + m_d + [2m_s + m_u + m_d])}{3} + 3A = \frac{1}{3}(m_\rho + 2m_{K^*}) + 3A, \end{aligned} \tag{5.35}$$

assuming again that $m_u = m_d$. We now introduce m_1, m_8 and m_{81} into the relations (5.17) between the eigenvalues $\lambda_1 = m_\phi$ and $\lambda_2 = m_\omega$ to obtain

$$\begin{aligned} m_\phi + m_\omega &= 3A + 2m_{K^*}, \\ m_\phi m_\omega &= 2m_{K^*}m_\rho - m_\rho^2 + A(4m_{K^*} - m_\rho). \end{aligned} \tag{5.36}$$

For $A = 0$ we recover ideal mixing (5.27). This suggests an intuitive reason for the strong deviation from ideal mixing in the pseudoscalar nonet: the perturbation shown in Fig. 5.2a involves two gluons (the exchange of a single (coloured) gluon is forbidden by colour conservation). For the ideally mixed vector mesons at least three gluons are required by virtue of the Fermi-Yang theorem [4] which forbids a spin-1 state to decay into two spin-1 gluons (Fig. 5.2b).[2] Hence the strength of the perturbation A is reduced for vectors, but enhanced for pseudoscalars. On the other hand, for orbital excitations $\ell > 0$ two-gluon exchange is allowed, but the centrifugal barrier suppresses the $q\overline{q}$ annihilation, which also leads to a reduced perturbation (Fig. 5.2c).

Eliminating A in (5.36) leads to the Schwinger sum rule [5]

$$\boxed{\Delta^2 \equiv (m_\omega + m_\phi)(4m_{K^*} - m_\rho) - 3m_\omega m_\phi + 8m_{K^*}m_\rho - 3m_\rho^2 - 8m_{K^*}^2 = 0}\,, \tag{5.37}$$

which should be fulfilled by all meson nonets. By introducing the measured values and ignoring mass uncertainties one finds that

$$\Delta(1^{--}) \sim 6\,\text{MeV}, \;\; \Delta(2^{++}) \sim 30\,\text{MeV}, \;\; \Delta(3^{--}) \sim 10\,\text{MeV}, \tag{5.38}$$

in contrast to

$$\Delta(0^{-+}) \sim 525\,\text{MeV}. \tag{5.39}$$

The sum rule (5.37) is poorly satisfied in the 0^{-+} nonet, which shows the limitations of this very simple model.

5.3 Okubo-Zweig-Iizuka Rule

The Okubo-Zweig-Iizuka (OZI) rule states that strong interaction processes described by Feynman graphs, which can be split into hadrons without cutting any quark line, are suppressed. The reaction may still occur, albeit with low probability, comparable to that of electromagnetic interactions. The OZI rule is best explained with a few examples.

Figure 5.3a shows the decay of the $f_2'(1525)$ meson which, according to Table 5.2, is a nearly pure $s\overline{s}$ state. The s and $\overline{s}$ quarks are transferred to the final states K^+K^- or $K^0\overline{K}^0$. The Feynman diagram cannot be split into hadrons without cutting any quark line. In contrast, the diagram in Fig. 5.3b can be split by the vertical dashed line into two parts without crossing any quark line, the $f_2'(1525)$ on the left and the two pions on the right. This process is OZI suppressed: the branching ratio for the decay into two pions is 8.2×10^{-3}, while $K\overline{K}$ decay occurs with a probability of 89%. The remaining 11% are attributed to the OZI allowed $f_2'(1525) \to \eta\eta$ decay, where the two η mesons are made of $s\overline{s}$ pairs, see (5.12) for

[2]Note that the Landau-Yang theorem also forbids the decay of 3^- mesons into two gluons, as well as all odd spin negative parity mesons (Appendix A).

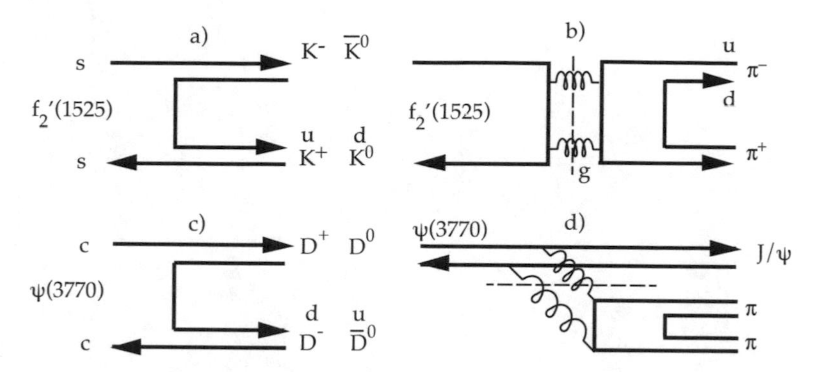

Fig. 5.3 OZI-allowed (**a**, **c**) and OZI-suppressed decays (**b**, **d**) of the $f_2'(1525)$ and $\psi(3770)$ mesons

the wavefunction of the η. The exchanged light quark in Fig. 5.3a is replaced by a strange quark.

A further example of OZI suppression is the decay of the $\phi(s\bar{s})$ meson into $\pi^+\pi^-\pi^0$, which occurs with a branching fraction of 15%, compared to 83% for the OZI allowed $K\overline{K}$ decay. Here the $\pi^+\pi^-\pi^0$ mode is enhanced by the larger phase space (Q value of $\simeq 600$ MeV for 3π, compared to only $\simeq 30$ MeV for $K\overline{K}$). In fact, the decay of the f_2' and the ϕ into pions occurs through an OZI allowed process via the small admixture of $u\bar{u}$ and $d\bar{d}$ in their wavefunctions.

The OZI rule suppresses the production of s quarks in proton-antiproton annihilation, since neither proton nor antiproton contain valence s or $\bar{s}$ quarks. This allows the pseudoscalar mixing angle to be determined from the measured annihilation rates into two mesons (Problem 5.1).

Figure 5.3b shows as a further example the decay of the $\psi(3770)$ into $D\overline{D}$ mesons (branching fraction close to 100%). The decay $\psi(3770) \rightarrow J/\psi\,\pi\pi$ is OZI suppressed with a branching fraction of $\sim 3\times 10^{-3}$. Note that the $\psi(2S)$ is too light to decay into $D\overline{D}$ pairs and decays with $\simeq$53% probability into $J/\psi\,\pi\pi$. This an OZI suppressed decay which competes with electromagnetic (radiative) decays, in particular γ transitions to the charmonium P states ($\simeq$29%, Sect. 9.1).

References

1. Yang, K.C.: Nucl. Phys. B 776, 187 (2007)
2. Amsler, C., Masoni, A.: In: Tanabashi, M., et al. (Particle Data Group): Phys. Rev. D 98, 030001 (2018), p. 665
3. Tanabashi, M., et al. (Particle Data Group): Phys. Rev. D 98, 030001 (2018)
4. Landau, L.D.: Dokl. Akad. Nauk. 60, 207 (1948); Yang, C.N.: Phys. Rev. 77, 242 (1950)
5. Schwinger, J.: Phys. Rev. Lett. 12, 237 (1964)

Chapter 6
SU(2)

6.1 Rotation Matrices

As already mentioned in Sect. 3.1 an active rotation of the physical system is represented by the operator[1]

$$U(\vec{J}) = e^{-i\vec{J}\cdot\vec{\alpha}}. \tag{6.1}$$

In this and the following chapters we shall use the lower case letters j and i for the spin and isospin to avoid confusion with the operators $\vec{J}$ and $\vec{I}$. We will also use natural units, e.g. setting $\hbar = 1$ throughout.

The unitary operator (6.1) belongs to the SU(2) group. SU(2) symmetry means that this operator commutes with the Hamiltonian, see (3.5), and thus refers to invariance under rotations in coordinate space when dealing with spin, and to charge independence when dealing with isospin. The three generators J_1, J_2 and J_3 (the components of $\vec{J}$) obey the commutation relation

$$\boxed{[J_i, J_j] = i\epsilon_{ijk} J_k}, \tag{6.2}$$

where the structure constants ϵ_{ijk} are given in Table 6.1. They are the elements of the antisymmetric unit tensor. The relation (6.2) is a property of angular momentum and is fulfilled for any value of the spin or isospin.

Now, depending on the value of j or i the spinors or isospinors span a $2j + 1$ or $2i + 1$ dimensional space. The operator (6.1) needs to be represented by square

[1] For a discussion on active and passive rotations see Fig. 18.1 in Chap. 18.

© The Author(s) 2018

C. Amsler, *The Quark Structure of Hadrons*, Lecture Notes in Physics 949,
https://doi.org/10.1007/978-3-319-98527-5_6

Table 6.1 SU(2) structure constants

ijk	123	213	312	321	231	132
ϵ_{ijk}	1	−1	1	−1	1	−1

Further combinations with repeated indices vanish

Table 6.2 Matrix elements of the SU(2) generators; m and m' vary between $-j$ and $+j$; omitted matrix elements vanish

m	$\langle j\,m\vert O\vert j\,m'\rangle$	O	
$m'+1$	$\sqrt{(j-m')(j+m'+1)}$	J_+	For
	$\frac{1}{2}\sqrt{(j-m')(j+m'+1)}$	J_1	$-j\leq m'\leq j-1$
	$-\frac{i}{2}\sqrt{(j-m')(j+m'+1)}$	J_2	Ditto
m'	m'	J_3	$-j\leq m'\leq j$
$m'-1$	$\sqrt{(j+m')(j-m'+1)}$	J_-	For
	$\frac{1}{2}\sqrt{(j+m')(j-m'+1)}$	J_1	$-j+1\leq m'\leq j$
	$\frac{i}{2}\sqrt{(j+m')(j-m'+1)}$	J_2	Ditto

Operating on $\vert jm'\rangle$ gives $O\vert jm'\rangle=\sum_m \vert jm\rangle\langle j\,m\vert O\vert j\,m'\rangle$

matrices with the corresponding dimensions. Let us consider active rotations by the angle θ about the y-axis and define the Wigner rotation matrix:

$$d^j_{mm'}(\theta)=\langle m\vert\,\overbrace{\mathrm{e}^{-iJ_2\theta}}^{U}\,\vert m'\rangle \tag{6.3}$$

Expanding the operator U gives

$$U=1-iJ_2\theta+\frac{1}{2!}(-iJ_2\theta)^2+\dots \tag{6.4}$$

We need the matrix elements of $(J_2)^n$. Table 6.2 lists the matrix elements of the three generators and of the ladder operators $J_\pm=J_1\pm iJ_2$, which will be used in later chapters. They follow from the commutation rules (6.2), see Appendix C. For example, for spin-$\frac{1}{2}$ the rotations are represented by 2-dimensional matrices and we need the fundamental representation of the SU(2) group, which is described by the three matrices

$$(J_1)=\langle m\vert J_1\vert m'\rangle=\frac{1}{2}\begin{pmatrix}0&1\\1&0\end{pmatrix}=\frac{1}{2}(\sigma_1),$$

$$(J_2)=\langle m\vert J_2\vert m'\rangle=\frac{1}{2}\begin{pmatrix}0&-i\\i&0\end{pmatrix}=\frac{1}{2}(\sigma_2),$$

$$(J_3)=\langle m\vert J_3\vert m'\rangle=\frac{1}{2}\begin{pmatrix}1&0\\0&-1\end{pmatrix}=\frac{1}{2}(\sigma_3). \tag{6.5}$$

One recognizes the Pauli matrices (σ_i). Applying J_2 twice gives

$$\langle m|(J_2)^2|m'\rangle = \sum_k \langle m|J_2|k\rangle\langle k|J_2|m'\rangle = \frac{1}{4}\begin{pmatrix}0 & -i\\ i & 0\end{pmatrix}\begin{pmatrix}0 & -i\\ i & 0\end{pmatrix} = \frac{1}{4}\begin{pmatrix}1 & 0\\ 0 & 1\end{pmatrix}. \tag{6.6}$$

The expansion (6.4) in matrix form is then equal to

$$\begin{aligned} d^{\frac{1}{2}}_{mm'}(\theta) &= \langle m|\mathrm{e}^{-iJ_2\theta}|m'\rangle \\ &= \begin{pmatrix}1 & 0\\ 0 & 1\end{pmatrix} - i\begin{pmatrix}0 & -i\\ i & 0\end{pmatrix}\frac{\theta}{2} - \frac{1}{2}\begin{pmatrix}1 & 0\\ 0 & 1\end{pmatrix}\left(\frac{\theta}{2}\right)^2 + \frac{i}{6}\begin{pmatrix}0 & -i\\ i & 0\end{pmatrix}\left(\frac{\theta}{2}\right)^3 + \ldots \\ &= \begin{pmatrix}1 - \frac{1}{2}\left(\frac{\theta}{2}\right)^2 + \ldots, & -\frac{\theta}{2} + \frac{1}{6}\left(\frac{\theta}{2}\right)^3 + \ldots \\ \frac{\theta}{2} - \frac{1}{6}\left(\frac{\theta}{2}\right)^3 + \ldots, & 1 - \frac{1}{2}\left(\frac{\theta}{2}\right)^2 + \ldots\end{pmatrix} = \begin{pmatrix}\cos\frac{\theta}{2} & -\sin\frac{\theta}{2}\\ \sin\frac{\theta}{2} & \cos\frac{\theta}{2}\end{pmatrix}. \end{aligned} \tag{6.7}$$

For $j = 1$ the matrix representation of SU(2) is given by 3-dimensional Pauli matrices which can be derived by using Table 6.2. For example, the matrix elements of J_2 are needed for rotations:

$$(J_2) \equiv \langle m|J_2|m'\rangle = \frac{1}{\sqrt{2}}\begin{pmatrix}0 & -i & 0\\ i & 0 & -i\\ 0 & i & 0\end{pmatrix}. \tag{6.8}$$

For the sake of completion we also list below the Wigner d matrices (6.3) for spins 1, $\frac{3}{2}$ and 2. They will be used later in the chapter on angular distribution. The following formula is useful for computations and to get higher spin matrices:

$$\begin{aligned} d^j_{mm'}(\theta) &= \sqrt{(j+m)!(j-m)!(j+m')!(j-m')!} \\ &\times \sum_\chi \frac{(-1)^{m-m'+\chi}}{(j+m'-\chi)!(j-m-\chi)(m-m'+\chi)!\chi!} \\ &\times \left(\cos\frac{\theta}{2}\right)^{2j+m'-m-2\chi} \times \left(\sin\frac{\theta}{2}\right)^{m-m'+2\chi}, \end{aligned} \tag{6.9}$$

where $\chi = 0, 1, 2\ldots$ and summation terms with negative factorials are ignored. The labels j and m stand for the spin and its projection, or equivalently for the isospin i and i_3. The d matrices satisfy the following symmetry relations:

$$\boxed{d^j_{m'm}(\theta) = (-1)^{m-m'} d^j_{mm'}(\theta) \quad \text{and} \quad d^j_{m'm}(\theta) = d^j_{-m-m'}(\theta)}\,. \tag{6.10}$$

Table 6.3 Wigner d-functions for spin $j = \frac{1}{2}, 1, \frac{3}{2}$ and 2

$$d^{\frac{1}{2}}_{mm'}(\theta) =$$

$$\begin{pmatrix} \cos\frac{\theta}{2} & -\sin\frac{\theta}{2} \\ \sin\frac{\theta}{2} & \cos\frac{\theta}{2} \end{pmatrix}$$

$$d^{1}_{mm'}(\theta) =$$

$$\begin{pmatrix} \frac{1+\cos\theta}{2} & -\frac{\sin\theta}{\sqrt{2}} & \frac{1-\cos\theta}{2} \\ \frac{\sin\theta}{\sqrt{2}} & \cos\theta & -\frac{\sin\theta}{\sqrt{2}} \\ \frac{1-\cos\theta}{2} & \frac{\sin\theta}{\sqrt{2}} & \frac{1+\cos\theta}{2} \end{pmatrix}$$

$$d^{\frac{3}{2}}_{mm'}(\theta) =$$

$$\begin{pmatrix} \frac{1+\cos\theta}{2}\cos\frac{\theta}{2} & -\sqrt{3}\left[\frac{1+\cos\theta}{2}\right]\sin\frac{\theta}{2} & \sqrt{3}\left[\frac{1-\cos\theta}{2}\right]\cos\frac{\theta}{2} & -\frac{1-\cos\theta}{2}\sin\frac{\theta}{2} \\ \sqrt{3}\left[\frac{1+\cos\theta}{2}\right]\sin\frac{\theta}{2} & \frac{3\cos\theta-1}{2}\cos\frac{\theta}{2} & -\frac{3\cos\theta+1}{2}\sin\frac{\theta}{2} & \sqrt{3}\left[\frac{1-\cos\theta}{2}\right]\cos\frac{\theta}{2} \\ \sqrt{3}\left[\frac{1-\cos\theta}{2}\right]\cos\frac{\theta}{2} & \frac{3\cos\theta+1}{2}\sin\frac{\theta}{2} & \frac{3\cos\theta-1}{2}\cos\frac{\theta}{2} & -\sqrt{3}\left[\frac{1+\cos\theta}{2}\right]\sin\frac{\theta}{2} \\ \frac{1-\cos\theta}{2}\sin\frac{\theta}{2} & \sqrt{3}\left[\frac{1-\cos\theta}{2}\right]\cos\frac{\theta}{2} & \sqrt{3}\left[\frac{1+\cos\theta}{2}\right]\sin\frac{\theta}{2} & \frac{1+\cos\theta}{2}\cos\frac{\theta}{2} \end{pmatrix}$$

$$d^{2}_{mm'}(\theta) =$$

$$\begin{pmatrix} =\left[\frac{1+\cos\theta}{2}\right]^2 & -\frac{1+\cos\theta}{2}\sin\theta & \frac{\sqrt{6}}{4}\sin^2\theta & -\frac{1-\cos\theta}{2}\sin\theta & \left[\frac{1-\cos\theta}{2}\right]^2 \\ \frac{1+\cos\theta}{2}\sin\theta & \frac{1+\cos\theta}{2}(2\cos\theta-1) & -\sqrt{\frac{3}{2}}\sin\theta\cos\theta & \frac{1-\cos\theta}{2}(2\cos\theta+1) & -\frac{1-\cos\theta}{2}\sin\theta \\ \frac{\sqrt{6}}{4}\sin^2\theta & \sqrt{\frac{3}{2}}\sin\theta\cos\theta & \frac{3}{2}\cos^2\theta-\frac{1}{2} & -\sqrt{\frac{3}{2}}\sin\theta\cos\theta & \frac{\sqrt{6}}{4}\sin^2\theta \\ \frac{1-\cos\theta}{2}\sin\theta & \frac{1-\cos\theta}{2}(2\cos\theta+1) & \sqrt{\frac{3}{2}}\sin\theta\cos\theta & \frac{1+\cos\theta}{2}(2\cos\theta-1) & -\frac{1+\cos\theta}{2}\sin\theta \\ \left[\frac{1-\cos\theta}{2}\right]^2 & \frac{1-\cos\theta}{2}\sin\theta & \frac{\sqrt{6}}{4}\sin^2\theta & \frac{1+\cos\theta}{2}\sin\theta & \left[\frac{1+\cos\theta}{2}\right]^2 \end{pmatrix}$$

The indices m and m' are understood to decrease from left to right and from top to bottom. For example, for $j = 1$ the matrix elements are

$$\begin{pmatrix} d^1_{11}(\theta) & d^1_{10}(\theta) & d^1_{1-1}(\theta) \\ d^1_{01}(\theta) & d^1_{00}(\theta) & d^1_{0-1}(\theta) \\ d^1_{-11}(\theta) & d^1_{-10}(\theta) & d^1_{-1-1}(\theta) \end{pmatrix}. \tag{6.11}$$

The d-functions are listed in Table 6.3 for $j \leq 2$.

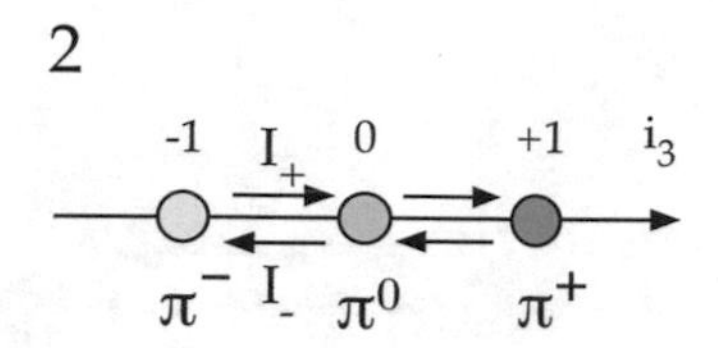

Fig. 6.1 Weight diagram of the pion (3-dimensional representation of SU(2)). The ladder operator I_+ (I_-) increases (decreases) i_3 by one unit

As a first example, let us apply SU(2) rotations to the isospinors of the pion. Among the three generators of SU(2), only I_3 is diagonal (corresponding to i_3 conservation). The operator I_3 is represented by the matrix

$$(I_3) \equiv \langle m|I_3|m'\rangle = \begin{pmatrix} 1 & 0 & 0 \\ 0 & 0 & 0 \\ 0 & 0 & -1 \end{pmatrix}, \tag{6.12}$$

see also Problem 6.1. Figure 6.1 shows the eigenvalues i_3 of I_3 (weight diagram of the 3-dimensional representation of SU(2)). Let us now perform a rotation by 180° about the isospin y-axis. We need the 3-dimensional representation of rotations applied on the π^+ isospinor. Thus, according to Table 6.3,

$$d^1_{mm'}|\pi^+\rangle = \begin{pmatrix} 0 & 0 & 1 \\ 0 & -1 & 0 \\ 1 & 0 & 0 \end{pmatrix} \begin{pmatrix} 1 \\ 0 \\ 0 \end{pmatrix} = \begin{pmatrix} 0 \\ 0 \\ 1 \end{pmatrix} = |\pi^-\rangle, \tag{6.13}$$

and similarly for $|\pi^-\rangle$: $d^1_{mm'}|\pi^-\rangle = |\pi^+\rangle$. For the π^0 one gets the opposite sign:

$$d^1_{mm'}|\pi^0\rangle = \begin{pmatrix} 0 & 0 & 1 \\ 0 & -1 & 0 \\ 1 & 0 & 0 \end{pmatrix} \begin{pmatrix} 0 \\ 1 \\ 0 \end{pmatrix} = -\begin{pmatrix} 0 \\ 1 \\ 0 \end{pmatrix} = -|\pi^0\rangle. \tag{6.14}$$

The last relation leads to a negative G parity for the π^0, since $C(\pi^0) = +1$. Charged pions have also negative G parity and therefore the plus sign in (6.13) requires that

$$C|\pi^\pm\rangle = -|\pi^\mp\rangle. \tag{6.15}$$

6.2 Isospinors of Quark and Antiquark

As an application of SU(2) let us derive expressions for the isospinors of the u and d quarks and their antiquark partners. The goal is to symmetrize the wavefunctions of the light mesons listed in Table 5.1 and to make the $i = 0$ and 1 wavefunctions

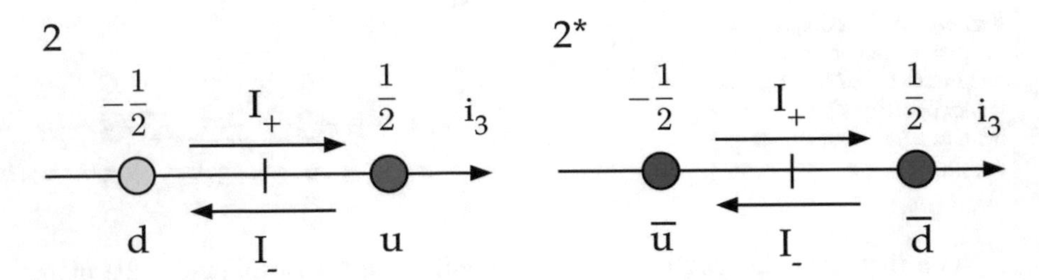

Fig. 6.2 Weight diagrams of the fundamental representation 2 of SU(2) and of the conjugate representation 2* for antiquarks

eigenstates of the G parity. Figure 6.2 shows the weight diagram of the fundamental representation of SU(2), and that of the conjugate representation. Let us apply a passive rotation about the y-axis of the quark isospinor with components u and d. According to Table 6.3 with $\theta \to -\theta$

$$\begin{pmatrix} u' \\ d' \end{pmatrix} = \begin{pmatrix} \cos\frac{\theta}{2} & \sin\frac{\theta}{2} \\ -\sin\frac{\theta}{2} & \cos\frac{\theta}{2} \end{pmatrix} \begin{pmatrix} u \\ d \end{pmatrix}, \tag{6.16}$$

hence

$$\begin{aligned} u' &= u\cos\frac{\theta}{2} + d\sin\frac{\theta}{2}, \\ d' &= -u\sin\frac{\theta}{2} + d\cos\frac{\theta}{2}. \end{aligned} \tag{6.17}$$

Charge conjugation transforms a quark q into its antiquark $\overline{q}$ with the definition

$$C|q\rangle = +|\overline{q}\rangle. \tag{6.18}$$

Charge conjugation $\overline{u} = Cu$ and $\overline{d} = Cd$ flips the sign of i_3 (Fig. 6.2). Reversing the order of the Eq. (6.17) and applying C gives

$$\begin{aligned} \overline{d}' &= -\overline{u}\sin\frac{\theta}{2} + \overline{d}\cos\frac{\theta}{2}, \\ \overline{u}' &= \overline{u}\cos\frac{\theta}{2} + \overline{d}\sin\frac{\theta}{2}, \end{aligned} \tag{6.19}$$

or in matrix form,

$$\begin{pmatrix} \overline{d}' \\ -\overline{u}' \end{pmatrix} = \begin{pmatrix} \cos\frac{\theta}{2} & +\sin\frac{\theta}{2} \\ -\sin\frac{\theta}{2} & \cos\frac{\theta}{2} \end{pmatrix} \begin{pmatrix} \overline{d} \\ -\overline{u} \end{pmatrix}. \tag{6.20}$$

The isospinor of the antiquark with components $(\overline{d}, -\overline{u})$ has the same transformation properties as those of the quark with components (u, d). Therefore we shall label the isospinor of the $\overline{u}$ quark as $|-\overline{u}\rangle$ (or $-|\overline{u}\rangle$).

Let us now construct the flavour wavefunctions of SU(2), first for the fundamental and conjugate representations. From the matrix elements of Table 6.2 one gets by applying the ladder operators I_+ and I_-

$$I_+|d\rangle = |u\rangle\langle u|I_+|d\rangle = |u\rangle, \quad I_-|u\rangle = |d\rangle\langle d|I_-|u\rangle = |d\rangle, \tag{6.21}$$

since $\langle d|I_+|d\rangle = \langle u|I_-|u\rangle = 0$. For the antiquarks one gets with the matrix representations (I_+) and (I_-) from Table 6.2

$$I_-|\overline{d}\rangle = \begin{pmatrix} 0 & 0 \\ 1 & 0 \end{pmatrix}\begin{pmatrix} 1 \\ 0 \end{pmatrix} = \begin{pmatrix} 0 \\ 1 \end{pmatrix} = -|\overline{u}\rangle \tag{6.22}$$

and

$$I_+|\overline{u}\rangle = \begin{pmatrix} 0 & 1 \\ 0 & 0 \end{pmatrix}\begin{pmatrix} 0 \\ -1 \end{pmatrix} = \begin{pmatrix} -1 \\ 0 \end{pmatrix} = -|\overline{d}\rangle\,. \tag{6.23}$$

The other combinations vanish, e.g. $I_+|u\rangle = I_-|\overline{u}\rangle = 0$, etc. For the isovectors, e.g. for the positive pion, operating with I_- on $|u\overline{d}\rangle$ gives

$$I_-|\pi^+\rangle = I_-|u\overline{d}\rangle = |d\overline{d} - u\overline{u}\rangle. \tag{6.24}$$

On the other hand, Table 6.2 gives with $i = 1$, $m' = 1$ and $m = 0$

$$I_-|\pi^+\rangle = \sqrt{2}|\pi^0\rangle \tag{6.25}$$

and therefore the π^0 isospinor reads

$$|\pi^0\rangle = \frac{1}{\sqrt{2}}|d\overline{d} - u\overline{u}\rangle. \tag{6.26}$$

Applying I_- a second time leads to the negative pion:

$$I_-|\pi^0\rangle = \sqrt{2}|\pi^-\rangle = \frac{1}{\sqrt{2}}I_-|d\overline{d} - u\overline{u}\rangle = \frac{1}{\sqrt{2}}|-d\overline{u} - d\overline{u}\rangle, \tag{6.27}$$

hence $|\pi^-\rangle = -|d\overline{u}\rangle$ (note the minus sign).[2]

[2] The components of the antiquark isospinor are also written $(-\overline{d}, \overline{u})$ in the literature. The π^+, π^0 and π^- wavefunctions are then given by $-|u\overline{d}\rangle$, $\frac{1}{\sqrt{2}}|u\overline{u} - d\overline{d}\rangle$ and $+|d\overline{u}\rangle$, respectively.

We now proceed to the symmetrization of these wavefunctions. The G parity operation (3.18) on the u and d quarks gives

$$C\begin{pmatrix} 0 & 1 \\ -1 & 0 \end{pmatrix}\begin{pmatrix} u \\ d \end{pmatrix} = \begin{pmatrix} \bar{d} \\ -\bar{u} \end{pmatrix} \text{ and } C\begin{pmatrix} 0 & 1 \\ -1 & 0 \end{pmatrix}\begin{pmatrix} \bar{d} \\ -\bar{u} \end{pmatrix} = \begin{pmatrix} -u \\ -d \end{pmatrix}, \tag{6.28}$$

leading to the transformations

$$\boxed{Gu = \bar{d},\ G\bar{u} = d,\ Gd = -\bar{u},\ G\bar{d} = -u}\,. \tag{6.29}$$

Hence G reverses the sign when applied to a d or $\bar{d}$ quark. We also find that $G^2u = -u$ and $G^2d = -d$, due to the fact that a rotation by $2 \times 180°$ flips the sign of the isospinor. For example, the following combinations are eigenstates of the G parity:

$$G(|u\bar{d}\rangle + |\bar{d}u\rangle) = -(|u\bar{d}\rangle + |\bar{d}u\rangle), \tag{6.30}$$

$$G(|u\bar{d}\rangle - |\bar{d}u\rangle) = +(|u\bar{d}\rangle - |\bar{d}u\rangle), \tag{6.31}$$

with negative and positive G parities, respectively. We have seen that the pion has negative G parity. Therefore, the symmetrized SU(2) wavefunctions are obtained by adding the permuted pairs. With proper normalisation,

$$|\pi^+\rangle = \frac{1}{\sqrt{2}}(|u\bar{d}\rangle \oplus |\bar{d}u\rangle). \tag{6.32}$$

We have encircled the plus sign to emphasize that the isospin wavefunctions are symmetric under permutations of the quark and antiquark. The wavefunction of the negative pion is

$$|\pi^-\rangle = -\frac{1}{\sqrt{2}}(|d\bar{u}\rangle \oplus |\bar{u}d\rangle). \tag{6.33}$$

For the neutral pion it is easy to verify that the orthogonal and normalized function

$$|\pi^0\rangle = \frac{1}{2}(|d\bar{d} - u\bar{u}\rangle \oplus |\bar{d}d - \bar{u}u\rangle) \tag{6.34}$$

also has negative G parity. On the other hand, for a spin-0 meson the spin wavefunction is antisymmetric, see (2.30), and for a pseudoscalar ($\ell = 0$) the orbital wavefunction Φ is symmetric. Hence the total wavefunction is antisymmetric, e.g.

$$|\pi^+\rangle = \Phi(\ell = 0) \cdot \frac{1}{\sqrt{2}}(|u\bar{d}\rangle + |\bar{d}u\rangle) \cdot \frac{1}{\sqrt{2}}(|\uparrow\downarrow\rangle - |\downarrow\uparrow\rangle). \tag{6.35}$$

In contrast, the ρ mesons have positive G parity. The isospin wavefunctions are antisymmetric and read

$$|\rho^+\rangle = \frac{1}{\sqrt{2}}(|u\bar{d}\rangle \ominus |\bar{d}u\rangle),$$
$$|\rho^-\rangle = -\frac{1}{\sqrt{2}}(|d\bar{u}\rangle \ominus |\bar{d}u\rangle),$$
$$|\rho^0\rangle = \frac{1}{2}(|d\bar{d} - u\bar{u}\rangle \ominus |\bar{d}d - \bar{u}u\rangle). \tag{6.36}$$

The spin-1 wavefunctions are symmetric, see (2.29). Hence the total wavefunctions are again antisymmetric, e.g.

$$|\rho^+\rangle = \Phi(\ell = 0) \quad \cdot \frac{1}{\sqrt{2}}(|u\bar{d}\rangle - |\bar{d}u\rangle) \quad \cdot \begin{bmatrix} |\uparrow\uparrow\rangle \\ \frac{1}{\sqrt{2}}(|\uparrow\downarrow\rangle + |\downarrow\uparrow\rangle) \\ |\downarrow\downarrow\rangle \end{bmatrix}. \tag{6.37}$$

Table 6.4 shows the symmetrized flavour wavefunctions of the pseudoscalar and vector mesons for the $i = 1$ and $i = 0$ isospin multiplets. For completeness we also list the wavefunctions of the two kaon isodoublets which are discussed below. We will see in Chap. 7 that the SU(2) multiplets can in turn be grouped into nonets belonging to the higher symmetry group SU(3). The 0^{-+} SU(3) wavefunctions are symmetric, the 1^{--} ones antisymmetric. The $i = 1$ wavefunctions have negative G parities for 0^{-+} and positive G parities for 1^{--}. The $i = 0$ octet $|8\rangle$ and singlet $|1\rangle$

Table 6.4 Symmetrized flavour wavefunctions of the light-quark ground state $^1S_0$ and $^3S_1$ mesons

i	$1^1S_0\ (0^{-+})$			$1^3S_1\ (1^{--})$		
1	$\|\pi^+\rangle$	=	$\frac{1}{\sqrt{2}}\|u\bar{d} + \bar{d}u\rangle$	$\|\rho^+\rangle$	=	$\frac{1}{\sqrt{2}}\|u\bar{d} - \bar{d}u\rangle$
	$\|\pi^0\rangle$	=	$\frac{1}{2}(\|d\bar{d} - u\bar{u}\rangle + \|\bar{d}d - \bar{u}u\rangle)$	$\|\rho^0\rangle$	=	$\frac{1}{2}(\|d\bar{d} - u\bar{u}\rangle - \|\bar{d}d - \bar{u}u\rangle)$
	$\|\pi^-\rangle$	=	$-\frac{1}{\sqrt{2}}\|d\bar{u} + \bar{u}d\rangle$	$\|\rho^-\rangle$	=	$-\frac{1}{\sqrt{2}}\|d\bar{u} - \bar{u}d\rangle$
0	$\|8\rangle$	=	$\frac{1}{2\sqrt{3}}(\|u\bar{u} + d\bar{d} - 2s\bar{s}\rangle$ $+\|\bar{u}u + \bar{d}d - 2\bar{s}s\rangle)$	$\|8\rangle$	=	$\frac{1}{2\sqrt{3}}(\|u\bar{u} + d\bar{d} - 2s\bar{s}\rangle$ $-\|\bar{u}u + \bar{d}d - 2\bar{s}s\rangle)$
	$\|1\rangle$	=	$\frac{1}{\sqrt{6}}(\|u\bar{u} + d\bar{d} + s\bar{s}\rangle$ $+\|\bar{u}u + \bar{d}d + \bar{s}s\rangle)$	$\|1\rangle$	=	$\frac{1}{\sqrt{6}}(\|u\bar{u} + d\bar{d} + s\bar{s}\rangle$ $-\|\bar{u}u + \bar{d}d + \bar{s}s\rangle)$
$\frac{1}{2}$	$\|K^+\rangle$	=	$\frac{1}{\sqrt{2}}\|u\bar{s} + \bar{s}u\rangle$	$\|K^{*+}\rangle$	=	$\frac{1}{\sqrt{2}}\|u\bar{s} - \bar{s}u\rangle$
	$\|K^0\rangle$	=	$\frac{1}{\sqrt{2}}\|d\bar{s} + \bar{s}d\rangle$	$\|K^{*0}\rangle$	=	$\frac{1}{\sqrt{2}}\|d\bar{s} - \bar{s}d\rangle$
$\frac{1}{2}$	$\|\overline{K}^0\rangle$	=	$-\frac{1}{\sqrt{2}}\|s\bar{d} + \bar{d}s\rangle$	$\|\overline{K}^{*0}\rangle$	=	$-\frac{1}{\sqrt{2}}\|s\bar{d} - \bar{d}s\rangle$
	$\|K^-\rangle$	=	$-\frac{1}{\sqrt{2}}\|s\bar{u} + \bar{u}s\rangle$	$\|K^{*-}\rangle$	=	$-\frac{1}{\sqrt{2}}\|s\bar{u} - \bar{u}s\rangle$

wavefunctions will be derived in Chap. 7. According to (4.4) the opposite holds for them: positive G parity for 0^{-+} and negative G parity for 1^{--}.

Let us now add the strange quark. The kaons build two doublets (K^+, K^0) and ($\overline{K}^0$, K^-) with isospinors

$$|K\rangle = \begin{pmatrix} u\overline{s} \\ d\overline{s} \end{pmatrix}, |\overline{K}\rangle = \begin{pmatrix} s\overline{d} \\ -s\overline{u} \end{pmatrix}. \tag{6.38}$$

The minus sign for the K^- takes into account the transformation properties (6.20) of an antidoublet. Applying the ladder operators gives

$$\begin{aligned}
I_+|K^0\rangle &= \begin{pmatrix} 0\,1 \\ 0\,0 \end{pmatrix}\begin{pmatrix} 0 \\ 1 \end{pmatrix} = \begin{pmatrix} 1 \\ 0 \end{pmatrix} = |K^+\rangle, \\
I_-|K^+\rangle &= \begin{pmatrix} 0\,0 \\ 1\,0 \end{pmatrix}\begin{pmatrix} 1 \\ 0 \end{pmatrix} = \begin{pmatrix} 0 \\ 1 \end{pmatrix} = |K^0\rangle, \\
I_+|K^-\rangle &= \begin{pmatrix} 0\,1 \\ 0\,0 \end{pmatrix}\begin{pmatrix} 0 \\ -1 \end{pmatrix} = \begin{pmatrix} -1 \\ 0 \end{pmatrix} = -|\overline{K}^0\rangle, \\
I_-|\overline{K}^0\rangle &= \begin{pmatrix} 0\,0 \\ 1\,0 \end{pmatrix}\begin{pmatrix} 1 \\ 0 \end{pmatrix} = \begin{pmatrix} 0 \\ 1 \end{pmatrix} = -|K^-\rangle.
\end{aligned} \tag{6.39}$$

The kaons are not eigenstates of the G parity. For the K^+ we choose the symmetrized isospinor in Table 6.4 with the overall plus sign. The isospinor of the K^0 is then obtained by applying I_- on the u quark with $I_-|u\rangle = |d\rangle$. The sign of the C parity operation is arbitrary. We choose the minus sign, $C|K^0\rangle = -|\overline{K}^0\rangle$,[3] which fixes with $C|d\rangle = |\overline{d}\rangle$ the overall sign of $|\overline{K}^0\rangle$ in Table 6.4, and with (6.39) that of K^-. With the transformations (6.29) we can derive the G parity transformations of the kaons, for example,

$$|K^+\rangle = \frac{1}{\sqrt{2}}|u\overline{s} + \overline{s}u\rangle \underbrace{\Rightarrow}_{G} \frac{1}{\sqrt{2}}|\overline{d}s + s\overline{d}\rangle = -|\overline{K}^0\rangle. \tag{6.40}$$

The kaon states transform under G as

$$G|K^+\rangle = -|\overline{K}^0\rangle,\ G|K^0\rangle = +|K^-\rangle,\ G|\overline{K}^0\rangle = +|K^+\rangle,\ G|K^-\rangle = -|K^0\rangle. \tag{6.41}$$

Under C they transform as

$$C|K^0\rangle = -|\overline{K}^0\rangle,\ C|\overline{K}^0\rangle = -|K^0\rangle,\ C|K^\pm\rangle = -|K^\mp\rangle. \tag{6.42}$$

[3]The $CP = +1$ eigenstate is then $K_1 = \frac{1}{\sqrt{2}}(|K^0\rangle + |\overline{K}^0\rangle)$.

For the K^* mesons it is easy to show with Table 6.4 that G and C lead to the opposite signs:

$$G|K^{*+}\rangle = +|\overline{K}^{*0}\rangle,\ G|K^{*0}\rangle = -|K^{*-}\rangle,\ G|\overline{K}^{*0}\rangle = -|K^{*+}\rangle,\ G|K^{*-}\rangle = +|K^{*0}\rangle, \tag{6.43}$$

and

$$C|K^{*0}\rangle = +|\overline{K}^{*0}\rangle,\ C|\overline{K}^{*0}\rangle = +|K^{*0}\rangle,\ C|K^{*\pm}\rangle = +|K^{*\mp}\rangle. \tag{6.44}$$

The eigenstates of G are obtained by symmetrizing the $\overline{K}K$ superpositions

$$\frac{1}{\sqrt{2}}(|\overline{K}^0K^0\rangle + |K^-K^+\rangle) \text{ and } \frac{1}{\sqrt{2}}(|\overline{K}^0K^0\rangle - |K^-K^+\rangle). \tag{6.45}$$

which correspond to the $i = 0$ and $i = 1$ eigenstates, respectively (Problem 6.2). Depending on the angular momentum ℓ carried by the $\overline{K}K$ pair, the eigenvalue is given by $G = (-1)^{i+\ell}$. For $\ell = 0$ (hence positive G-parity for $i = 0$ and negative G-parity for $i = 1$) the eigenstates of G are obtained by adding to (6.45) the charge conjugated states. This is for example the case for the $J^P = 0^+$ isoscalar $f_0(980)$ and the isovector $a_0(980)$ which have the (not yet symmetrized) eigenfunctions (6.45), respectively.[4]

In contrast, for the $\ell = 1$ (hence negative G-parity for $i = 0$ and positive G-parity for $i = 1$) the charge conjugated states must be subtracted from (6.45), see Problem 6.2.

6.3 Young Tableaux

We have constructed the isospin wavefunctions in Table 6.4 with the help of the ladder operators $I_\pm$. For isospin $\frac{1}{2}$ the fundamental representation 2 is combined with its conjugate 2*, leading to an isotriplet and an isosinglet. This is illustrated in Fig. 6.3 for the vector ground states, where the ρ^0 and ω have both $i_3 = 0$. The isospin wavefunction of the $(i = 0)$ ω cannot be derived from those of the $(i = 1)$ ρ with ladder operators, because we are dealing with two irreducible representations $2 \times 2^* = 1 + 3$. The same applies to spin-$\frac{1}{2}$ where the fundamental representation 2 is combined with itself: the wavefunctions (2.29) and (2.30) belong to two irreducible representations, a spin triplet and a spin singlet, respectively.

[4]This is true in the absence of isospin mixing. The decays of both mesons into $\overline{K}K$ induces isospin breaking transitions from one state to the other via $\overline{K}K$ loops. There is experimental evidence that the mixing angle differs significantly from the $\cos 45° = \frac{1}{\sqrt{2}}$ assumed in the superpositions (6.45) [1].

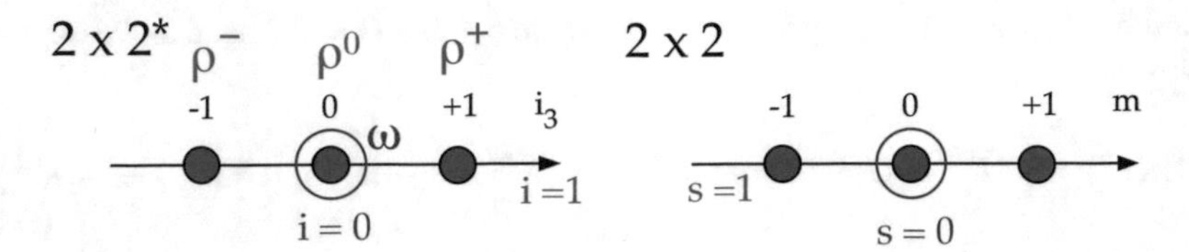

Fig. 6.3 Weight diagram of SU(2) × SU(2) for isospin $\frac{1}{2}$ (left) and for spin $\frac{1}{2}$ (right). The states $i_3 = 0$ and $m = 0$ are doubly occupied

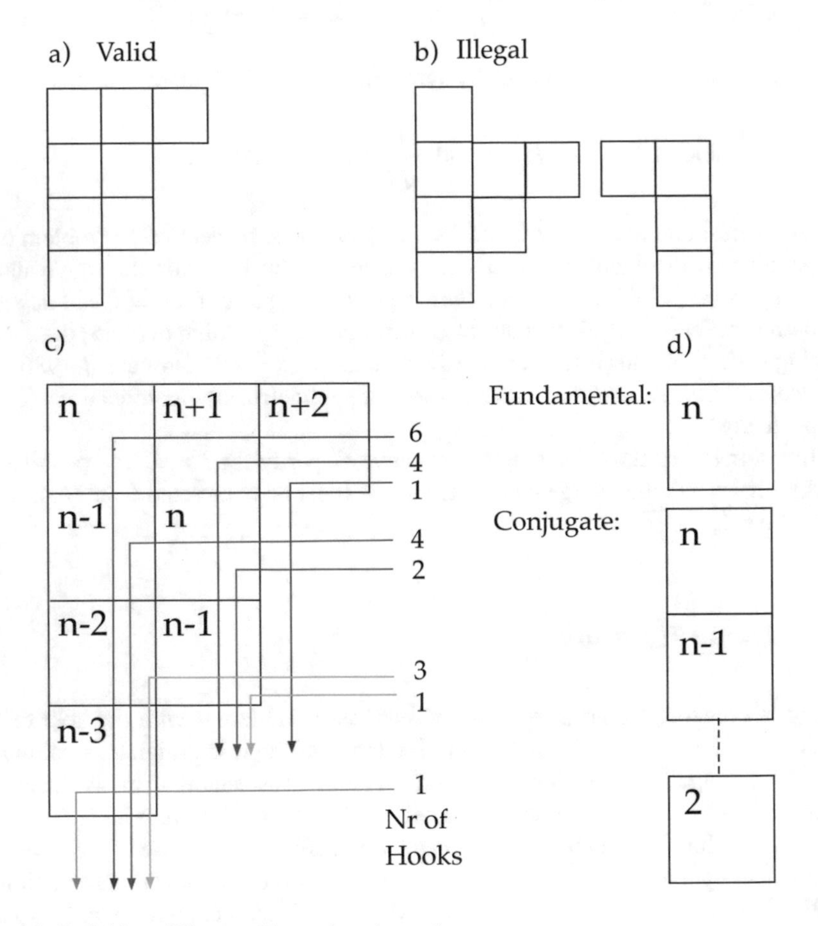

Fig. 6.4 Definition of Young tableaux (see the text)

The method of the Young tableaux is a convenient way to find the multiplet structure of SU(2)× SU(2), SU(2)× SU(2) × SU(2), etc., as well as couplings of higher symmetry groups SU(n) which will be needed in the following chapters. The procedure has been described in [2] and we shall give the recipe here without proof.

The definitions are shown in Fig. 6.4. To be valid a Young tableau should be left and top rectified (a). Tableaux such as the one shown in (b) are not allowed. The

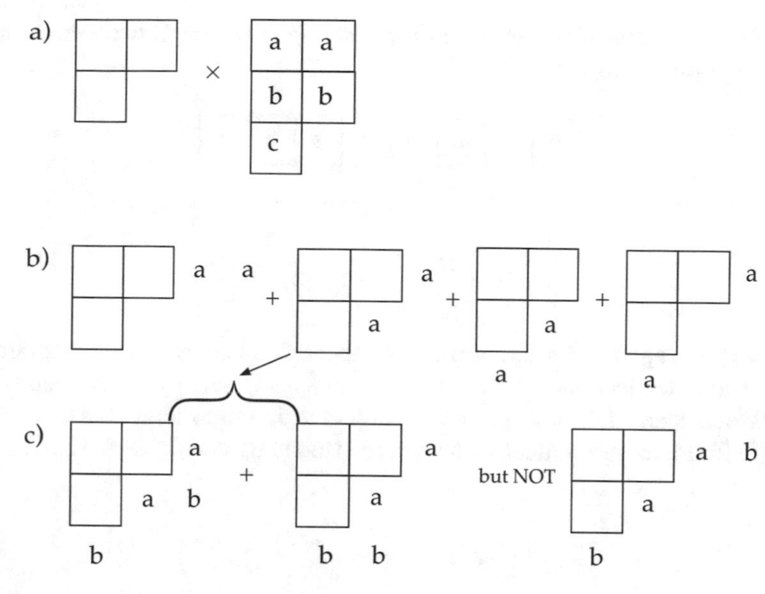

Fig. 6.5 (**a**) Coupling of two Young tableaux; (**b**) adding the a's; (**c**) adding the b's to the second tableau in (**b**), see also the text

dimension of a tableau is calculated as follows: first introduce the dimension n of SU(n) into the top left box and then subtract one unit in each box while descending to the bottom of the tableau (Fig. 6.4c). Add one unit in each box while moving from left to right. The number in the box is called the "value". Now draw lines crossing all boxes, starting from the right-hand side and moving to the bottom. A "hook" is defined as the number of crossed boxes. The dimension of the tableau is given by the product of values divided by the product of hooks. Figure 6.4d shows the tableau of the fundamental representation n (dimension n) and of the conjugated representation n^* with dimension equal to $n(n-1)\ldots2/(n-1)(n-2)\ldots1 = n$.

The couplings of two SU(n) groups is explained in Fig. 6.5. First insert a series of a's, b's, c's in each row of the second tableau (Fig. 6.5a). Then complete the first tableau from the right by the a's to obtain a valid tableau with no more than one a in each column, as shown in Fig. 6.5b. Repeat the procedure with b's, but there should not be more b's than a's when counting from right lo left and from top to bottom. An example is shown in (c) when using the second tableau in (b). Repeat with c's and then calculate the dimensions.

Let us illustrate the procedure by a simple example and couple two spin-$\frac{1}{2}$ quarks. The tableau for SU(2) $\times$ SU(2) is

$$\begin{array}{|c|}\hline 2 \\ \hline \end{array} \times \begin{array}{|c|}\hline 2 \\ \hline \end{array} = \begin{array}{|c|}\hline 2 \\ \hline 1 \\ \hline \end{array} + \begin{array}{|c|c|}\hline 2 & 3 \\ \hline \end{array}$$

$$= \frac{2}{2} + \frac{6}{3} = 1 + 3, \tag{6.46}$$

giving a spin singlet and a spin triplet, as expected. Since the Young diagrams for 2 and 2* are identical, this decomposition also applies to a $q\overline{q}$ meson made of the light flavours u and d. The next example shows the decomposition of SU(2)$\times$ SU(2) $\times$ SU(2) for three spin-$\frac{1}{2}$ quarks which is relevant to baryons in Sect. 13.2:

$$\begin{array}{|c|}\hline 2 \\ \hline \end{array} \times \begin{array}{|c|}\hline 2 \\ \hline \end{array} \times \begin{array}{|c|}\hline 2 \\ \hline \end{array} = \left(\begin{array}{|c|}\hline \\ \hline \\ \hline \end{array} + \begin{array}{|c|c|}\hline & \\ \hline \end{array} \right) \times \begin{array}{|c|}\hline \\ \hline \end{array}$$

$$= \begin{array}{|c|}\hline 2 \\ \hline 1 \\ \hline 0 \\ \hline \end{array} + \begin{array}{|c|c}\hline 2 & \multicolumn{1}{c|}{3} \\ \hline 1 & \\ \cline{1-1} \end{array} + \begin{array}{|c|c}\hline 2 & \multicolumn{1}{c|}{3} \\ \hline 1 & \\ \cline{1-1} \end{array} + \begin{array}{|c|c|c|}\hline 2 & 3 & 4 \\ \hline \end{array}$$

$$= 0 + \frac{6}{3} + \frac{6}{3} + \frac{24}{6} = 2 + 2 + 4, \tag{6.47}$$

giving two doublets and one quadruplet. See Problems 10.1 and 10.2 for further examples.

The connection between a Young tableau and the corresponding weight diagram is explained in detail in Ref. [3] and in Appendix D for the most interesting case $n = 3$.

References

1. Close, F.E., Kirk, A.: Phys. Lett. B 515, 13 (2001)
2. Hammermesh, M.: Group Theory and Its Application to Physical Problems. Addison-Wesley, Reading (1962)
3. Tanabashi, M., et al. (Particle Data Group): Phys. Rev. D 98, 030001 (2018)

Chapter 7
SU(3)

7.1 Fundamental Representation

Let us now extend the isospin symmetry group to the three light quarks u, d and s. Apart from I_3, strangeness S is also a conserved quantum number in strong interactions. It is often more convenient to use the hypercharge Y as a conserved quantity, with quantum number y related to S through (1.4):

$$y = B + S = \frac{1}{3} + S. \tag{7.1}$$

We are therefore seeking a group of unitary operators which commute with the Hamiltonian, but for which two of the generators are simultaneously diagonal:

$$[U, H] = 0 \Rightarrow [I_3, H] = 0, \quad [Y, H] = 0 \text{ and } [I_3, Y] = 0. \tag{7.2}$$

The symmetry group is SU(3) with the unitary operators

$$U = e^{i \sum_{i=1}^{8} G_i \alpha_i}, \tag{7.3}$$

where the eight generators G_i obey the commutation relations

$$\boxed{[G_i, G_j] = i f_{ijk} G_k}. \tag{7.4}$$

The structure constants f_{ijk} are given in Table 7.1. Summing over k, which occurs for the commutators $[G_4, G_5]$ and $[G_6, G_7]$ is implicitly assumed. For example,

$$[G_4, G_5] = \frac{i}{2} G_3 + i \frac{\sqrt{3}}{2} G_8. \tag{7.5}$$

© The Author(s) 2018

C. Amsler, *The Quark Structure of Hadrons*, Lecture Notes in Physics 949,
https://doi.org/10.1007/978-3-319-98527-5_7

Table 7.1 SU(3) structure constants

ijk	123	147	156	246	257	345	367	458	678
f_{ijk}	1	$\frac{1}{2}$	$-\frac{1}{2}$	$\frac{1}{2}$	$\frac{1}{2}$	$\frac{1}{2}$	$-\frac{1}{2}$	$\frac{\sqrt{3}}{2}$	$\frac{\sqrt{3}}{2}$

The tensor f_{ijk} is antisymmetric. Combinations with repeated indices vanish

SU(3) symmetry means that all masses in the multiplets are equal, which is not the case, the s quark being much heavier than the u and d quarks. Nonetheless this broken flavour symmetry, SU(3)$_f$, leads to useful relations within hadron multiplets. In Chap. 10 we will discuss SU(3) applied to colour, SU(3)$_c$, which is an exact symmetry.

The fundamental representation of SU(3) is described by the following 3×3 matrices which fulfil the commutation rules (7.4) (the proof is left a simple exercise):

$$(G_1) = \frac{1}{2}\begin{pmatrix} 0 & 1 & 0 \\ 1 & 0 & 0 \\ 0 & 0 & 0 \end{pmatrix}, \quad (G_2) = \frac{1}{2}\begin{pmatrix} 0 & -i & 0 \\ i & 0 & 0 \\ 0 & 0 & 0 \end{pmatrix}, \quad (G_3) = \frac{1}{2}\begin{pmatrix} 1 & 0 & 0 \\ 0 & -1 & 0 \\ 0 & 0 & 0 \end{pmatrix},$$

$$(G_4) = \frac{1}{2}\begin{pmatrix} 0 & 0 & 1 \\ 0 & 0 & 0 \\ 1 & 0 & 0 \end{pmatrix}, \quad (G_5) = \frac{1}{2}\begin{pmatrix} 0 & 0 & -i \\ 0 & 0 & 0 \\ i & 0 & 0 \end{pmatrix}, \quad (G_6) = \frac{1}{2}\begin{pmatrix} 0 & 0 & 0 \\ 0 & 0 & 1 \\ 0 & 1 & 0 \end{pmatrix},$$

$$(G_7) = \frac{1}{2}\begin{pmatrix} 0 & 0 & 0 \\ 0 & 0 & -i \\ 0 & i & 0 \end{pmatrix}, \quad (G_8) = \frac{1}{2\sqrt{3}}\begin{pmatrix} 1 & 0 & 0 \\ 0 & 1 & 0 \\ 0 & 0 & -2 \end{pmatrix}, \tag{7.6}$$

with $(G_i) \equiv \frac{1}{2}(\lambda_i)$, where the Gell-Mann matrices (λ_i) replace the Pauli matrices of SU(2). The diagonal operators are I_3 and Y with

$$I_3 \equiv G_3 \quad \text{and} \quad Y \equiv \frac{2}{\sqrt{3}} G_8. \tag{7.7}$$

The quantum numbers i_3 and y for the three light quarks can be read off the diagonal elements of the corresponding matrices (I_3) and (Y). The weight diagram of the fundamental representation is shown in Fig. 7.1 (left). The SU(2) matrices (I_1) and (I_2) (6.5) are embedded in G_1 and G_2, and occur also in G_4, G_5 (first and third rows and columns) and in G_6 ,G_7 (second and third rows and columns): SU(3) therefore includes three SU(2) subgroups, those of the I-spin, and the so-called V-spin and U-spin, respectively. The corresponding ladder operators consist of our already familiar

$$\boxed{I_\pm = G_1 \pm i G_2}, \tag{7.8}$$

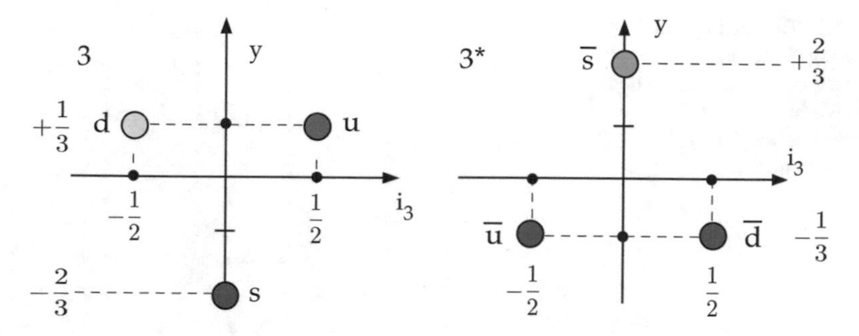

Fig. 7.1 Left: weight diagram of the fundamental representation (3) of SU(3). Right: weight diagram of the conjugate representation (3*)

plus the new ones

$$\boxed{V_{\pm} = G_4 \pm iG_5}\,, \quad \boxed{U_{\pm} = G_6 \pm iG_7}\,. \tag{7.9}$$

The generators G_i are Hermitian ($G_i^{\dagger} = G_i$) and therefore

$$I_{\pm} = I_{\mp}^{\dagger}, \quad V_{\pm} = V_{\mp}^{\dagger}, \quad U_{\pm} = U_{\mp}^{\dagger}. \tag{7.10}$$

The three flavour vectors

$$u = \begin{pmatrix} 1 \\ 0 \\ 0 \end{pmatrix}, \quad d = \begin{pmatrix} 0 \\ 1 \\ 0 \end{pmatrix}, \quad s = \begin{pmatrix} 0 \\ 0 \\ 1 \end{pmatrix} \tag{7.11}$$

are eigenstates of the matrices (I_3) and (Y), see (7.6). By acting on the wavefunctions the ladder operators increment or decrement the quantum numbers i_3 and y by one unit, as illustrated in Fig. 7.2. For example, V_+ transforms an s quark into a u quark: applying the matrix (V_+) on an s quark indeed gives

$$V_+|s\rangle = \begin{pmatrix} 0 & 0 & 1 \\ 0 & 0 & 0 \\ 0 & 0 & 0 \end{pmatrix} \begin{pmatrix} 0 \\ 0 \\ 1 \end{pmatrix} = \begin{pmatrix} 1 \\ 0 \\ 0 \end{pmatrix} = |u\rangle. \tag{7.12}$$

However, when using the commutators (7.4), the demonstration becomes also valid for representations of higher dimensions:

$$\begin{aligned} I_3 V_+ &= I_3(G_4 + iG_5) = I_3 G_4 + i I_3 G_5 = \frac{i}{2} G_5 + G_4 I_3 + i\left(-\frac{i}{2} G_4 + G_5 I_3\right) \\ &= \frac{1}{2}(G_4 + iG_5) + (G_4 + iG_5) I_3 = V_+ \left(I_3 + \frac{1}{2}\right). \end{aligned} \tag{7.13}$$

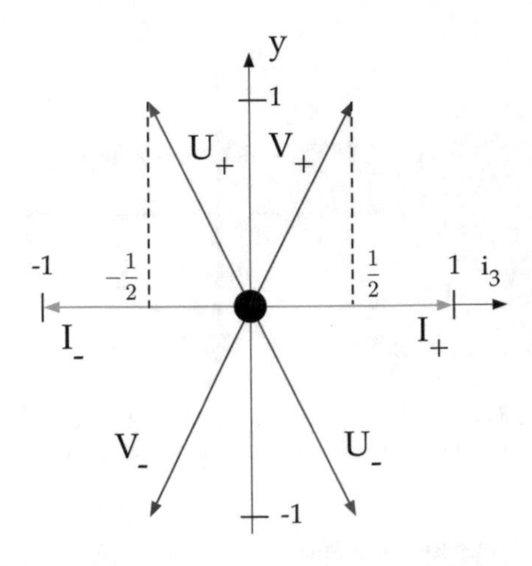

Fig. 7.2 The raising and lowering operators $I_\pm$ increase or decrease i_3 by one unit, while $V_\pm$ and $U_\pm$ increase or decrease i_3 by $\frac{1}{2}$ and y by one unit

Table 7.2 Commutation rules for the SU(3) ladder operators

$[Y, I_\pm] = 0$	$[Y, U_\pm] = \pm U_\pm$	$[Y, V_\pm] = \pm V_\pm$
$[I_3, I_\pm] = \pm I_\pm$	$[I_3, U_\pm] = \mp\frac{1}{2}U_\pm$	$[I_3, V_\pm] = \pm\frac{1}{2}V_\pm$
$[I_+, I_-] = 2I_3$	$[U_+, U_-] = \frac{3}{2}Y - I_3$	$[V_+, V_-] = \frac{3}{2}Y + I_3$
$[I_+, V_-] = -U_-$	$[I_+, U_+] = V_+$	$[U_+, V_-] = I_-$
$[I_+, V_+] = 0$	$[I_+, U_-] = 0$	$[U_+, V_+] = 0$

$I_3 V_+$ applied on $|i_3 y\rangle$ gives

$$I_3 V_+ |i_3 y\rangle = V_+ \left(i_3 + \frac{1}{2}\right) |i_3 y\rangle = \left(i_3 + \frac{1}{2}\right) V_+ |i_3 y\rangle. \tag{7.14}$$

On the other hand,

$$\begin{aligned} Y V_+ &= \frac{2}{\sqrt{3}} G_8 (G_4 + i G_5) = \frac{2}{\sqrt{3}} \left(G_4 G_8 + i \frac{\sqrt{3}}{2} G_5 \right) + \frac{2}{\sqrt{3}} \left(i G_5 G_8 + \frac{\sqrt{3}}{2} G_4 \right) \\ &= \frac{2}{\sqrt{3}} G_4 G_8 + i G_5 + i \frac{2}{\sqrt{3}} G_5 G_8 + G_4 \\ &= G_4 Y + i G_5 Y + G_4 + i G_5 = V_+ (Y + 1). \end{aligned} \tag{7.15}$$

Therefore

$$Y V_+ |i_3 y\rangle = V_+ (Y + 1) |i_3 y\rangle = (y + 1) V_+ |i_3 y\rangle. \tag{7.16}$$

Thus V_+ raises i_3 by $+\frac{1}{2}$ and y by $+1$, e.g. V_+ transforms an s quark into a u quark. Likewise $U_+|s\rangle = |d\rangle$ and $I_+|d\rangle = |u\rangle$. The commutation rules in Table 7.2 can be derived from the structure constants in Table 7.1 (Problem 7.1). Further commutators follow from (7.10).

7.2 Conjugate Representation

What about the antiquarks? The Young tableaux of the fundamental and conjugate representations of SU(3) are different,

$$3 = \square \qquad 3^* = \begin{array}{c}\square\\\square\end{array}$$

and so are the weight diagrams (Fig. 7.1). Let us write for the antiquark states the column vectors

$$\overline{u} = \begin{pmatrix}1\\0\\0\end{pmatrix}, \quad \overline{d} = \begin{pmatrix}0\\1\\0\end{pmatrix}, \quad \overline{s} = \begin{pmatrix}0\\0\\1\end{pmatrix}. \tag{7.17}$$

The matrices (7.6) do not represent the conjugate representation. In particular, (G_3) and (G_8) should have the opposite signs for i_3 and y. The group elements of 3^* are obtained by complex conjugating the unitary operator U (7.3), i.e. by replacing the generators G_i by $G'_i = -G^*_i$:

$$\begin{aligned}
G'_1 &= -G_1,\ \ G'_2 = G_2,\ \ G'_3 = -G_3,\\
G'_4 &= -G_4,\ \ G'_5 = G_5,\\
G'_6 &= -G_6,\ \ G'_7 = G_7,\\
G'_8 &= -G_8,
\end{aligned} \tag{7.18}$$

which also fulfil the commutation rules (7.4). For example, according to Table 7.1 the commutator between G'_1 and G'_5 should read

$$[G'_1, G'_5] = -\frac{i}{2}G'_6 = \frac{i}{2}G_6, \tag{7.19}$$

which is correct since

$$[G'_1, G'_5] = G'_1G'_5 - G'_5G'_1 = G_5G_1 - G_1G_5 = \frac{i}{2}G_6. \tag{7.20}$$

For antiquarks the ladder operators $V_\pm$ and $U_\pm$ are represented by the matrices

$$\begin{aligned}
(I_\pm)' &= (G'_1) \pm i(G'_2) = -(G_1) \pm i(G_2) = -(I_\mp),\\
(V_\pm)' &= (G'_4) \pm i(G'_5) = -(G_4) \pm i(G_5) = -(V_\mp),\\
(U_\pm)' &= (G'_6) \pm i(G'_7) = -(G_6) \pm i(G_7) = -(U_\mp).
\end{aligned} \tag{7.21}$$

Therefore $I_\pm$, $V_\pm$ and $U_\pm$ flip signs when applied on antiquarks (Problem 7.2), as was the case for $I_\pm$ in SU(2), see (6.22,6.23):

$$\begin{aligned} I_+|\overline{u}\rangle &= -|\overline{d}\rangle, \quad I_-|\overline{d}\rangle = -|\overline{u}\rangle, \\ V_+|\overline{u}\rangle &= -|\overline{s}\rangle, \quad V_-|\overline{s}\rangle = -|\overline{u}\rangle, \\ U_+|\overline{d}\rangle &= -|\overline{s}\rangle, \quad U_-|\overline{s}\rangle = -|\overline{d}\rangle. \end{aligned} \tag{7.22}$$

From the 3* weight diagram (7.1) one immediately sees that all other combinations must vanish (e.g. $V_-|\overline{u}\rangle = 0$).[1]

We have now acquired all the tools needed to construct the SU(3) wavefunctions of $q\overline{q}$ mesons. Coupling 3 with 3* gives

$$\boxed{3} \times \begin{array}{|c|}\hline 3 \\ \hline 2 \\ \hline \end{array} = \begin{array}{|c|c|}\hline 3 & 4 \\ \hline 2 & \\ \cline{1-1} \end{array} + \begin{array}{|c|}\hline 3 \\ \hline 2 \\ \hline 1 \\ \hline \end{array}$$

$$= 3 \times 3^* = 8 + 1, \tag{7.23}$$

an octet and a singlet. The weight diagram is easily obtained by superimposing the three triangles of the conjugate representation to the corners of the fundamental one, as shown in Fig. 7.3 for pseudoscalar mesons, i_3 and y being additive quantum numbers. The center of the hexagon ($i_3 = y = 0$) is occupied by three states, the $i = 1$ and the two $i = 0$ states (5.3).

Let us derive the wavefunctions of the two isoscalar states, first for the octet member. By applying U_+ on the π^+ and then V_- (green arrows in Fig. 7.3) one obtains the state

$$|\varphi\rangle = V_-U_+|\pi^+\rangle = V_-U_+|u\overline{d}\rangle = -V_-|u\overline{s}\rangle = -|s\overline{s}\rangle + |u\overline{u}\rangle. \tag{7.24}$$

On the other hand, operating with I_- on $|\pi^+\rangle$ leads to $|\varphi'\rangle = |d\bar{d}\rangle - |u\bar{u}\rangle$, the (unnormalized) π^0. The state $|\psi\rangle$ orthogonal to $|\varphi'\rangle$ is found by the standard procedure illustrated in Fig. 7.4. Defining

$$|\psi\rangle = |\varphi'\rangle - \alpha|\varphi\rangle = |d\bar{d}\rangle - |u\bar{u}\rangle + \alpha(|s\overline{s}\rangle - |u\overline{u}\rangle) \tag{7.25}$$

such that $\langle\varphi'|\psi\rangle = 0$, requires $\alpha = -2$, hence

$$|\psi\rangle = |d\bar{d}\rangle - |u\bar{u}\rangle - 2(|s\bar{s}\rangle - |u\bar{u}\rangle) = |u\bar{u}\rangle + |d\bar{d}\rangle - 2|s\bar{s}\rangle. \tag{7.26}$$

[1]Note that one needs to move twice around the periphery of the 3* triangle to retrieve the original state, e.g. when moving clockwise, $\overline{u} \to -\overline{s} \to \overline{d} \to -\overline{u} \to \overline{s} \to -\overline{d} \to \overline{u}$.

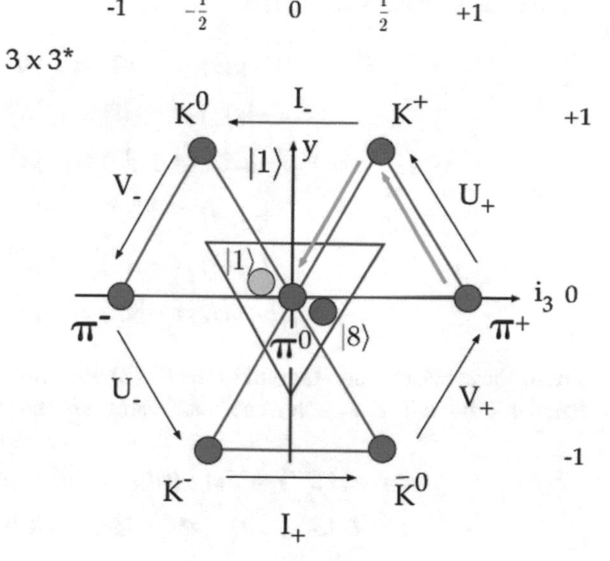

Fig. 7.3 The nonet of the 0^{-+} mesons ($3 \times 3^*$ representation) decomposes into an octet and a singlet

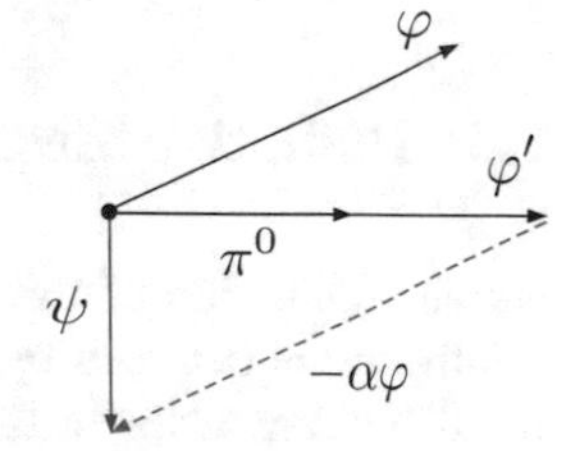

Fig. 7.4 Gram-Schmidt procedure to find the state orthogonal to φ'

The normalized state is

$$\boxed{|8\rangle = \frac{1}{\sqrt{6}}(|uu\rangle + |d\bar{d}\rangle - 2|s\bar{s}\rangle)} \; . \tag{7.27}$$

The $q\overline{q}$ combination orthogonal to all octet states is the singlet state

$$\boxed{|1\rangle = \frac{1}{\sqrt{3}}(|u\bar{u}\rangle + |d\bar{d}\rangle + |s\bar{s}\rangle)} \; , \tag{7.28}$$

as advertised earlier (5.3). With the ladder operators shown in Fig. 7.3 one can construct the wavefunctions of the mesons along the periphery of the hexagon. Again, one needs to circle twice to retrieve the sign of the original state. However, the eigenstates of I_3 and Y are defined up to phase factors $e^{i\phi}$ and we wish to retain within isospin multiplets the SU(2) symmetries discussed in Sect. 6.2. Starting with $|\pi^+\rangle = |u\overline{d}\rangle$ and moving counterclockwise around the hexagon we define the meson wavefunctions as follows:

$$\begin{aligned}
U_+|u\bar{d}\rangle &= -|u\bar{s}\rangle \equiv -|K^+\rangle, \\
I_-(-|u\bar{s}\rangle) &= -|d\bar{s}\rangle \equiv -|K^0\rangle, \\
V_-(-|d\bar{s}\rangle) &= +|d\bar{u}\rangle \equiv -|\pi^-\rangle, \\
U_-|d\bar{u}\rangle &= +|s\bar{u}\rangle \equiv -|K^-\rangle, \\
I_+|s\bar{u}\rangle &= -|s\bar{d}\rangle \equiv +|\overline{K}^0\rangle, \\
V_+(-|s\bar{d}\rangle) &= -|u\bar{d}\rangle \equiv -|\pi^+\rangle.
\end{aligned} \tag{7.29}$$

These phase choices correspond to the SU(3) wavefunctions before symmetrization listed in Table 6.4. The kaon wavefunctions satisfy the relations (6.39). For example,

$$\begin{aligned}
I_+|K^-\rangle &= I_+(-|s\bar{u}\rangle) = |s\bar{d}\rangle = -|\overline{K}^0\rangle, \\
I_-|K^+\rangle &= I_-|u\bar{s}\rangle = |d\bar{s}\rangle = |K^0\rangle.
\end{aligned} \tag{7.30}$$

7.3 Radiative Meson Decays

As an application of the SU(3) let us predict with the flavour wavefunctions the relative rates between radiative decays of vector mesons. We shall first compute the branching ratios of the decays $\omega \to \pi^0\gamma$, $\rho^0 \to \pi^0\gamma$ and $\rho^\pm \to \pi^\pm\gamma$, and then compare with experimental results. In these processes the mesons in the initial and final states have the same (negative) parities. The photon removes one unit of angular momentum and flips the spin of one the quarks. Hence these decays are M1 transitions involving the interaction of the quark magnetic moments $\vec{\mu}_k$ with the field of the photon. The transition operator is given by

$$M = \sum_1^2 \vec{\mu}_k \cdot \vec{\epsilon} = \sum_1^2 \frac{g e_k}{2m_k} \vec{s}_k \cdot \vec{\epsilon} = \sum_1^2 e_k \xi_k \vec{s}_k \cdot \vec{\epsilon}, \tag{7.31}$$

where $g = 2$ and $\xi_k \equiv \frac{1}{m_k}$. The sum extends over the quark and antiquark masses m_k, charges e_k and spins $\vec{s}_k$. Let us choose the quantization axis along the flight direction of the photon and assume +100% polarization of the vector meson (Fig. 7.5). The photon is then right-circularly polarized [1] and the electric field is proportional to

$$\vec{\epsilon} = \frac{1}{\sqrt{2}}(1, i, 0), \tag{7.32}$$

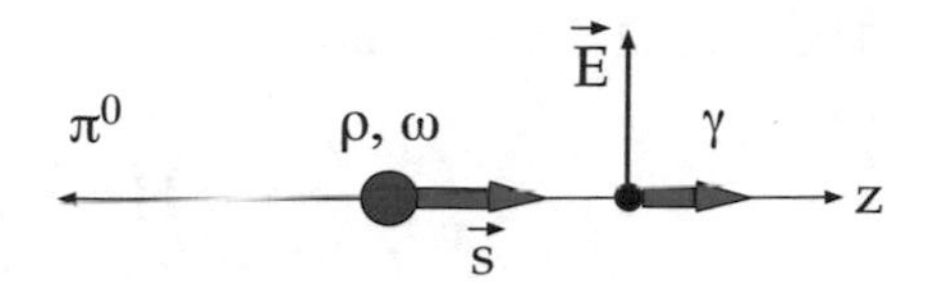

Fig. 7.5 Radiative decay of a 100% polarized vector meson

hence

$$\vec{s}_k \cdot \vec{\epsilon} = \frac{1}{\sqrt{2}} s_{+k}, \tag{7.33}$$

where s_{+k} is the SU(2) raising ladder operator $(s_1 + is_2)_k$.

For $\omega \to \pi^0\gamma$ the transition operator is represented by the matrix element

$$M_\omega = \frac{\xi}{\sqrt{2}} \times$$
$$\left\langle \frac{(d\bar{d} + u\bar{u}) - (\bar{d}d + \bar{u}u)}{2}[\uparrow\uparrow] \middle| e_1 s_{+1} + e_2 s_{+2} \middle| \frac{(d\bar{d} - u\bar{u}) + (\bar{d}d - \bar{u}u)}{2}\left[\frac{\uparrow\downarrow - \downarrow\uparrow}{\sqrt{2}}\right] \right\rangle . \tag{7.34}$$

We have introduced the SU(3) wavefunctions from Table 6.4 (ideally mixed ω) and the SU(2) spin functions (2.29, 2.30). We have also assumed that $m_u = m_d$. The first and second terms of the ladder operators give

$$\frac{\xi e}{4\sqrt{2}}\left(-\frac{1}{\sqrt{2}}\right)\left(-\frac{1}{3} - \frac{2}{3} - \frac{1}{3} - \frac{2}{3}\right) = \frac{\xi e}{4} \tag{7.35}$$

and

$$\frac{\xi e}{4\sqrt{2}}\left(\frac{1}{\sqrt{2}}\right)\left(\frac{1}{3} + \frac{2}{3} + \frac{1}{3} + \frac{2}{3}\right) = \frac{\xi e}{4}, \tag{7.36}$$

hence

$$M_\omega = \frac{\xi e}{2}. \tag{7.37}$$

We repeat the calculation for $\rho^0 \to \pi^0\gamma$ by replacing the positive signs by negative signs in the left-hand side of (7.34):

$$M_{\rho^0} = \frac{\xi}{\sqrt{2}} \times$$
$$\left\langle \frac{(d\bar{d} - u\bar{u}) - (\bar{d}d - \bar{u}u)}{2}[\uparrow\uparrow] \middle| e_1 s_{+1} + e_2 s_{+2} \middle| \frac{(d\bar{d} - u\bar{u}) + (\bar{d}d - \bar{u}u)}{2}\left[\frac{\uparrow\downarrow - \downarrow\uparrow}{\sqrt{2}}\right] \right\rangle . \tag{7.38}$$

The first and second terms give

$$\frac{\xi e}{4\sqrt{2}}\left(-\frac{1}{\sqrt{2}}\right)\left(-\frac{1}{3}+\frac{2}{3}-\frac{1}{3}+\frac{2}{3}\right)=-\frac{\xi e}{12} \tag{7.39}$$

and

$$\frac{\xi e}{4\sqrt{2}}\left(\frac{1}{\sqrt{2}}\right)\left(\frac{1}{3}-\frac{2}{3}+\frac{1}{3}-\frac{2}{3}\right)=-\frac{\xi e}{12}, \tag{7.40}$$

therefore

$$M_{\rho^0}=-\frac{\xi e}{6}. \tag{7.41}$$

The ω and ρ masses are almost equal and hence also the final state phase space factors. The ratio of partial decay widths is

$$\frac{\Gamma(\omega\to\pi^0\gamma)}{\Gamma(\rho^0\to\pi^0\gamma)}\simeq\left|\frac{M_\omega}{M_{\rho^0}}\right|^2=9. \tag{7.42}$$

Finally for $\rho^\pm\to\pi^\pm\gamma$, e.g. ρ^- decay:

$$M_{\rho^-}=\frac{\xi}{\sqrt{2}}\left\langle\frac{d\bar{u}-\bar{u}d}{\sqrt{2}}\,[\uparrow\uparrow]\,\middle|\,e_1s_{+1}+e_2s_{+2}\,\middle|\,\frac{d\bar{u}+\bar{u}d}{\sqrt{2}}\left[\frac{\uparrow\downarrow-\downarrow\uparrow}{\sqrt{2}}\right]\right\rangle. \tag{7.43}$$

The first and second terms give

$$\frac{\xi e}{2\sqrt{2}}\left(-\frac{1}{\sqrt{2}}\right)\left(-\frac{1}{3}+\frac{2}{3}\right)=-\frac{\xi e}{12}\ \text{and}\ \frac{\xi e}{2\sqrt{2}}\left(\frac{1}{\sqrt{2}}\right)\left(-\frac{2}{3}+\frac{1}{3}\right)=-\frac{\xi e}{12}, \tag{7.44}$$

hence

$$M_{\rho^-}=-\frac{\xi e}{6}, \tag{7.45}$$

equal to M_{ρ^0}, as expected from charge invariance.

Let us now compare these predictions with data. The most accurate branching ratios for $\omega\to\pi^0\gamma$ and $\rho^0\to\pi^0\gamma$ have been obtained by comparing the cross sections for ω and ρ^0 production and decay in e^+e^- annihilation. Figure 7.6 (left) shows the cross section measured at the VEPP-2M e^+e^- collider in Novosibirsk, which is dominated by the narrow ω signal. The contribution from the very broad and much weaker ρ^0 signal is not directly visible, but can be taken into account by

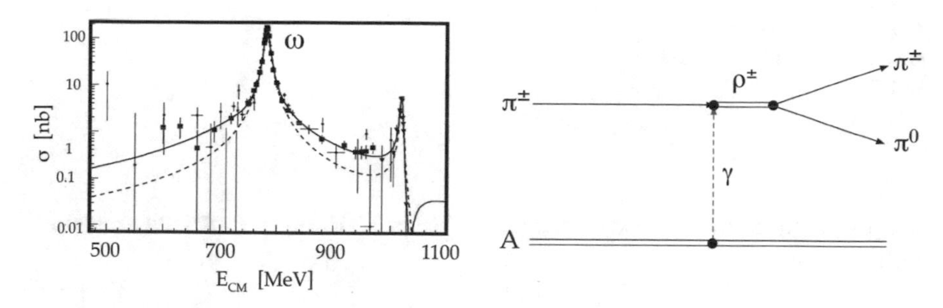

Fig. 7.6 Left: the cross section for $e^+e^- \to \pi^0\gamma$ is dominated by ω production. The full (dashed) curve shows the fit including (neglecting) $\rho^0 \to \pi^0\gamma$ interference [2]. The dispersive shape at the upper energy end is due to the interference between $e^+e^- \to \omega \to \pi^0\gamma$ and $e^+e^- \to \phi \to \pi^0\gamma$. Right: Primakoff production of the ρ meson on a heavy target A

a fit using the vector dominance model (VDM) [2]. With the recommended values for the decay branching ratios f [3] one finds the partial widths $\Gamma = f \times \Gamma_T$

$$\Gamma(\omega \to \pi^0\gamma) = 0.0840 \times 8.49\,\text{MeV} = 713\,\text{keV},$$
$$\Gamma(\rho^0 \to \pi^0\gamma) = 4.7 \times 10^{-4} \times 149.1\,\text{MeV} = 70\,\text{keV}. \tag{7.46}$$

Taking into account the experimental errors (mainly from ρ^0 decay) one obtains the experimental ratio

$$\frac{\Gamma(\omega \to \pi^0\gamma)}{\Gamma(\rho^0 \to \pi^0\gamma)} = 10.2 \pm 1.3, \tag{7.47}$$

which is in agreement with the prediction (7.42).

The decay $\rho^\pm \to \gamma\pi^\pm$ is difficult to measure owing to the smallness of the branching ratio. It was investigated with the inverse reaction $\gamma\pi^\pm \to \rho^\pm$. In the Primakoff process a high energy (e.g. 200 GeV) charged pion is scattered in the Coulomb field of a heavy target, such as Cu or Pb (Fig. 7.6, right). Photon exchange dominates nuclear processes at small momentum transfer and the cross section for $\rho^\pm \to \gamma\pi^+$ can be calculated [4]. The recommended value for the partial width is [3]

$$\Gamma(\rho^\pm \to \pi^\pm\gamma) = 67.1 \pm 7.4\,\text{keV}, \tag{7.48}$$

which agrees with the neutral mode (7.46). Other radiative decays such as $K^{*+} \to K^+\gamma$ and $K^{*0} \to K^0\gamma$ can be calculated with the same procedure and compared to experimental data from Primakoff scattering (Problem 7.3).

We have seen in Sect. 5.2 how the pseudoscalar mixing angle can be obtained from the 0^{-+} masses. The following application of SU(3) shows how this angle is determined directly from the measured partial widths for π^0, η, and $\eta' \to \gamma\gamma$ decays. Let us write the photon wavefunction as a superposition of $q\overline{q}$ pairs with

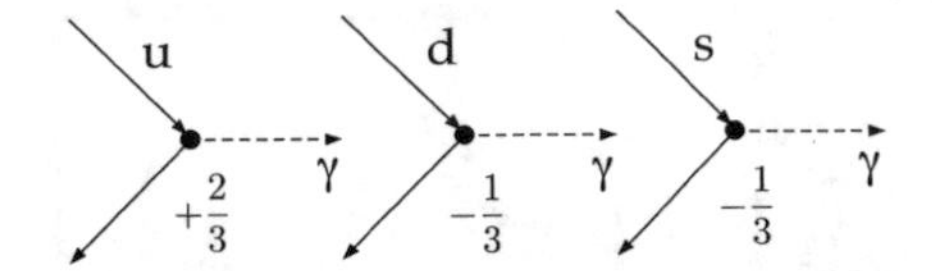

Fig. 7.7 The photon as a superposition of $u\overline{u}$, $d\overline{d}$, and $s\overline{s}$ pairs

couplings equal to the quark charges Q_k (Fig. 7.7). The normalized wavefunction

$$|\gamma\rangle = \frac{1}{\sqrt{2}}\sqrt{\frac{3}{2}}\left[\left(\frac{2}{3}u\bar{u} - \frac{1}{3}d\bar{d} - \frac{1}{3}s\bar{s}\right) - \left(\frac{2}{3}\bar{u}u - \frac{1}{3}\bar{d}d - \frac{1}{3}\bar{s}s\right)\right]. \tag{7.49}$$

is chosen to be antisymmetric like that of the ω since the photon is also a vector boson. In fact, the symmetric wavefunction—with the plus sign—would lead to a vanishing π^0 decay rate.

The matrix element for radiative π^0 decay is obtained by replacing the wavefunction for the ω in (7.34) by (7.49):

$$M_{\pi^0} = \frac{\xi}{\sqrt{2}}\frac{1}{\sqrt{2}}\sqrt{\frac{3}{2}} \times$$
$$\left\langle\left[\tfrac{2}{3}u\bar{u} - \tfrac{1}{3}d\bar{d} - \tfrac{1}{3}s\bar{s} - \left(\tfrac{2}{3}\bar{u}u - \tfrac{1}{3}\bar{d}d - \tfrac{1}{3}\bar{s}s\right)\right](\uparrow\uparrow)\left|e_1 s_{+1} + e_2 s_{+2}\right|\tfrac{(d\bar{d}-u\bar{u})+(\bar{d}d-\bar{u}u)}{2}\tfrac{(\uparrow\downarrow-\downarrow\uparrow)}{\sqrt{2}}\right\rangle. \tag{7.50}$$

The s_{+1} term gives

$$\frac{\xi e}{\sqrt{2}}\frac{1}{\sqrt{2}}\sqrt{\frac{3}{2}\frac{1}{2}}\frac{1}{\sqrt{2}}\left(\frac{1}{9} - \frac{4}{9} + \frac{1}{9} - \frac{4}{9}\right) = \xi e\frac{\sqrt{3}}{8}\left(-\frac{2}{3}\right) = -\frac{\xi e}{4\sqrt{3}} \tag{7.51}$$

and the s_{+2} term

$$\frac{\xi e}{\sqrt{2}}\frac{1}{\sqrt{2}}\sqrt{\frac{3}{2}\frac{1}{2}}\left(-\frac{1}{\sqrt{2}}\right)\left(-\frac{1}{9} + \frac{4}{9} - \frac{1}{9} + \frac{4}{9}\right) = \xi e\frac{\sqrt{3}}{8}\left(-\frac{2}{3}\right) = -\frac{\xi e}{4\sqrt{3}}, \tag{7.52}$$

hence

$$M_{\pi^0} = -\frac{\xi e}{2\sqrt{3}}. \tag{7.53}$$

The calculation can be repeated for the octet $|8\rangle$ and the singlet $|1\rangle$ by replacing the π^0 in (7.50) by (5.3), which we leave to the reader as an exercise. To simplify, we shall assume exact SU(3), that is $m_s = m_u = m_d$ (for $m_s > m_u$ see [5]). The results for $|8 \to \gamma\gamma\rangle$ and $|1 \to \gamma\gamma\rangle$ are:

$$M_8 = \frac{1}{\sqrt{3}} M_{\pi^0},$$
$$M_1 = 2\sqrt{\frac{2}{3}} M_{\pi^0}. \tag{7.54}$$

The η and η' wavefunctions are given by the rotated superpositions (5.11). For the η the matrix element is

$$M_\eta = M_8 \cos\theta_P - M_1 \sin\theta_P = \left(\frac{1}{\sqrt{3}} \cos\theta_P - 2\sqrt{\frac{2}{3}} \sin\theta_P \right) M_{\pi^0}. \tag{7.55}$$

The mass differences between η to π^0 have to be taken into account to derive the ratio of $\gamma\gamma$ partial widths. The 0^{-+} mesons decay into two 1^{--} photons by conserving P and C parity and hence one unit of angular momentum ℓ is removed from the system. The two-body phase space factor W is usually written as

$$W = pF_\ell^2(p), \tag{7.56}$$

where p is the break-up momentum in the rest frame of the decaying particle, and $F_\ell(p)$ is a damping factor which suppresses high angular momenta ℓ for small p [6]. This factor is determined by the range of the interaction, typically equal to 1 fm (corresponding to 197 MeV/c). Convenient expressions for $F_\ell(p)$ are given in Table 7.3. For large daughter momenta, $F_\ell(p) = 1$, while for momenta that are not significantly larger than 197 MeV/c the phase space factor W is proportional to $p^{2\ell+1}$. Following [5] we adopt the latter. With the photons carrying half of the meson masses one then gets the ratio

$$\frac{\Gamma_{\eta\to\gamma\gamma}}{\Gamma_{\pi^0\to\gamma\gamma}} = \left(\frac{m_\eta}{m_\pi}\right)^3 \frac{1}{3} \left(\cos\theta_P - 2\sqrt{2}\sin\theta_p\right)^2. \tag{7.57}$$

Table 7.3 Damping factors (squared) with $z = (p/p_R)^2$ and $p_R = 197$ MeV/c

ℓ	$F_\ell^2(p)$
0	1
1	$\frac{2z}{z+1}$
2	$\frac{13z^2}{(z-3)^2+9z}$
3	$\frac{277z^3}{z(z-15)^2+9(2z-5)^2}$
4	$\frac{12\,746z^4}{(z^2-45z+105)^2+25z(2z-21)^2}$

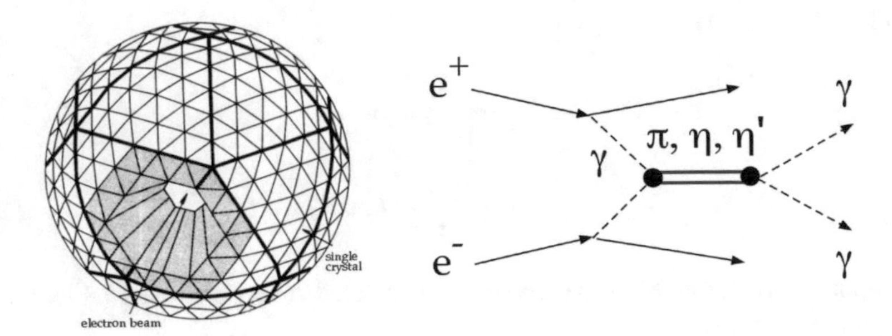

Fig. 7.8 Left: the Crystal Ball detector [5]. The shaded area shows the 30 crystals adjacent to the beam pipe which are vetoed to reduce background (see also Fig. 9.2 in Sect. 9.1). Right: production of pseudoscalar mesons with quasireal photons. The leptons are scattered under small angles and are not detected

The analogous calculation for the η' gives[2]

$$\frac{\Gamma_{\eta' \to \gamma\gamma}}{\Gamma_{\pi^0 \to \gamma\gamma}} = \left(\frac{m_\eta}{m_\pi}\right)^3 \frac{1}{3}\left(\sin\theta_P + 2\sqrt{2}\cos\theta_p\right)^2 . \tag{7.58}$$

The 2γ decays have been measured with the Crystal Ball detector at the DORIS e^+e^- storage ring at DESY. The beams were tuned to a colliding energy of about 10 GeV to also study the heavy quark resonances in the Υ region (Sect. 9.2). The detector was a ball of 672 NaI(T1) crystals covering a solid angle of $93\% \times 4\pi$ (Fig. 7.8). The arrangement was based on an icosahedron with each face subdivided into 36 triangles. The electron and positron beams entered the detector through openings on either side. The light was detected by one photomultiplier at the rear of each crystal. Charged particles could be vetoed with proportional tubes. The energy resolution on the photon was typically $\frac{\sigma}{E_\gamma} = 0.027$ $E[\mathrm{GeV}]^{-1/4}$ and the photon direction could be determined with an uncertainty of about 2°.

The measurement were performed without detecting the scattered electron and positron which flew through the forward and backward holes (Fig. 7.8, right). The photons were therefore quasireal ($m_\gamma \simeq 0$). The small momentum transfer to the photon ensured the production of low mass mesons and also suppressed spin-1 contributions by virtue of the Landau-Yang theorem.

[2] A more involved calculation lifting the assumption $m_s = m_u$ leads to the results (7.57, 7.58), but with the trigonometric functions multiplied by the factors $\frac{f_\pi}{f_8} \simeq 0.8$ and $\frac{f_\pi}{f_1} \simeq 0.95$, respectively, where f_π is the pion decay constant and f_8, f_1 are the octet and singlet decay constants [5].

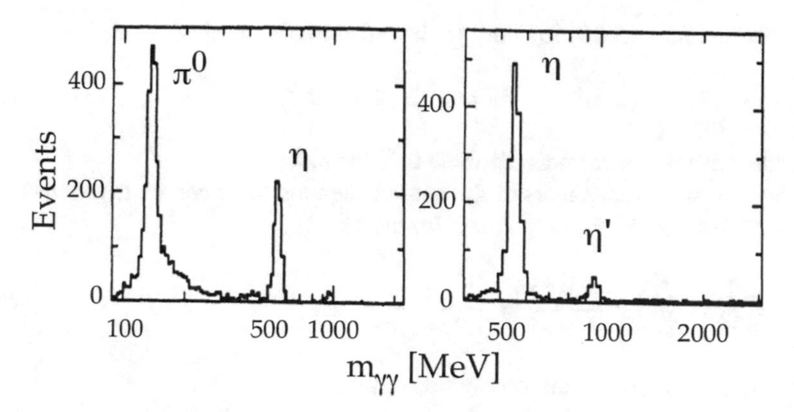

Fig. 7.9 2γ invariant mass distributions. Note the different bin sizes in the two mass ranges and the logarithmic scale [5]

Figure 7.9 shows the 2γ signals for the three pseudoscalars. The observed widths are determined by the energy resolution of the Crystal Ball and are much larger than their natural widths Γ_P. The partial width is given by

$$\Gamma(P \to \gamma\gamma) = f(P \to \gamma\gamma)\Gamma_P, \tag{7.59}$$

where f stands for the decay branching ratio. The pseudoscalar production cross section depends on the partial width and is given in good approximation by

$$\sigma(\gamma\gamma \to P) = \frac{16\pi^2 q}{m_P^2} \cdot \Gamma(P \to \gamma\gamma) \tag{7.60}$$

for small momentum transfers q to the photons [5]. From the number N_P of events observed in Fig. 7.9 one can determine the pseudoscalar partial widths from

$$N_P = \mathcal{L}\epsilon \cdot \sigma(\gamma\gamma \to P) f(P \to \gamma\gamma) = \mathcal{L}\epsilon \cdot \frac{16\pi^2 q}{m_P^2}\,\Gamma(P \to \gamma\gamma) f(P \to \gamma\gamma), \tag{7.61}$$

by using the known decay branching ratios $f(P \to \gamma\gamma)$. The collider integrated luminosity is $\mathcal{L}$ and the detection efficiency ϵ. The results for the partial widths are [5]

$$\begin{aligned}
\Gamma(\pi^0 \to \gamma\gamma) &= 7.7 \pm 0.7\,\text{eV}, \\
\Gamma(\eta \to \gamma\gamma) &= 514 \pm 39\,\text{eV}, \\
\Gamma(\eta' \to \gamma\gamma) &= 4.7 \pm 0.7\,\text{keV}.
\end{aligned} \tag{7.62}$$

Substituting into (7.57, 7.58) leads to the pseudoscalar mixing angle

$$\theta_P = (-22.4 \pm 1.2)^\circ, \tag{7.63}$$

in accord with the linear mass formula in Table 5.2.

The very short mean lives of the neutral pseudoscalars can be determined from the processes $\gamma\gamma \to P_S \to \gamma\gamma$, e.g. for the π^0

$$\tau = \frac{1}{\Gamma} = \frac{f(\pi^0 \to \gamma\gamma)}{\Gamma(\pi^0 \to \gamma\gamma)} = \frac{1}{7.7 \times 10^{-6}}\,\text{MeV}^{-1} = 8.5 \times 10^{-17}\,\text{s}, \tag{7.64}$$

where we have used the unit conversion (1.9).

References

1. Amsler, C.: Nuclear and Particle Physics, pp. 11-14. IOP Publishing, Bristol (2015)
2. Achasov, M.N., et al.: Phys. Lett. B 559, 171 (2003)
3. Patrignani, C., et al. (Particle Data Group): Chin. Phys. C 40, 100001 (2016); 2017 update
4. Capraro, L., et al.: Nucl. Phys. 288, 659 (1987)
5. Williams, D.A., et al.: Phys. Rev. D 38, 1365 (1988)
6. Blatt, J., Weisskopf, V.: Theoretical Nuclear Physics, p. 361. Wiley, New York (1952)

Chapter 8
Heavy Quark Mesons

8.1 Charm Quark

The J/ψ resonance was discovered at Brookhaven National Laboratory (BNL) and at SLAC in November 1974. The 30 GeV protons from BNL's alternating-gradient synchrotron impinged on a beryllium target and new particles decaying into e^+e^- pairs were searched with two magnetic spectrometers. Electrons and positrons were selected by Čerenkov hydrogen gas counters. A narrow peak—dubbed J—was observed around the e^+e^- mass of 3100 MeV in the reaction $p\mathrm{Be} \to JX$, $J \to e^+e^-$ (Fig. 8.1, left).

At SLAC electrons were collided with positrons in the SPEAR storage ring. The outgoing particles were detected in the cylindrical magnetic spectrometer MARK I, equipped with spark chambers and shower counters. A very sharp resonance—dubbed ψ—was observed to decay into hadrons or $\mu^+\mu^-$ pairs (Fig. 8.1, right) around the collision energy E of 3100 MeV. At resonance the cross section rose by about two orders of magnitude.

The ψ is produced by the emitted photon from e^+e^- annihilation. However, the spin-1 assignment is not straightforward since the photon is virtual.[1] The assignment $J^{PC} = 1^{--}$ to the ψ follows from the dispersive line shape observed in the $\mu^+\mu^-$ cross section (inset in Fig. 8.1, right): the line shape is distorted by the interference between $e^+e^- \to J/\psi \to \mu^+\mu^-$ and $e^+e^- \to \gamma \to \mu^+\mu^-$. Indeed, interference between two channels leading to the same final state arises when the two processes have the same quantum numbers. Figure 8.2 (left) shows a Feynman diagram of the production and decay of the ψ, established as a pair of bound charm-anticharm quarks, to be explained below. The graph of the inverse reaction observed at BNL is depicted in Fig. 8.2 (right).

[1] Another example of spin non-conservation in virtual processes is the spin-0 π^+ which decays into $\mu^+\nu_\mu$ via the emission of a spin-1 virtual W^+ boson.

© The Author(s) 2018

C. Amsler, *The Quark Structure of Hadrons*, Lecture Notes in Physics 949,
https://doi.org/10.1007/978-3-319-98527-5_8

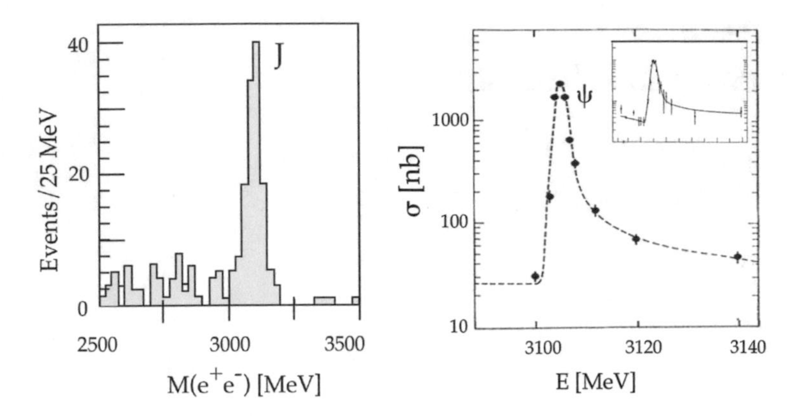

Fig. 8.1 Left: the $J \to e^+e^-$ resonance observed at BNL in 1974 [1]. Right: the ψ resonance produced in e^+e^- collisions and decaying into hadrons. The dashed curve shows the expected signal for a zero-width resonance folded by the experimental resolution. The tail is due to radiative processes in e^+e^- [2]. The inset shows the dispersive shape of the resonance line in $e^+e^- \to \psi \to \mu^+\mu^-$ [3]

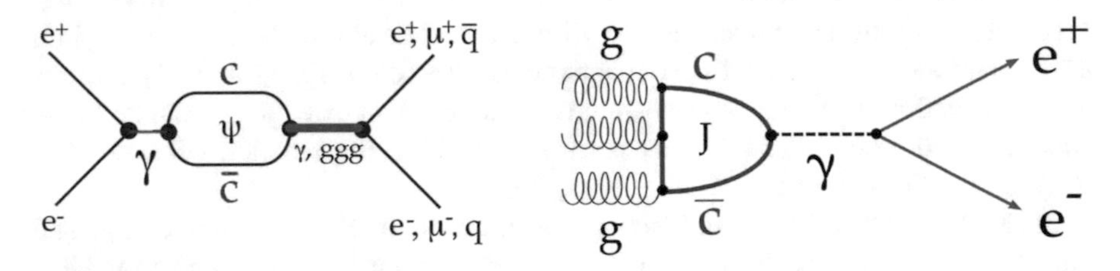

Fig. 8.2 Production and decay of the J/ψ resonance produced in e^+e^- collisions (left) and in hadronic collisions, the latter decaying for example into an electron-positron pair (right)

The measured width of the resonance was determined by detector resolution in the BNL experiment and by the energy spread of the colliding beams at SLAC. The measured full width of the ψ was measured to be less than 1.3 MeV [2]. The natural width can be estimated by comparing the integrated signal (Fig. 8.1, right) with the integral of a Breit-Wigner cross section σ centered at the mass M of the ψ (Problem 8.1):

$$\begin{aligned}\int_{-\infty}^{\infty} \sigma dE &= \int_{-\infty}^{\infty} \frac{2j+1}{(2s+1)(2s+1)} \cdot \frac{4\pi}{E^2} \cdot \frac{\Gamma_{ee}\Gamma_h}{(E-M)^2+\Gamma^2/4} dE \\ &= \frac{3}{4} \cdot \frac{8\pi^2}{M^2} \left(\frac{\Gamma_{ee}\Gamma_h}{\Gamma}\right) = \frac{3}{4} \cdot \frac{8\pi^2}{M^2} \left(\frac{\Gamma_{ee}}{\Gamma}\right)\left(\frac{\Gamma_h}{\Gamma}\right)\Gamma \\ &= \frac{3}{4} \cdot \frac{8\pi^2}{M^2} f(e^+e^-) f(h)\Gamma. \end{aligned} \tag{8.1}$$

We have made the approximation $4\pi/E^2 \simeq 4\pi/M^2$, valid for a narrow resonance, and used for the integration the formula

$$\int_{-\infty}^{\infty} \frac{1}{(E-M)^2+\Gamma^2/4} dE = \frac{2\pi}{\Gamma}. \tag{8.2}$$

The resonance spin is $j = 1$ the spins of the leptons are $s = \frac{1}{2}$. Γ_{ee} and Γ_h are the partial widths for $\psi \to e^+e^-$ and $\psi \to$ hadrons, respectively. The decay branching ratio $f(e^+e^-)$ is equal to $\simeq$6%, that for decays into hadrons $f(h) \simeq 88\%$.

The natural width of the ψ (Γ = 93 keV) turned out to be surprisingly small. The observation of this sharp resonance (now called $J/\psi(1S)$ or simply J/ψ) is sometimes referred to as the "November revolution" which spurred a flood of possible theorctical explanations (Fig. 8.3, left). Only 10 days after the initial discovery, a second narrow state, the ψ'—now called $\psi(2S)$—was observed in e^+e^- collisions about 600 MeV above the J/ψ. Figure 8.3 (right) shows a pictorial event display from the MARK I experiment at SPEAR of the ψ' decaying into $J/\psi\,\pi^+\pi^-$, with $J/\psi \to e^+e^-$. The observation of the ψ' was soon followed by the discovery of higher lying states at SPEAR and at DORIS/DESY (for an excellent review on the history of the charm quark discovery see [5]).

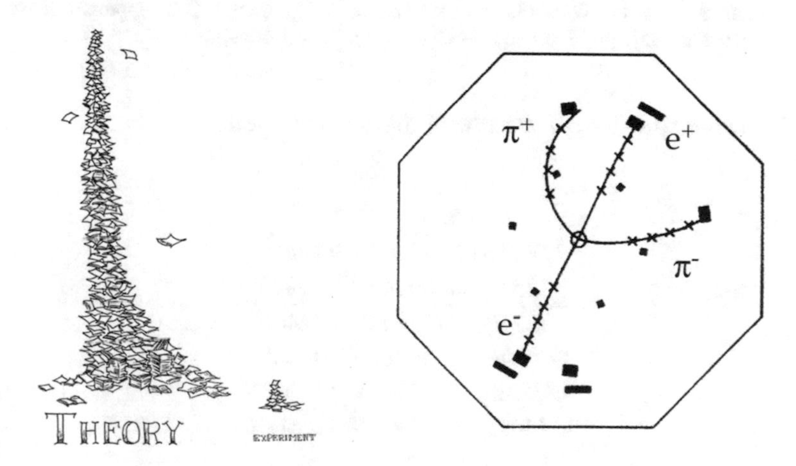

Fig. 8.3 Left: The J/ψ discovery triggered a flood of theoretical speculations (sketch by J.D. Jackson, credit CERN courier, April 1975). Right: event display from the MARK I detector at SPEAR showing the ψ' decaying into $J/\psi(\to e^+e^-)\pi^+\pi^-$ [4]

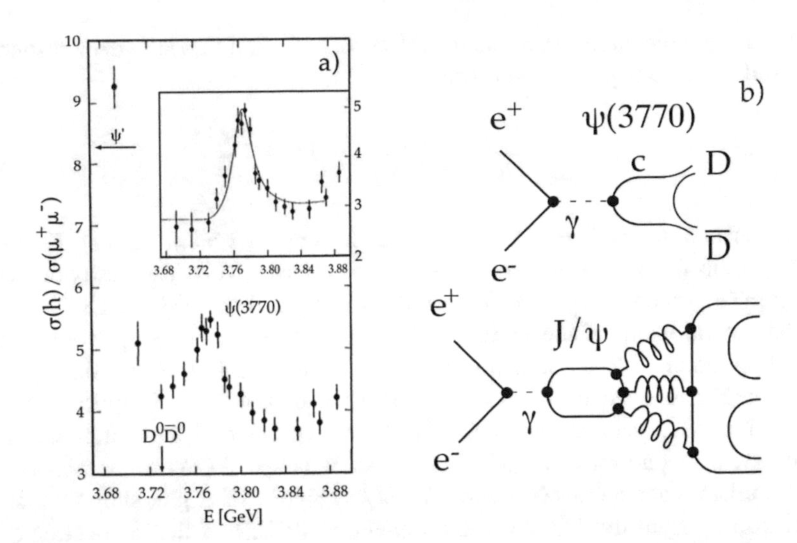

Fig. 8.4 (**a**) Observation of the $\psi(3770)$ in the cross section for $e^+e^- \rightarrow$ hadrons just above $D^0\overline{D}^0$ production threshold (see the text, adapted from [6]). The inset shows the fitted signal after subtracting the radiative processes from the ψ'; (**b**) excitation and decay of the $\psi(3770)$ resonance into $D\overline{D}$ ($D^0\overline{D}^0$ or D^+D^-). D decay contributes to (and dominates) the hadronic final state. The J/ψ and ψ' lie below threshold and decay hadronically through OZI-suppressed decays (or through electromagnetic processes such as e^+e^-, $\mu^+\mu^-$, or γ + hadrons)

The following spin-1 resonances with masses and widths have been observed to date:

	M	Γ
$J/\psi(1S)$	3097 MeV	**93 keV**
$\psi(2S)$	3686 MeV	**303 keV**
$\psi(3770)$	3773 MeV	27 MeV
$\psi(4040)$	4039 MeV	80 MeV
$\psi(4160)$	4153 MeV	103 MeV
$\psi(4415)$	4421 MeV	62 MeV.

(8.3)

The reason for the astonishing much smaller widths of the two lowest states was settled with the discovery of the much broader $\psi(3770)$. Figure 8.4a shows the cross section for hadron production measured at SPEAR, normalized to the calculated cross section for non-resonant $\mu^+\mu^-$ production (i.e. via γ exchange only) [6]. The fast rise towards low energy is due to the radiative tail of the ψ'. The inset shows the data after radiative corrections and a Breit-Wigner fit, leading to a width of 27 MeV, much larger than that of the J/ψ.

Under the assumption that the ψ resonances are made of pairs $c\overline{c}$ of a new type of heavy quark, they would decay through the emission of open charmed mesons ($c\overline{q}+$

$\bar{c}q$ pairs), the D mesons. However, the mass of the lightest pair of charmed mesons ($D^0\overline{D}^0$) is 3730 MeV (shown by the vertical arrow in Fig. 8.4a), which is larger than the masses of the J/ψ and ψ'. The J/ψ and the ψ' therefore disintegrate through OZI-hindered processes, the former into light hadrons and the latter into $J/\psi\,\pi\pi$ (or electromagnetically), which reduces their widths and increases their lifetimes accordingly. The OZI suppression of the J/ψ and ψ' decays was the indisputable evidence for the existence of the charm quark.

The J/ψ is the vector (1^{--}) ground state of the $c\bar{c}$ system ($\ell = 0, s = 1$). The branching ratio into $\rho\pi$ is 1.7%, hence its G parity is negative and from (4.4) its isospin is $i = 0$. The $\psi' = \psi(2S)$, $\psi(4040)$ and $\psi(4415)$ mesons are radially excited vector mesons.

The $\psi(3770)$ is a 1^{--} orbital excitation with two units of relative angular momentum (1^3D_1 state), as can be deduced from its small e^+e^- partial width: the decay rate into electron-positron pairs is proportional to the overlap probability of the quark and antiquark wavefunctions, which decreases with excitation energy and becomes very small for $\ell > 0$ (Van Royen-Weisskopf formula [7], Problem 8.2). The measured partial widths, plotted in Fig. 8.5 (left) show that the $\psi(3770)$ state does not fit in the $\ell = 0$ sequence but is the 1^3D_1 state. The $\psi(4160)$ has also been suggested to be the 2^3D_1 state. The $\psi(2S)$ decays OZI suppressed into $J/\psi\,\pi\pi$ or electromagnetically, e.g. to the lower lying $c\bar{c}$ states with $\ell = 1$ and $j = 0, 1$ or 2, as we shall discuss in the next section.

The states above $D\overline{D}$ threshold decay into mesons with "open" charm ($c\bar{q}$ or $q\bar{c}$, $q \neq c$). In the next section we shall extend the 1^1S_0 nonet of the light quarks to a 16-plet by including the c quark. The "hidden" charm isosinglet state $\eta_c = |c\bar{c}\rangle$ with $y = 0$ and mass 2983 MeV is the lightest $c\bar{c}$ hadron. It decays OZI-suppressed, e.g. into

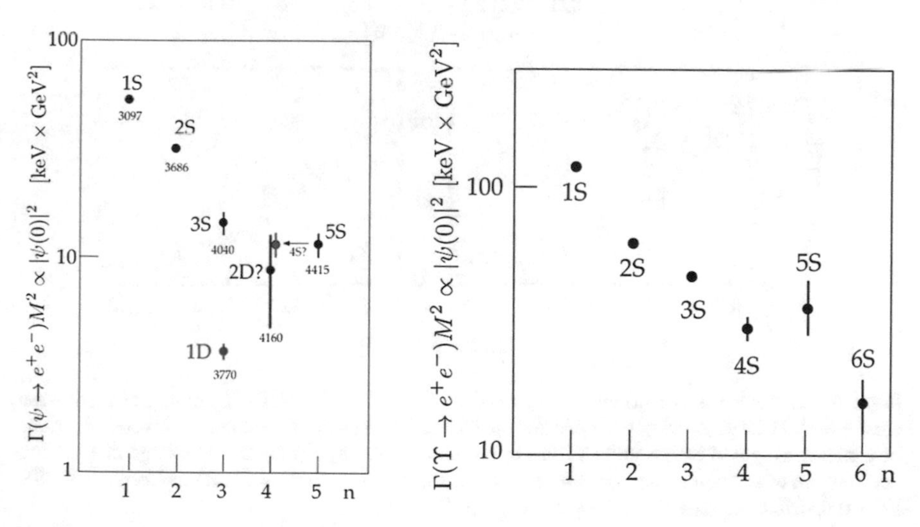

Fig. 8.5 Left: e^+e^- partial widths of 1^{--} $c\bar{c}$ mesons. The $\psi(3770)$, and possibly the $\psi(4160)$, are 1^3D_1 and 2^3D_1 orbital excitations. Right: e^+e^- partial widths of the 1^{--} $b\bar{b}$ mesons

$K\overline{K}\pi$ or into multipion final states.The 1^1S_0 D mesons form the two isodoublets

$$D^+ = |c\overline{d}\rangle,\ D^0 = |c\overline{u}\rangle \text{ and } \overline{D}^0 = |\overline{c}u\rangle,\ D^- = |\overline{c}d\rangle, \tag{8.4}$$

with hypercharge $y = -\frac{1}{3}$ and $+\frac{1}{3}$, respectively, and the two isosinglets

$$D_s^+ = |c\overline{s}\rangle,\ D_s^- = |s\overline{c}\rangle. \tag{8.5}$$

with $y = \pm\frac{2}{3}$.

The D mesons decay mostly into kaonic final states, which reflects the Cabibbo favoured $c \to sW^+$ over $c \to dW^+$ transition. For example, the branching ratio $f(D^0 \to K^-\pi^+)$ is 3.9×10^{-2}, while $f(D^0 \to \pi^+\pi^-) = 1.4 \times 10^{-3}$. Also, the "wrong" charge channel $D^0 \to K^+\pi^-$ is suppressed ($c \not\to \overline{s}$) with $f = 1.4 \times 10^{-4}$. Thus the charge of the kaon determines the flavour (c or $\overline{c}$) of the D meson (Fig. 8.6, top). The lifetime of the D^+ meson is 1 ps, that of the D^0 is 0.4 ps, which are long enough to permit the identification of the decay vertex in high energy experiments.

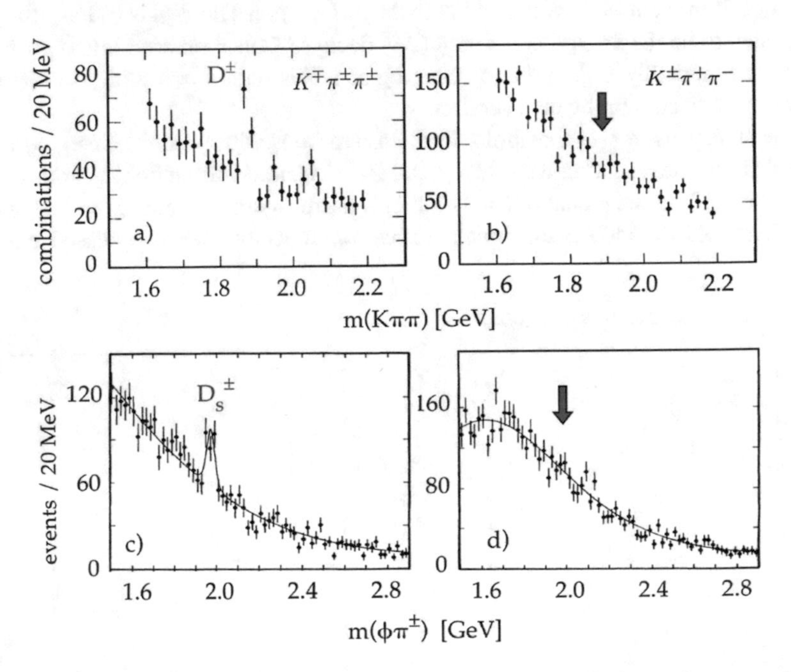

Fig. 8.6 (**a**) $K\pi\pi$ mass distribution measured by MARK I with SPEAR running at a collision energy of 4.03 GeV, showing evidence for the charged D meson [8]. Plotted are all combinations of particles weighted by the probabilities to be pions or kaons; (**b**) the wrong charge distribution does not show any signal; (**c**) first observation of the $D_s^\pm \to \phi\pi^\pm$ by CLEO at Cornell's CESR [9]; (**d**) distribution outside the $\phi \to K^+K^-$ signal

Table 8.1 D^* decay channels and branching ratios

D^{*0}	$D^0\pi^0$	0.647			$D^0\gamma$	0.353
$\overline{D}^{*0}$	$\overline{D}^0\pi^0$	Ditto			$\overline{D}^0\gamma$	Ditto
D^{*+}	$D^0\pi^+$	0.677	$D^+\pi^0$	0.307	$D^+\gamma$	0.016
D^{*-}	$\overline{D}^0\pi^-$	Ditto	$D^-\pi^0$	Ditto	$D^-\gamma$	Ditto

The decays $D^{*0} \rightarrow D^+\pi^-$ and $\overline{D}^{0*} \rightarrow D^-\pi^+$ are kinematically forbidden

Table 8.2 Quark model assignments for the charm and bottom mesons with open flavours and known quantum numbers [10]

		$i=\frac{1}{2}$	$i=0$	$i=\frac{1}{2}$	$i=0$	$i=0$
		$c\overline{u}, c\overline{d}$	$c\overline{s}$	$\overline{b}u, \overline{b}d$	$\overline{b}s$	$\overline{b}c$
$n^{2s+1}\ell_J$	J^{PC}	$u\overline{c}, d\overline{c}$	$s\overline{c}$	$b\overline{u}, b\overline{d}$	$b\overline{s}$	$b\overline{c}$
1^1S_0	0^{-+}	$D(1869.6)^{\pm}$	$D_s(1968.3)^{\pm}$	$B(5279.3)^{\pm}$	$B_s(5366.8)^0$	$B_c(6275)^{\pm}$
		$D(1864.8)^0$		$B(5279.6)^0$		
1^3S_1	1^{--}	$D^*(2010.3)^{\pm}$	$D_s^*(2112.1)^{\pm}$	$B^*(5324.7)^{\pm}$	$B_s^*(5415.4)^0$	
		$D^*(2006.9)^0$		$B^*(5324.6)^0$		
1^1P_1	1^{+-}	$D_1(2420)$	$D_{s1}(2536)^{\pm}$	$B_1(5721)$	$B_{s1}(5830)^0$	
1^3P_0	0^{++}	$D_0^*(2400)$	$D_{s0}^*(2317)^{\pm a}$			
1^3P_1	1^{++}	$D_1(2430)$	$D_{s1}(2460)^{\pm a}$			
1^3P_2	2^{++}	$D_2^*(2460)$	$D_{s2}^*(2573)^{\pm}$	$B_2^*(5747)$	$B_{s2}^*(5840)^0$	
1^3D_1	1^{--}		$D_{s1}^*(2860)^{\pm b}$			
1^3D_3	3^{--}	$D_3^*(2750)$	$D_{s3}^*(2860)^{\pm}$			
2^1S_0	0^{-+}	$D(2550)$				$B_c(2S)^{\pm}$
2^3S_1	1^{--}		$D_{s1}^*(2700)^{\pm b}$			

The C parity is that of the corresponding light quark isoscalar. The 1^{+-} and 1^{++} mesons are mixtures of the $1^{1\pm}$ states

[a]The masses are much smaller than most theoretical predictions. They have alternatively been interpreted as four-quark states

[b]These mesons are mixtures of the 1^3D_1 and 2^3S_1 states

The corresponding 1^3S_1 open charm mesons are labelled D^*, D_s^* and the $J/\psi(1S)$ is the hidden charm $y=0$ isosinglet. The D^* mesons decay rapidly into D mesons (Table 8.1).

Table 8.2 lists all the open charm mesons observed so far [10]. The mass difference between D and D^* is of the order of the pion mass. The identification of the $D^{*\pm}$ charge is achieved by measuring the charge of the very slow pion in the $D^{*\pm} \rightarrow D^0(\overline{D}^0)\pi^{\pm}$ rest frame.[2] The mass difference between the $D_s^{*\pm}$ and the $D_s^{\pm}$ is also close to the pion mass and hence the $D_s^{*\pm}$ decays dominantly into $D_s^{\pm}\gamma$ ($f \simeq 94\%$), while $f(D_s^{\pm}\pi^0) \simeq 6\%$. The $D_s^{\pm}$ has many decay channels [10], among

[2]This heavy quark "tagging" is useful when studying $D^0 - \overline{D}^0$ oscillations, since the pion charge determines the flavour of the initial heavy quark (c or $\overline{c}$).

them $\phi\pi^{\pm}$, which was the discovery channel (Fig. 8.6, bottom), and $\tau^{\pm}\nu_{\tau}$ which has been used as a source of τ-neutrinos [11]. Orbital excitations have also been reported.

8.2 Bottom Quark

The bottom quark was discovered at Fermilab in 1978 when an enhancement was observed in the $\mu^+\mu^-$ mass spectrum of the reaction p(Pt or Cu)$\rightarrow$ $\mu^+\mu^- X$ with 400 GeV protons [12]. The muons were analyzed by a two-arm magnetic spectrometer. Figure 8.7 (left) shows the signal from $\Upsilon(1S) \rightarrow \mu^+\mu^-$. A few years later the $\Upsilon(1S)$ and radial excitations at higher masses were observed with the CUSB detector in e^+e^- collisions at the Cornell's CESR[13] and also at DORIS (DESY) [14]. Figure 8.7 (right) shows the three lower Υ states observed recently by CMS with the LHC running at a center-of-mass energy of 7 TeV [15]. The masses and widths of the (1^{--}) Υ observed so far are

	M	Γ
$\Upsilon(1S)$	9460 MeV	**54 keV**
$\Upsilon(2S)$	10,023 MeV	**32 keV**
$\Upsilon(3S)$	10,355 MeV	**20 keV**
$\Upsilon(4S)$	10,579 MeV	21 MeV
$\Upsilon(5S)$	10,891 MeV	54 MeV
$\Upsilon(6S)$	10,987 MeV	61 MeV.

(8.6)

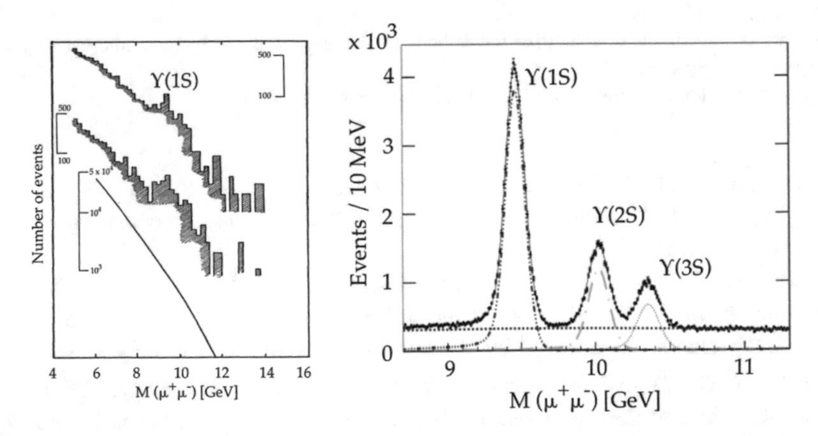

Fig. 8.7 Left: the $\Upsilon(1S)$ decaying into a muon pair was first reported at Fermilab. The graph shows the event distribution for two magnetic field settings in the pair spectrometer [12]. The line shows the distribution of accidental pairs (simulated with muons from different events). Right: the $\Upsilon(1S, 2S, 3S)$ signals observed by CMS at the LHC [15]

The existence of a fifth quark, the b quark, is responsible for the narrow widths of the $\Upsilon(1S, 2S, 3S)$ mesons. The kinematic threshold decaying into a pair of open bottom ($B\overline{B}$) mesons lies at 10,558 MeV, just below the mass of the $\Upsilon(4S)$. The $\Upsilon(1S, 2S, 3S)$ states have substantially longer lifetimes than the states above threshold because they have to decay through OZI-suppressed processes or electromagnetically. Figure 8.5 (right) shows the measured e^+e^- partial widths for 1^3S_1 $b\overline{b}$ states.

In the next section we shall add to the pseudoscalar 16-plet another 9 mesons containing the b quark. The $i = \frac{1}{2}$ B mesons form the two isodoublets

$$\overline{B}^0 = |b\overline{d}\rangle,\ B^- = |b\overline{u}\rangle \text{ and } B^+ = |u\overline{b}\rangle,\ B^0 = |d\overline{b}\rangle, \tag{8.7}$$

with hypercharge $y = -\frac{1}{3}$ and $+\frac{1}{3}$, respectively. The two isosinglets

$$\overline{B}_s^0 = |b\overline{s}\rangle,\ B_s^0 = |s\overline{b}\rangle \tag{8.8}$$

have $y = \pm\frac{2}{3}$, and the two isosinglets containing two heavy quarks, the

$$B_c^+ = |c\overline{b}\rangle,\ B_c^- = |b\overline{c}\rangle, \tag{8.9}$$

have $y = 0$. The spin and parity of the B mesons are assumed from the quark model predictions.

The hidden bottom isosinglet state $\eta_b = |b\overline{b}\rangle$ with $y = 0$ and mass 9399 MeV is the lightest $b\overline{b}$ hadron, decaying OZI-suppressed into multihadron final states. The B mesons decay mostly into states involving D mesons (and hence ultimately kaonic final states), since the coupling $b \to cW^-$ is favoured over $b \to uW^+$. The lifetime of the B meson ($\simeq$1.5 ps) is large, comparable to that of the D^+. This is due to the dominant but kinematically forbidden bottom–top coupling. Thanks to its long mean life, the B meson can be identified by its decay vertex in high energy experiments.

The B_s^0 decays dominantly into D_s^- + anything. The event display of the ALEPH experiment at LEP (Fig. 8.8) shows one of the first observed B_s^0 mesons decaying into $\psi(2S)\phi$, a rare decay channel with a branching ratio of 5×10^{-4}, but convenient to identify with the $\mu^+\mu^-$ pair from J/ψ decay.

The $B_c^\pm$ is difficult to observe due to its small production cross section. A hint was reported by OPAL at LEP in 1997, soon confirmed by CDF at Fermilab which measured later its mass and lifetime [16]. Figure 8.9 shows the signal from $B_c^\pm \to J/\psi\,\pi^\pm$ near 6.3 GeV. Note that the branching ratio for the Cabibbo favoured $B_c^\pm \to J/\psi\,\pi^\pm$ is much larger than that for $B_c^\pm \to J/\psi\,K^\pm$, while the opposite holds for $B^\pm \to J/\psi\,\pi^\pm$ and the Cabibbo favoured $B^\pm \to J/\psi\,K^\pm$ (Fig. 8.9, bottom) which enhances the signal of the $B^\pm$ relative to $B_c^\pm$ in the case of insufficient particle identification. The lifetime of the $B_c^\pm$ is 0.5 ps.

The corresponding 1^3S_1 vector mesons are labelled B^* and the 1^3S_1 $b\overline{b}$ ground state is the $\Upsilon(1S)$. The mass difference between B^* and B is only 48.6 MeV, hence

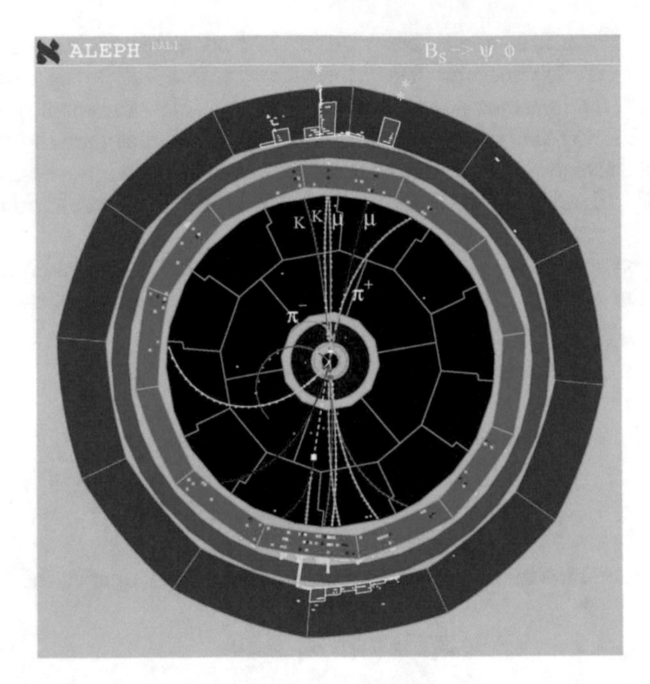

Fig. 8.8 Event display from the ALEPH detector at CERN-LEP showing one of the first observed B_s mesons decaying into $\psi(2S)\phi(1020)$ with the ϕ decaying into K^+K^- and the $\psi(2S)$ into $J/\psi\,\pi^+\pi^-$, followed by $J/\psi \to \mu^+\mu^-$ (photo CERN)

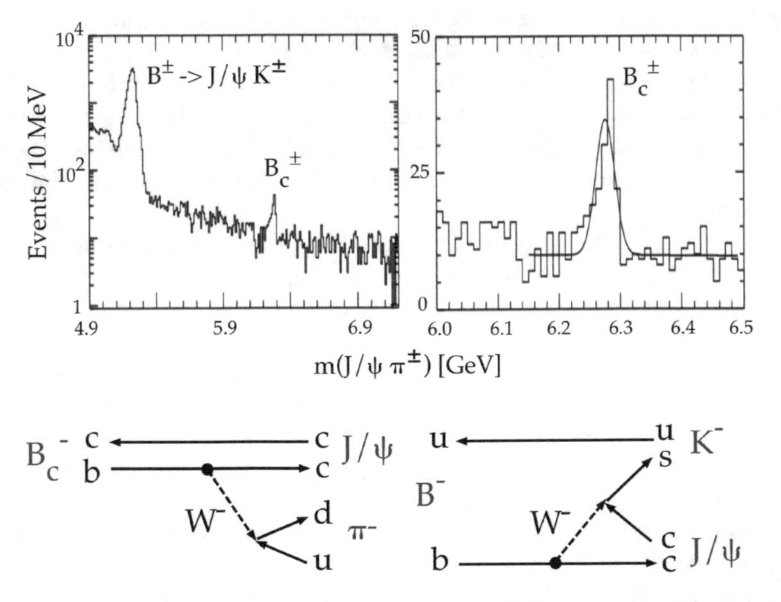

Fig. 8.9 Top: $J/\psi\,\pi^\pm$ mass distribution (left) and enhanced region around the $B_c^\pm$ signal (right). The peak around 5.2 GeV is due to the background channel $B^\pm \to J/\psi\,K^\pm$ with the pion mass incorrectly attributed to the kaon [16]. Bottom: Cabibbo favoured decays of the B_c^- and B^-

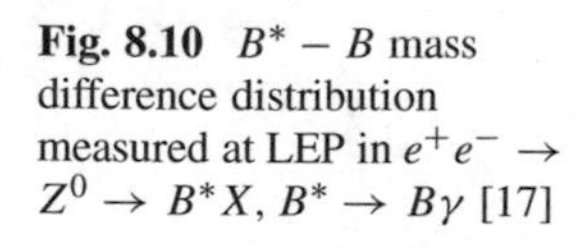
Fig. 8.10 $B^* - B$ mass difference distribution measured at LEP in $e^+e^- \to Z^0 \to B^*X$, $B^* \to B\gamma$ [17]

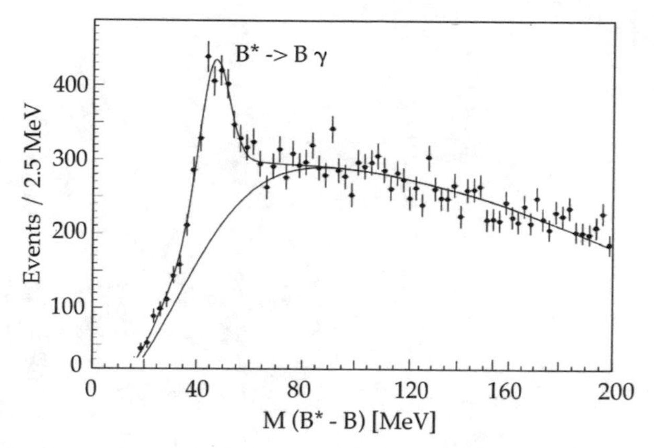

the B^* decays exclusively into $B\gamma$. Figure 8.10 shows the B^* signal observed by the OPAL experiment at LEP at the Z^0 peak. The $b\overline{b}$ pairs led to hadron jets and the energy of the decay γ was measured by a lead glass calorimeter.

Table 8.2 lists all the bottom mesons observed so far [10]. The $B_c^{*\pm}$ has not been observed yet. It is expected to decay exclusively into $B_c^{\pm}$ with the emission of a 70 MeV photon [18].

8.3 SU(4)

Let us now include the c quark in the flavour symmetry group. The quantum numbers conserved by the strong interaction are i_3, y and C, which are related to the electric charge through (1.2) and (1.4). For mesons

$$Q = i_3 + \frac{S+C}{2} \quad \text{and} \quad y = S - \frac{C}{3}. \tag{8.10}$$

Eliminating S gives the relation between the charge and the three quantum numbers

$$Q = i_3 + \frac{3y+4C}{6}. \tag{8.11}$$

Figure 8.11 shows the weight diagram of the fundamental representation of SU(4). Since the c quark is very heavy the symmetry group is badly broken and is therefore only of limited use. Nevertheless, SU(4) is convenient to classify the hadrons containing c quarks (see also Sect. 17.1 on charmed baryons). For mesons the 16-dimensional representation reduces into a singlet and a 15-plet:

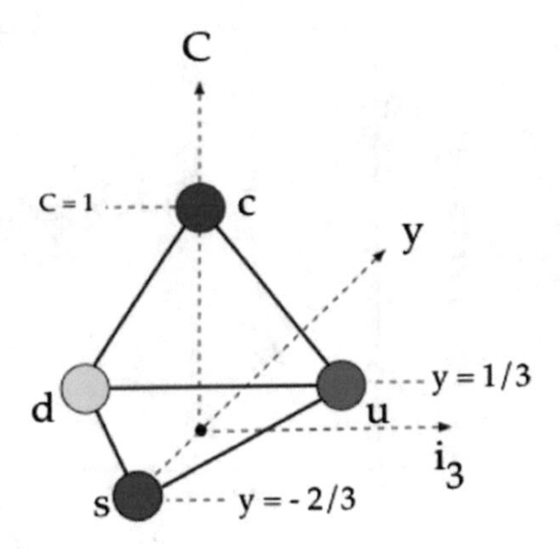

Fig. 8.11 Weight diagram of the fundamental representation of SU(4) including the charm quark

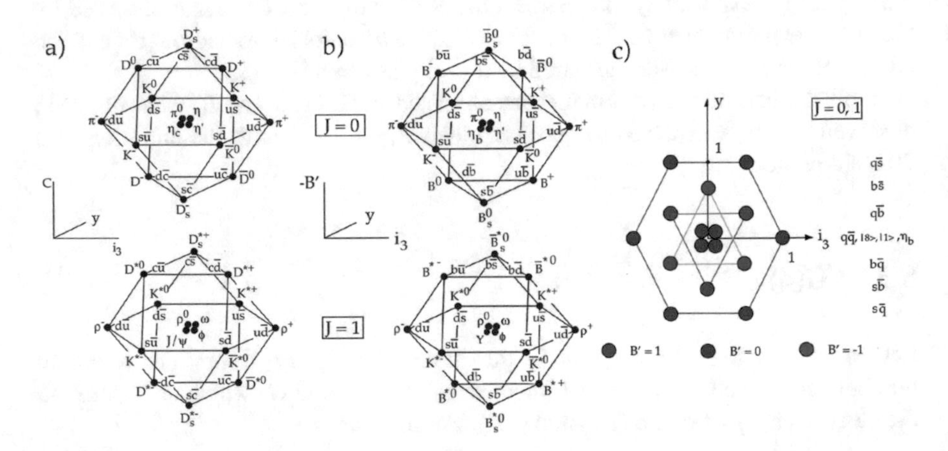

Fig. 8.12 (**a**) SU(4) weight diagrams of $q\overline{q}$ mesons made of four quarks, u, d, s, c [19]; (**b**) with u, d, s and b quarks; (**c**) weight diagram of bottom mesons projected onto the $i_3 - y$ plane [20]

$$\begin{array}{|c|}\hline 4\\\hline\end{array} \times \begin{array}{|c|}\hline 4\\\hline 3\\\hline 2\\\hline\end{array} = \begin{array}{|c|}\hline 4\\\hline 3\\\hline 2\\\hline 1\\\hline\end{array} + \begin{array}{|c|c|}\hline 4 & 5\\\hline 3 & \\\cline{1-1} 2 & \\\cline{1-1}\end{array}$$

$$= 4 \times 4 = \frac{24}{24} + \frac{120}{8} = 1 + 15. \tag{8.12}$$

The weight diagrams of pseudoscalar and vector mesons are shown in Fig. 8.12a. The nonets of SU(3) are now supplemented by a third isoscalar state, the η_c and

J/ψ, respectively. Due to the large mass differences the $c\bar{c}$ isoscalar is not expected to mix significantly with the octet and singlet states of SU(3). Alternatively, one can extend SU(3) to SU(4) by including the b quark and thus with $y = S + \frac{B'}{3}$ the electric charge is equal to

$$Q = i_3 + \frac{3y + 2B'}{6}. \tag{8.13}$$

The weight diagrams of pseudoscalar and vector mesons[3] are shown in Fig. 8.12b and their projections onto the $i_3 - y$ plane in Fig. 8.12c.

The SU(5) group would include all u, d, s, c and b quarks with a 4-dimensional weight diagram which for mesons decomposes into a singlet and a 24-plet. The 25 pseudoscalar mesons contain the 16-plet shown in Fig. 8.12b or c, the 6 D mesons in Fig. 8.12a, the η_c and the $B_c^{\pm}$. These states have all been identified experimentally. Among the corresponding 25 vector mesons only the $B_c^{*\pm}$ has not been observed yet.

References

1. Aubert, J.J., et al.: Phys. Rev. Lett. 33, 1404 (1974)
2. Augustin, J.-E., et al.: Phys. Rev Lett. 33, 1406 (1974)
3. Boyarski, A., et al.: Phys. Rev. Lett. 34, 1357 (1975)
4. Abrams, G.S., et al.: Phys. Rev. Lett. 34, 1181 (1974)
5. Cahn, R.N., Goldhaber, G.: The Experimental Foundations of Particle Physics, p. 257. Cambridge, New York (1989)
6. Rapidis, A., et al.: Phys. Rev. Lett. 39, 526 (1977)
7. Van Royen, H., Weisskopf, V.F.: Nuovo Cimento 50A, 617 (1967)
8. Peruzzi, I., et al.: Phys. Rev. Lett. 37, 569 (1976)
9. Chen, A., et al.: Phys. Rev. Lett. 51, 634 (1983)
10. Tanabashi, M., et al. (Particle Data Group): Phys. Rev. D 98, 030001 (2018)
11. Kodama, K., et al.: Phys. Lett. B 504, 218 (2001)
12. Herb, S.W., et al.: Phys. Rev. Lett. 39, 252 (1977)
13. Andrews, D., et al.: Phys. Rev. Lett. 44, 1108 (1980)
14. Bienlein, J.K., et al.: Phys. Lett. 78B, 360 (1978)
15. Khachatryan, V., et al.: Phys. Lett. B 749, 14 (2015)
16. Aaltonen, T., et al.: Phys. Rev. Lett. 100, 182002 (2008)
17. Ackerstaff, K., et al.: Z. Phys. C 74, 413 (1997)
18. Gershtein, S.S., et al.: Phys. Rev. D 51, 3613 (1995)
19. Amsler, C., DeGrand, T., Krusche, B.: In: Tanabashi, M., et al. (Particle Data Group): Phys. Rev. D 98, 030001 (2018), p. 287
20. Amsler, C.: Nuclear and Particle Physics. IOP Publishing, Bristol (2015)

[3]The hypercharge is sometimes defined as $y = B + S + B' + C + T$ so that relation (1.5) also applies to heavy quark states. However, this definition does not reproduce the elegant SU(4) weight diagrams in Fig. 8.12.

Chapter 9
Quarkonium

Quarkonia are mesons made of $q\bar{q}$ pairs of the same flavour, such as the $J/\psi(1S)$ ($c\bar{c}$ "charmonium"), the $\Upsilon(1S)$ ($b\bar{b}$ "bottomonium"), and even the ϕ ($s\bar{s}$ "strangeonium"), as well as their radial and orbital excitations. These states are bound by the strong interaction mediated by gluons. The term "quarkonium" has been coined in analogy to the electromagnetically bound e^+e^- system (positronium). In the previous chapter we have dealt with the $c\bar{c}$ and $b\bar{b}$ ground state ($\ell = 0$) vector mesons and their radial excitations (ψ and Υ sequences). The present chapter is devoted to their $\ell > 0$ orbital excitations.

Let us first examine the positronium system with the predicted (hydrogen-like) spectrum shown in Fig. 9.1. The hyperfine interaction in the ground state splits the $^1S_0$ (parapositronium) and the $^3S_1$ (orthopositronium) states by a trifling fraction of the positronium mass. Furthermore, the levels are almost degenerate. As we shall see, this contrasts with quarkonium for which the energy splittings are a substantial fraction of the mass. Parapositronium decays into 2γ with a mean life of 124 ps, orthopositronium into 3γ with a mean life of 142 ns. The first observation of positronium dates back to 1951, when positrons from a sodium source were stopped in a gas such as N_2 or O_2 [1]. Annihilation into 2γ was detected by observing the two 511 keV annihilation γ. Quenching of orthopositronium into parapositronium (leading to an increase of prompt decays) was demonstrated by adding a small quantity of nitric oxide gas.

To date the $1S$, $2S$ and $3S$ states have been observed, as well as the $n = 10 - 31$ (Rydberg) states [2]. Evidence for the 1^1P_1 singlet state has been reported in [3].

© The Author(s) 2018

C. Amsler, *The Quark Structure of Hadrons*, Lecture Notes in Physics 949,
https://doi.org/10.1007/978-3-319-98527-5_9

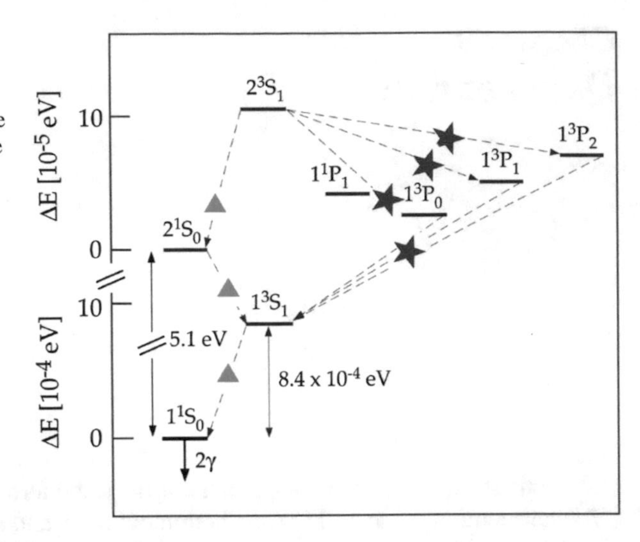

Fig. 9.1 The positronium (e^+e^-) spectrum. Note that the energy differences between radial excitations are much larger (~5 eV) than the hyperfine splitting of the $^3S_1$ and $^1S_0$ levels. The levels with equal n are nearly degenerated. The red stars label E1 and the green triangles M1 transitions

9.1 Charmonium

The $\psi(2S)$ decays OZI suppressed into $J/\psi(1S)\pi\pi$ or into light hadrons with branching fractions of 53% and 15%, respectively [4]. Radiative transitions into lower lying $c\bar{c}$ states ($\ell > 0$) also occur with a probability of about 30%. The $\psi(2S)$ decays electromagnetically into the χ_{cj} states with quantum numbers $\ell = 1$, $j = 0$, 1 or 2 (E1 parity changing transitions) or into the η_c with $\ell = 0$, $j = 0$ (M1 parity conserving transitions). These transitions have been observed first at SLAC by the Crystal Ball detector sketched in Fig. 9.2, see also Fig. 7.8.

The collision energies of the e^+ and e^- beams were tuned to 1843 MeV each to excite the $\psi(2S)$ resonance. The ball of 672 NaI (Tl) crystals recorded all decay photons, among them the monochromatic photons emitted by the transitions to the χ_{cj} (1^3P_j) and η_c (1^1S_0, 2^1S_0) states. The transition lines expected from the decays depicted in the right-hand side of the figure were indeed observed (Fig. 9.3). This is a direct and irrefutable proof that hadrons have an internal structure made of elementary constituents. The measured line widths were determined by detector resolution. The large background continuum stems from photons from π^0 decays that are emitted during the OZI-suppressed decays of the $\psi(2S)$ resonance.

Figure 9.4 shows the $c\bar{c}$ charmonium levels [4]. The first state above charm threshold which decays into $D\overline{D}$ mesons is the $\psi(3770)$, an orbitally excited vector meson already mentioned in Sect. 8.1. The 0^{-+} $\eta_c(2S)$ and the 1^{+-} spin singlet $h_c(1P)$ are now well established, but have been controversial for many years. The hyperfine splitting of the $2S$ level of 47 MeV is much smaller than initially reported by Crystal Ball (labelled 1 in Fig. 9.3).

The search and identification of the $h_c(1P)$ deserves a few more words. Radiative transitions from the (1^{--}) $\psi(2S)$ to the (1^{+-}) $h_c(1P)$ are forbidden by C

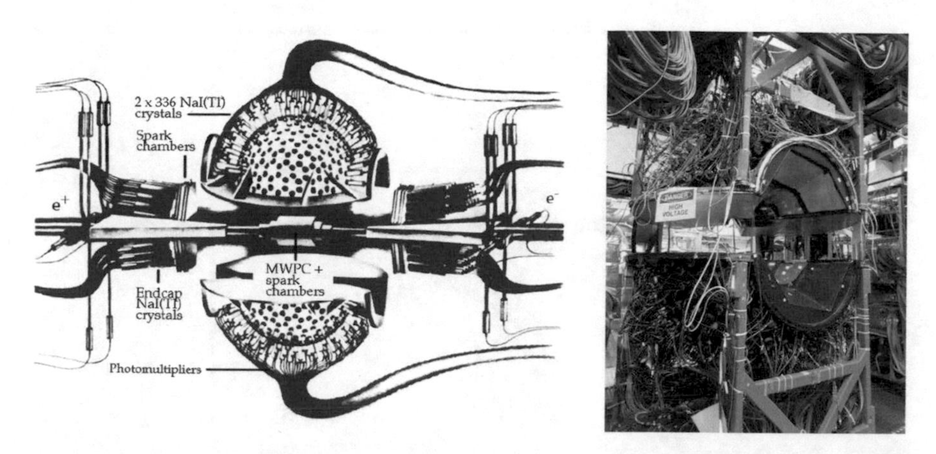

Fig. 9.2 The NaI(Tl) Crystal Ball (left) at the SPEAR storage ring covered a solid angle of 98% $\times 4\pi$ for photon detection [5]. Charged particles could be vetoed with a multiwire proportional chamber (MWPC) and spark chambers. The detector was moved to the DORIS ring at DESY, later to the AGS in Brookhaven and then finally to the MAMI microtron in Mainz (right, photographed by the author)

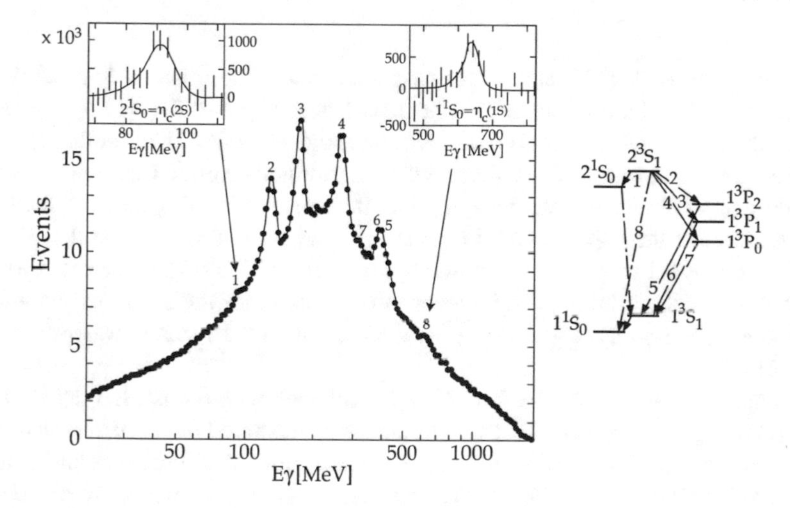

Fig. 9.3 Radiative decay spectrum from the $\psi(2S)$ level observed by the Crystal Ball in e^+e^- collisions. The monochromatic γ lines correspond to the transitions shown on the right (adapted from [6])

conservation and therefore the $h_c(1P)$ cannot be produced in e^+e^- annihilations. The first sign of this state emerged as a $\overline{p}p$ resonance, when protons were collided with antiprotons at the CERN ISR [7]. Since the radiative transitions $h_c(1P) \to \gamma J/\psi(1S)$ are forbidden by C conservation, one expects to see instead transitions to the $J/\psi(1S)$ associated with the emission of pions. Five events were observed

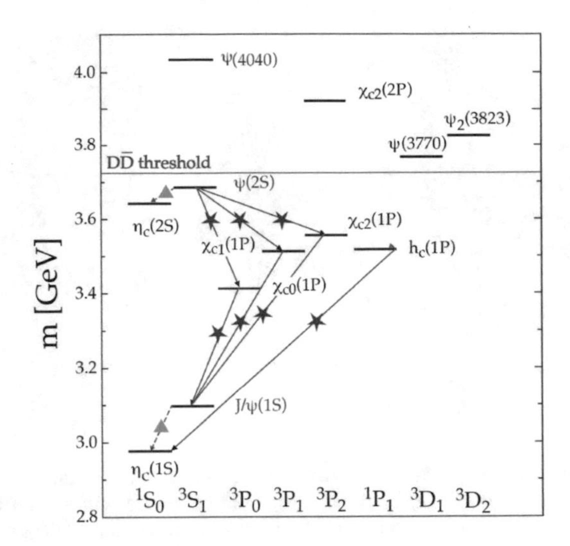

Fig. 9.4 Observed charmonium states and γ transitions. The red stars label E1 transitions and the green triangles M1 transitions. Radiative transitions $\psi(2S) \rightarrow \gamma h_c(1P)$ and $h_c(1P) \rightarrow \gamma J/\psi(1S)$ are forbidden by C parity conservation

from the putative $h_c(1P)$ state at the expected resonance mass of 3525 MeV (the center of gravity of the χ_{cj} states, see Sect. 9.4), decaying into $J/\psi(1S)(\rightarrow e^+e^-)+$ pions. Proton-antiproton annihilation was investigated by the E760 collaboration at the Fermilab antiproton accumulator, where antiprotons collided with protons from a hydrogen jet target. A state decaying into the isospin violating $J/\psi\,\pi^0$ final state was reported at the expected $h_c(1P)$ mass [8], but the claim was withdrawn when no $J/\psi\,\pi^0$ signal was observed with an upgraded version of the detector operated by the E785 collaboration [9]. However, when looking for the E1 transition into the final state $\eta_c(1S)\gamma$, a narrow Γ <1 MeV structure with 15 events was observed at 3526 MeV.

Comprehensive data on the $h_c(1P)$ were obtained with the BESIII detector [10] at the Beijing e^+e^- collider BEPC. The detector, illustrated in Fig. 9.5, consists of a superconducting 1T solenoidal magnet equipped with a drift chamber and a matrix of 6240 CsI(Tl) crystals. Charged particle identification is achieved by dE/dx and by time-of-flight from scintillation counters. Muons are detected by resistive plate chambers interleaved with the iron plates providing the magnetic flux return. Data were taken at the $\psi(2S)$ and π^0 transitions to the $h_c(1P)$ state were sought, with $h_c(1P)$ decaying into $\gamma\eta_c(1S)$, and where the $\eta_c(1S)$ was reconstructed in sixteen different decay channels, such as $\overline{p}p$, $K^+K^-\pi^+\pi^-$, $K^+K^-4\pi^\pm$ or $\eta 4\pi^\pm$ [12]. Note that he transition $\psi(2S) \rightarrow \pi^0 h_c(1P)$ is isospin violating and occurs with a branching ratio f of about 10^{-3}, while $f \simeq 50\%$ for the radiative E1 transition $h_c(1P) \rightarrow \gamma\eta_c(1S)$. The decay chain $\psi(2S) \rightarrow \pi^0 X$, $X \rightarrow \gamma\eta_c(1S)$, $\eta_c(1S) \rightarrow$ final state can be fully reconstructed from energy and momentum conservation. Figure 9.5 (right) shows the distribution of mass m_X recoiling against the π^0. The

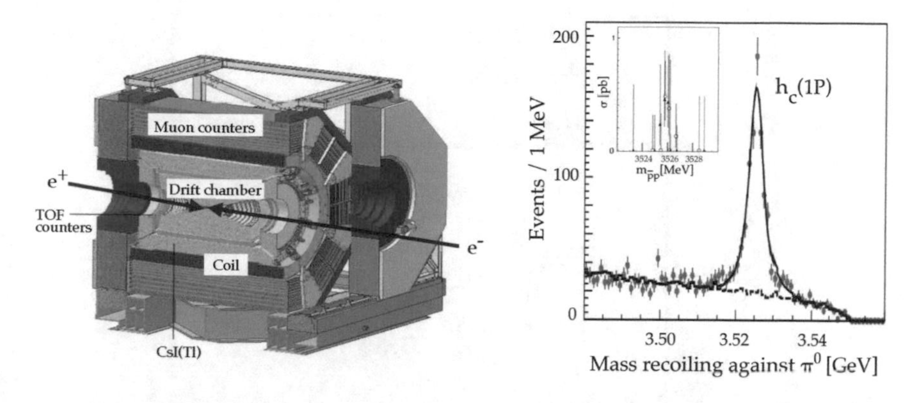

Fig. 9.5 Left: the BESIII detector at the electron-positron collider BEPC [11]. Right: the $h_c(1P)$ observed in $\psi(2S) \to \pi^0 h_c(1P)$, $h_c(1P) \to \gamma\eta_c(1S)$ (adapted from [9, 12]). The inset shows one of the first evidences for the $h_c(1P)$, observed earlier in $\overline{p}p$ annihilations at Fermilab [9]

0.7 MeV narrow signal peaks at a mass of 3525.3 ± 0.2 MeV, which lies exactly at the center-of-mass of the χ_{cj} spin triplet (Sect. 9.4).

Meson spectroscopy in the charm region (see also Sect. 16.2) will be pursued in 2025 by the PANDA experiment at the FAIR proton-antiproton facility near Darmstadt.

9.2 Bottomonium

Figure 9.6 shows the $b\overline{b}$ bottomonium levels that have been observed to date. The $\chi_{bj}(2P)$ orbitals were the first to be seen, produced by radiative decays from the $\Upsilon(3S)$ state [13]. The experiment was performed at the CESR e^+e^- ring by the CUSB detector, an array of 320 rectangular NaI(Tl) crystals in a square geometry, surrounded by lead glass counters [14]. The radiative transitions to the $\chi_{bj}(1P)$ orbitals, followed by the radiative transitions to the $\Upsilon(1S)$ ground state [15] (labelled 1, 2, 3 and 4, 5, 6, respectively in Fig. 9.6), were discovered by CUSP shortly afterwards by running CESR on the $\Upsilon(2S)$.

High statistics data are available from the CLEO III detector at CESR. A sketch of the apparatus, with design similar to BESIII which also features a large modular CsI(Tl) array, is shown in Fig. 9.7. The photon spectra obtained by CLEO III running on the $\Upsilon(2S)$ and $\Upsilon(3S)$ are displayed in Fig. 9.7. The top figure shows the three transition lines 1, 2, 3 on top of a large γ background. The peak around 400 MeV is due to the experimentally unresolved lines 4, 5 and 6. The bottom figure also shows the corresponding three lines from $\Upsilon(3S) \to \gamma\chi_b(2P)$. The peaks around 250 and 800 MeV stem from the subsequent transitions to the $\Upsilon(2S)$ and $\Upsilon(1S)$, respectively. The background was measured by running off the $\Upsilon(2S/3S)$ resonance energies (dashed lines) and by adding the contribution from the ground state measured directly by running on the $\Upsilon(1S)$ (dotted lines).

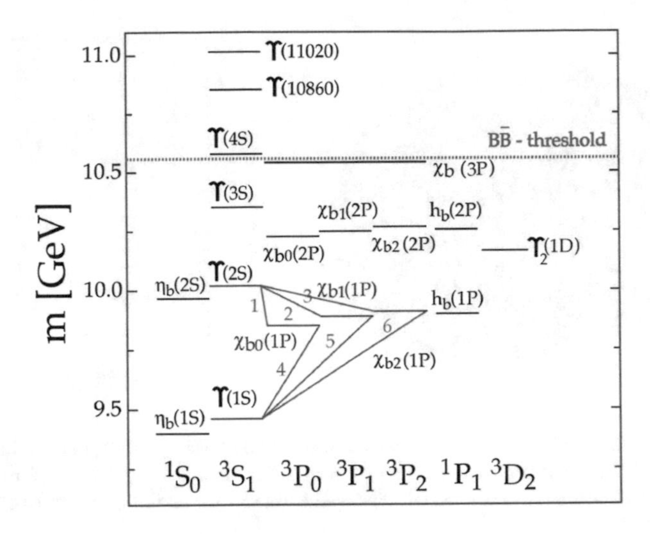

Fig. 9.6 Observed bottomonium states (see the text)

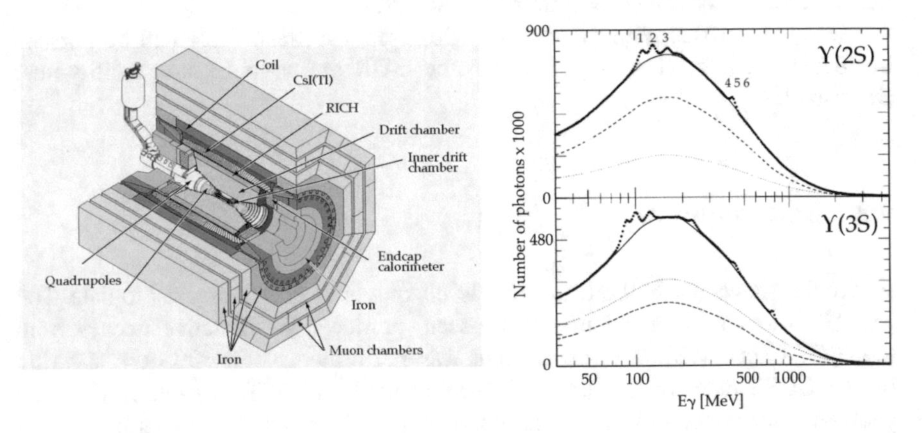

Fig. 9.7 Left: the CLEO III / CLEO-c detector at the Cornell electron-positron collider CESR [16]. Right: radiative transitions from the $\Upsilon(2S)$ and $\Upsilon(3S)$ states [17] (see the text and Fig. 9.6)

The $3P$ levels, which lie about 30 MeV below the open bottom threshold, have been reported by the ATLAS collaboration at the LHC running at a collision energy of 7 TeV [18]. Events with $\Upsilon(1S)$ or $\Upsilon(2S)$ were selected by their decays into $\mu^+\mu^-$ (Fig. 9.8, left), and γ transitions from the $1P, 2P, 3P$ levels to the $\Upsilon(1S)$ and from the $2P, 3P$ levels to the $\Upsilon(2S)$ were detected by the liquid argon calorimeter. The $3P$ levels are observed for the first time (Fig. 9.8, right). The experimental resolution of the calorimeter was not sufficient to resolve the fine splittings of the P levels. The mass splitting of $\sim$11 MeV between the $\chi_{b1}(3P)$ and $\chi_{b2}(3P)$ has been measured in CMS by using γ conversion into e^+e^- pairs [19].

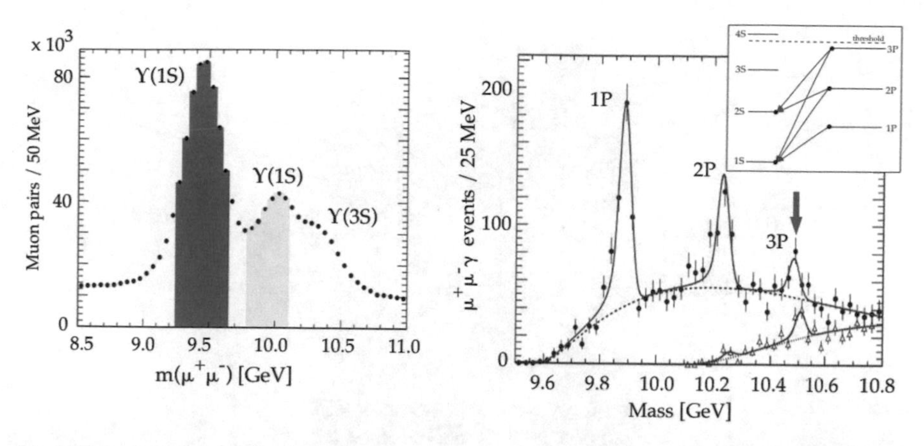

Fig. 9.8 Left: $\Upsilon(1S)$ and $\Upsilon(2S)$ selection windows (blue and yellow, respectively) for ATLAS events decaying into muon pairs. The latter window is asymmetric to reduce the contamination from the $\Upsilon(3S)$. Right: distribution of $\mu^+\mu^-\gamma$ events as a function of $m(\mu^+\mu^-\gamma) - m(\mu^+\mu^-) + m[\Upsilon(1S \text{ or } 2S)]$ (adapted from [18]). The peaks are due to the radiative transitions to the $\Upsilon(1S)$ (top red curve) and to the $\Upsilon(2S)$ (bottom pink curve). The inset shows the corresponding transitions

Table 9.1 Measured splittings (in MeV) between the S and P levels of charmonium and bottomonium below open threshold

$c\bar{c}$		$b\bar{b}$						
2^3S_1 -1^3S_1	2^3S_1 $-1P_j$	2^3S_1- -1^3S_1	2^3S_1 $-1P_j$	3^3S_1 -2^3S_1	3^3S_1 $-2P_j$	4^3S_1 -3^3S_1	4^3S_1 $-3P_j$	J^{PC}
590	271	563	164	332	123	224		0^{++}
	175		130		100		49[a]	1^{++}
	130		111		86			2^{++}
	161		124		95			1^{+-}

[a]Fine structure unresolved

Table 9.1 lists the measured level splittings between S and P levels in bottomonium, and also includes those of charmonium for comparison.

The lightest bottomonium state, the $\eta_b(1S)$, was established by the BaBar collaboration at SLAC's PEP-II electron-positron collider. It was observed in the radiative decay spectrum of the $\Upsilon(3S)$ [20]. The $\eta_b(2S)$, $h_b(1P)$ and $h_b(2P)$ were studied by the Belle collaboration at the KEKB electron-positron collider in Japan [21]. The $h_b(2P)$ and $h_b(1P)$ decayed into $\gamma\eta_b(1S)$ / $\gamma\eta_b(2S)$ and $\gamma\eta_b(1S)$, respectively. They were produced via the emission of two charged pions from

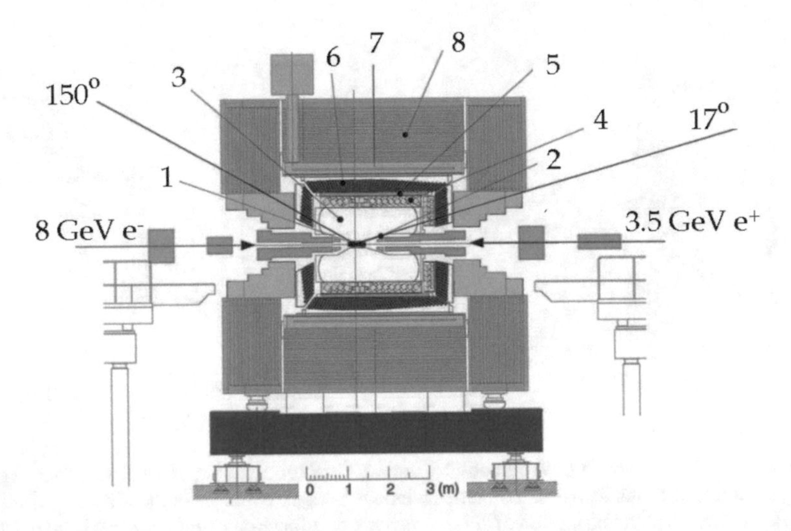

Fig. 9.9 The Belle detector at the KEKB asymmetric e^+e^- collider [22]. 1—silicon vertex detector, 2—forward BGO calorimeter, 3—drift chamber, 4—particle identifier (aerogel), 5—time-of-flight counters, 6—CsI(Tl) γ-calorimeter, 7—superconducting solenoid, 8—muon detector (resistive plate counters)

the $\Upsilon(5S)$. A drawing of the Belle detector is shown in Fig. 9.9. The lepton beams collide with different energies, hence the center of mass moves along the electron beam direction.[1]

9.3 Potential Models of Quarkonium

The measured excitation spectrum of quarkonium ($s\bar{s}$, $c\bar{c}$ and $b\bar{b}$) shown in Fig. 9.10 suggests that the mass differences between the radial excitations is roughly independent of quark flavour, e.g the $2S$ and $1S$ levels are separated by $\sim$600 MeV for all three flavours. Furthermore, the excitation energy is relatively small compared to at least the c and b quark masses, and hence one expects heavy quarkonia to be non-relativistic systems (this will be justified below), which can be described by a suitable flavour independent potential and by solving the Schrödinder equation. Also, the level spacings and splittings are much larger in quarkonium than in

[1]In symmetric e^+e^- colliders the electron and positron collide with the same energies, hence the B and $\overline{B}$ mesons produced at the $\Upsilon(4S)$ are emitted back-to-back and the collision point, where the B and $\overline{B}$ are produced, cannot be determined. In asymmetric colliders both B mesons are emitted into the same hemisphere. The production point can then be reconstructed to study oscillations and CP violation in the $B^0 - \overline{B}^0$ system (see e.g [23]).

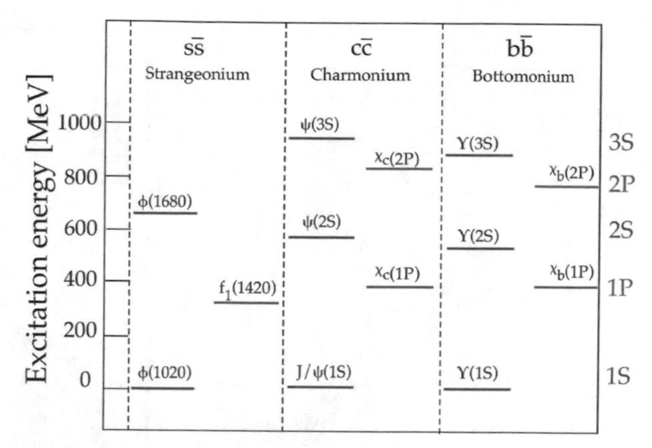

Fig. 9.10 Excitation energies of the quarkonia S and P levels

positronium (Fig. 9.1), due to the much stronger binding force mediated by gluon exchange.

The level spacings in quarkonium decrease much less rapidly with increasing radial number n than for positronium, where they decrease as $1/n^2$. Furthermore, in quarkonium the first ($1P$) and second ($2P$) orbitals lie approximately halfway between the S levels, in contrast to positronium, where S and P levels are nearly degenerate. Thus a heavy $q\bar{q}$ system resembles roughly the 3-dimensional harmonic oscillator [24] for which the potential levels are equidistant and equal to $\frac{3}{2}\hbar\omega$, $\frac{7}{2}\hbar\omega$, $\frac{9}{2}\hbar\omega, \ldots$ for $\ell = 0$), while the $\ell = 1$ orbitals lie in-between ($\frac{5}{2}\hbar\omega$, $\frac{9}{2}\hbar\omega$, ..., see Fig. 15.2 in Sect. 15.1). A reasonable ansatz for the potential energy binding the $q\bar{q}$ pair is therefore between $1/r$ (Coulomb potential) and r^2 (harmonic potential). The Cornell potential (energy) is of the form [25]

$$V(r) = -\frac{a}{r} + br, \tag{9.1}$$

where the "constant" a is related to the strong coupling α_s which decreases with energy. A fit to the charmonium spectrum gave $a = 0.52$ and $b = 0.925\,\text{GeV fm}^{-1}$ [25]. The potential energy, expressed in GeV with (1.7), is then approximately given by

$$V(r)[\text{GeV}] \sim -\frac{0.1}{r[\text{fm}]} + 0.9\,r[\text{fm}]. \tag{9.2}$$

The $1/r$ term, inspired by the $1/r$ Coulomb potential for photons, describes the exchange of a spin-1 gluon and dominates at short distances. The second term arises from the exchange of many gluons. The potential energy increases with r and reproduces the quark confinement $V(\infty) = \infty$. Translated into SI units the force becomes $b = 0.9\,\text{GeV fm}^{-1} = 1.4 \times 10^5\,\text{J m}^{-1} = 1.4 \times 10^5\,\text{N}$.

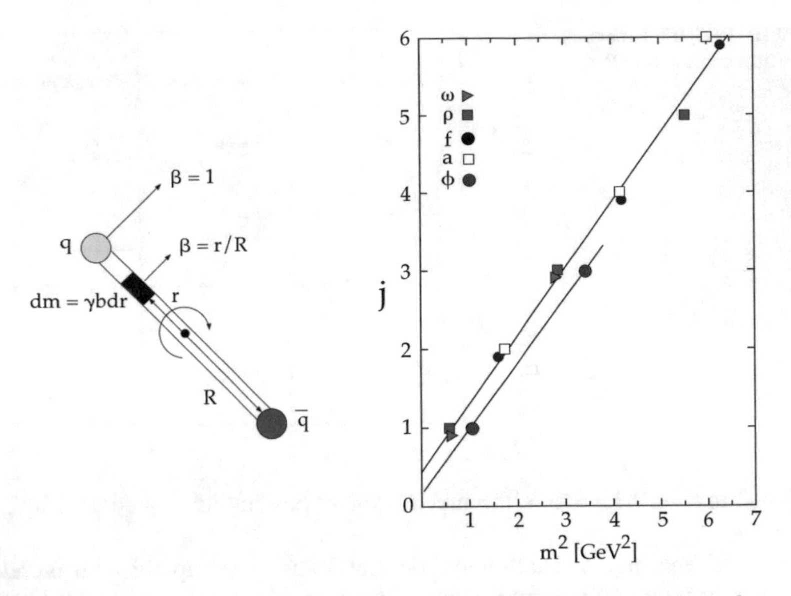

Fig. 9.11 Left: light quark meson modelled as a gluon tube with the (nearly) massless quarks rotating at the speed of light. Right: Chew-Frautschi plot of the Regge trajectories of the spin triplet mesons ($J^{PC} = 1^{--}, 2^{++}, 3^{--}, 4^{++}, 5^{--}, 6^{++}$)

Let us briefly return to the light quarks. Assuming a separation of 1 fm gives a potential energy of ~800 MeV, which is the typical mass of ground state mesons (such as the ρ). Hence the mass of light quark mesons is dominated by gluons. This contrasts with heavy mesons for which the main contribution stems from the bare masses of the c and b quarks. Naively, a $q\overline{q}$ pair could be viewed as consisting of a rotating pair connected by a gluon tube of length $2R$ (Fig. 9.11). For light mesons the quarks are almost massless and therefore rotate with nearly the speed of light. The contribution of the gluons to the meson mass m at a distance r from the center is $dm = \gamma b dr = b dr/\sqrt{1-\beta^2}$ with $b \approx 0.9\,\text{GeV fm}^{-1}$ and $\beta = r/R$.
The mass m and the angular momentum ℓ of the meson is then given by

$$m = 2\int_0^R \frac{b dr}{\sqrt{1-\beta^2}} = \pi R b \quad \text{and} \quad \ell = 2\int_0^R \frac{r \cdot \beta b dr}{\sqrt{1-\beta^2}} = \frac{\pi R^2 b}{2}. \tag{9.3}$$

By eliminating R one finds that $\ell = \frac{m^2}{2\pi b}$. By adding a constant which takes into account (i) the spin of the pair ($s = 0$ or 1) and (ii) the (small) quark masses one obtains the Regge trajectory j vs m^2:

$$j = \frac{m^2}{2\pi b} + \text{const} \Rightarrow \boxed{j \simeq 0.9\, m^2[\text{GeV}^2] + \text{const}}, \tag{9.4}$$

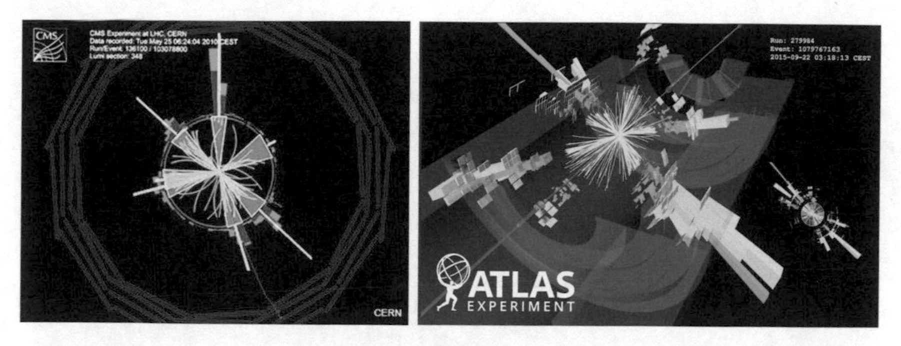

Fig. 9.12 Left: a 6-jet event recorded by the LHC/CMS experiment at 7 TeV collision energy. Right: a multijet event observed with the ATLAS detector at 13 TeV (image credits CMS and ATLAS/CERN)

where we have expressed b in GeV^2 by using the unit conversion (1.7). For example, the Regge trajectory of the (u, d) spin triplet mesons (Fig. 9.11) is indeed linear with a slope of 0.9 GeV^{-2}. That of the ϕ mesons is displaced due to the larger s mass.

When r increases the excitation energy pair is stored in $V(r)$ until the creation of a new quark-antiquark pair from vacuum becomes more economical. The two pairs then fly apart (and may in turn also split), leading to hadrons emitted within narrow cones (a process called fragmentation or hadronization). The observation of such jets at high collision energies (Fig. 9.12) is another direct evidence for the existence of the quark substructure in hadrons.

Let us now estimate the average kinetic energies of the quarks and the interquark distance. The virial theorem states that a bound system of N bodies has a total mean kinetic energy given by

$$\langle T \rangle = -\frac{1}{2}\langle \sum_{k=1}^{N} \vec{F}_k \cdot \vec{r}_k \rangle, \tag{9.5}$$

where $\vec{F}_k$ denotes the force acting on body k at the space coordinate $\vec{r}_k$. For a two-body system with separation distance r one gets

$$\langle T \rangle = \frac{1}{2}\left\langle r \frac{dV(r)}{dr} \right\rangle. \tag{9.6}$$

For a power law $V \propto r^n$ this becomes simply

$$\langle T \rangle = \frac{n}{2}\langle V \rangle. \tag{9.7}$$

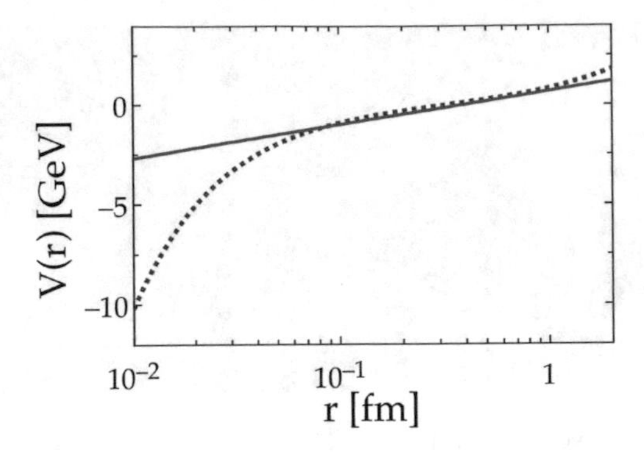

Fig. 9.13 Quarkonium potential energy V as a function on interquark distance r for the power law (9.1) (dotted curve) and logarithmic law (9.9) (straight line)

Hence for the Cornell potential (9.1) the quark-antiquark pair share the average total kinetic energy

$$\langle T \rangle = \frac{1}{2}\left(\frac{a}{\langle r \rangle} + b\langle r \rangle\right), \tag{9.8}$$

about 500 MeV for an average distance of 1 fm. An alternative potential which reproduces (9.1) in the relevant range is that from Ref. [26] (Fig. 9.13),

$$V(r) = \kappa \ln \frac{r}{r_0} \tag{9.9}$$

with $\kappa = 0.733$ GeV and $r_0 = 0.39$ fm. The virial theorem (9.6) then gives the average total kinetic energy

$$\langle T \rangle = \frac{\kappa}{2} = 365\,\text{MeV}, \tag{9.10}$$

independent of $\langle r \rangle$. The energy is shared between the quark and the antiquark which can be treated non-relativistically. Hence the energy levels of charmonium and bottomonium (and to a lesser extent strangeonium $s\bar{s}$) can be predicted with the potential $V(r)$ by solving the Schrödinger equation. This has been performed for various potentials such as (9.1) [25], (9.9) [26] or even with potentials of the form $A + Br^{\nu}$ with $\nu = 0.104$ [27]. The agreement with data is impressive for the ground states, as Fig. 9.14 testifies, as well as for the fine and hyperfine structures that are described in the next subsection. Relativistic effects have been taken into account in [30]. For details on theoretical and experimental aspects on quarkonia we refer to the homepage of the Quarkonium Working Group [31].

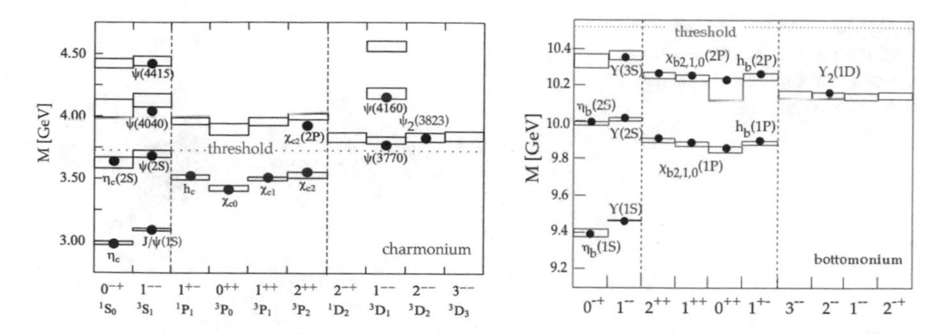

Fig. 9.14 Left: predicted charmonium spectrum (boxes) compared with the available data (black dots) [28]. Right: bottomonium spectrum (extracted from the predictions in Ref. [29])

Finally, let us estimate the interquark distance from the uncertainty relation:

$$\langle r\rangle \cdot \langle p\rangle = \langle r\rangle \cdot \sqrt{\langle T\rangle m} \simeq 1 \tag{9.11}$$

with $\langle T\rangle = 365\,\text{MeV}$ leads to $\langle r\rangle = 0.15\,\text{fm}$ for $b\bar{b}$ and $0.27\,\text{fm}$ for $c\bar{c}$, where we have used the quark masses $m_b = 4.86\,\text{GeV}$ and $m_c = 1.45\,\text{GeV}$ from the potential of Ref. [27].

9.4 Fine and Hyperfine Splitting

Let us now deal with the $\ell > 0$ orbitals, in particular the P levels, and start with a general theorem on the ordering of the energy levels $E(n, \ell)$ associated with solutions of the Schrödinger equation. The quantum number $n = 0, 1, 2, \ldots$ is the number of nodes of the wavefunction (the principal quantum number being $N = n + \ell + 1$). It has been shown [32] that, for potential energies $V(r)$ satisfying the following conditions for all value of $r > 0$,

$$\Delta V(r) = 0 \Rightarrow E(n, \ell) = E(n-1, \ell+1), \tag{9.12}$$

$$\Delta V(r) < 0 \Rightarrow E(n, \ell) < E(n-1, \ell+1), \tag{9.13}$$

$$\Delta V(r) > 0 \Rightarrow E(n, \ell) > E(n-1, \ell+1), \tag{9.14}$$

where Δ is the Laplace operator which, applied on spherically symmetric potentials reads

$$\Delta V(r) = \frac{d^2V}{dr^2} + \frac{2}{r}\frac{dV}{dr}. \tag{9.15}$$

For example, the potential energy of an electron in the Coulomb field satisfies $\Delta V(r) = -\frac{\rho(r)}{\epsilon_0}(-e)$, where $\rho(r)$ is the charge density (Poisson's equation). For the

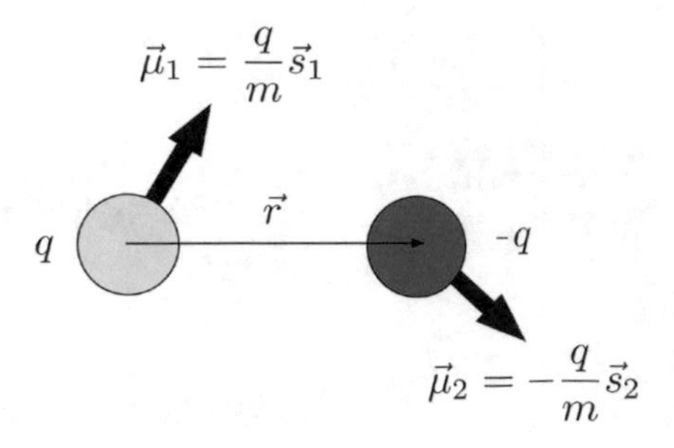

Fig. 9.15 The non-central tensor force acts between the quark and antiquark magnetic moments $\vec{\mu}_1$ and $\vec{\mu}_2$

hydrogen atom $\rho(r) = 0$, except in the proton (which is assumed here to be point-like). Hence condition (9.12) predicts the degeneracy of the energy levels. In alkaline atoms the single valence electron feels the negative charge of the inner electrons, hence from (9.13) the P levels are predicted correctly to lie above the S levels.

Returning to quarkonium, we find for the Cornell potential energy (9.1) that

$$\Delta V(r) = -\frac{2a}{r^3} + \frac{2}{r} \cdot \frac{a}{r^2} + \frac{2b}{r} = \frac{2b}{r} > 0. \tag{9.16}$$

The P states lie below the S states, which is verified experimentally, e.g. the $\chi_{c0,1,2}$ $(1P)$ levels lie below the $\psi(2S)$. The same conclusion also applies to the logarithmic potential (9.9) and to the 3-dimensional harmonic oscillator (Problem 9.1).

The level fine splitting is due to the interaction between the angular momentum $\vec{L}$ and the quark spins (LS coupling) and to the tensor interaction T between the quark and antiquark magnetic moments (Fig. 9.15), while the hyperfine splitting (ss) arises from the very short range interaction between the quark and antiquark spins. With the Coulomb-like potential $-a/r$ the treatment follows the same lines as for the hydrogen atom, but the additional scalar term br in (9.1) is included to take into account the long range confinement forces. This term contributes to the LS coupling only [33].

The interaction Hamiltonian between two spin-$\frac{1}{2}$ particles with equal masses m is written as (see e.g. [33])

$$H = H_{LS} + H_T + H_{ss}, \tag{9.17}$$

where

$$H_{LS} = -\frac{b}{2m^2 r}[\vec{L} \cdot \vec{S}] + \frac{2\alpha_s}{m^2 r^3}[\vec{L} \cdot \vec{S}],$$

$$H_T = \frac{\alpha_s}{3m^2 r^3}[3(\vec{\sigma}_1 \cdot \vec{n})(\vec{\sigma}_2 \cdot \vec{n}) - \vec{\sigma}_1 \cdot \vec{\sigma}_2],$$

$$H_{ss} = \frac{8\pi\alpha_s}{9m^2}[\vec{\sigma}_1 \cdot \vec{\sigma}_2]\,\delta^3(\vec{r}), \tag{9.18}$$

with the unit vector $\vec{n} = \frac{\vec{r}}{r}$, $\vec{S} = \vec{s}_1 + \vec{s}_2$ and $\vec{\sigma}_{1,2} = 2\vec{s}_{1,2}$.

These contributions have been derived in first order perturbation (Breit-Fermi Hamiltonian), therefore large QCD corrections are expected since α_s is large. The level spacings decrease with quark mass. Let us now calculate the expectation values of the operators (9.18) and absorb some of the constant terms and the unknown radial distributions into proportionality constants:

$$\Delta E_{LS} \propto \langle \vec{L} \cdot \vec{S} \rangle, \tag{9.19}$$

$$\Delta E_T \propto \langle 3(\vec{\sigma}_1 \cdot \vec{n})(\vec{\sigma}_2 \cdot \vec{n}) - \vec{\sigma}_1 \cdot \vec{\sigma}_2 \rangle, \tag{9.20}$$

$$\Delta E_{ss} \propto \frac{1}{m^2} |\psi(0)|^2, \tag{9.21}$$

where $\psi(0)$ is the wavefunction at $r = 0$, which vanishes for $\ell > 0$. Thus this term contributes only to the $\ell = 0$ states. With

$$2\langle \vec{L} \cdot \vec{S} \rangle = \langle \vec{J}^{\,2} - \vec{L}^{\,2} - \vec{S}^{\,2} \rangle = [j(j+1) - \ell(\ell+1) - s(s+1)] \tag{9.22}$$

one finds for $\ell = s = 1$

$$\Delta E_{LS} \propto \begin{pmatrix} -4 \\ -2 \\ +2 \end{pmatrix} \text{ for } j = \begin{pmatrix} 0 \\ 1 \\ 2 \end{pmatrix}, \tag{9.23}$$

while LS coupling does not contribute to $\ell = 0$ nor to spin singlet states. The tensor force also vanishes for spin singlet states, since for $\vec{S} = 0$, $\vec{\sigma}_1 = -\vec{\sigma}_2$ and with (9.20)

$$(\vec{\sigma}_1 \cdot \vec{n})^2 = 1 \ , \ (\vec{\sigma}_1)^2 = 3 \Rightarrow \Delta E_T = 0. \tag{9.24}$$

The expectation values of the operator (9.20) for spin triplet states are given by [34]

$$\Delta E_T \propto \begin{pmatrix} \frac{-2(\ell+1)}{2\ell-1} \\ 2 \\ \frac{-2\ell}{2\ell+3} \end{pmatrix} \text{ for } j = \begin{pmatrix} \ell - 1 \\ \ell \\ \ell + 1 \end{pmatrix}, \tag{9.25}$$

hence for P states

$$\Delta E_T \propto \begin{pmatrix} -4 \\ 2 \\ -\frac{2}{5} \end{pmatrix} \text{ for } j = \begin{pmatrix} 0 \\ 1 \\ 2 \end{pmatrix}. \tag{9.26}$$

Table 9.2 shows the 1^3P_j mass splittings as a function of two proportionality constants A and B, compared to the masses of the $\chi_{cj}(1P)$ levels. Since the spin singlet $h_c(1P)$ does not experience any spin force, one expects its mass to lie at the

Table 9.2 1^3P_J mass splittings in the charmonium sector, due to the LS coupling and the tensor force T, compared with experimental data [4]

j	LS	T	$M_{\chi_{cj}}$ [MeV]
0	$-2A$	$-4B$	3414.7
1	$-A$	$2B$	3510.7
2	A	$-\frac{2}{5}B$	3556.2

center of gravity of the triplet states, where

$$5 \times \left(A - \frac{2B}{5}\right) + 3 \times (-A + 2B) + 1 \times (-2A - 4B) = 0, \tag{9.27}$$

taking into account the spin multiplicities. The equation is fulfilled and the mass differences reproduced with $A = 35.0\,\text{MeV}$ and $B = 10.2\,\text{MeV}$. The predicted $h_c(1P)$ mass is

$$M(c\bar{c}) = \frac{5M(\chi_{c2}) + 3M(\chi_{c1}) + M(\chi_{c0})}{9} = 3525.3\,\text{MeV}, \tag{9.28}$$

in impressive agreement with the measured value (3525.3 ± 0.2 MeV, Sect. 9.1). The 2^{++} state is correctly predicted to be the least bound state.

Finally, let us show that the size of bottomonium is much smaller than that of charmonium. Experimentally, the hyperfine splitting of the $1S$ level is 114 MeV for charmonium ($J/\psi(1S) - \eta_c(1S)$) and 61 MeV for bottomonium ($\Upsilon(1S) - \eta_b(1S)$) [4]. From the contact interaction (9.21) one then estimates that

$$\frac{m_b^2}{m_c^2}\frac{|\psi(0)|^2_\psi}{|\psi(0)|^2_\Upsilon} \sim 2, \tag{9.29}$$

where we assume α_s in (9.18) to be equal ($\simeq 0.5$) for charmonium and bottomonium [33]. Using the bare quark masses one then finds the ratio of overlap probabilities at the origin

$$\frac{|\psi(0)|^2_\psi}{|\psi(0)|^2_\Upsilon} \sim 0.17, \tag{9.30}$$

which demonstrates that the $b\bar{b}$ mesons are much smaller than the $c\bar{c}$ ones.

References

1. Deutsch, M.: Phys. Rev. 82, 455 (1951)
2. Aghion, S., et al.: Phys. Rev. A 94, 012507 (2016), and references therein
3. Conti, R.S., et al.: Phys. Lett. A 177, 43 (1993)
4. Tanabashi, M., et al. (Particle Data Group): Phys. Rev. D 98, 030001 (2018)
5. Bloom, E.D.: International Symposium on Lepton and Photon Interaction at High Energy, Batavia, SLAC-PUB-2425 (1979)
6. Partridge, R., et al.: Phys. Rev. Lett. 45, 1150 (1980)
7. Baglin, C., et al.: Phys. Lett. B 171, 135 (1986)
8. Armstrong, T.A., et al.: Phys. Rev. Lett. 69, 2337 (1992)
9. Andreotti, M., et al.: Phys. Rev. D 72, 032001 (2005)
10. Ablikim, M., et al.: Nucl. Instrum. Methods Phys. Res. A 614, 345 (2010)
11. Dobbs, S.: Hadron 2017 Conference, Salamanca, Spain (2017)
12. Ablikim, M., et al.: Phys. Rev. D 86, 092009 (2012)
13. Han, K., et al.: Phys. Rev. Lett. 49, 1612 (1982)
14. Lee-Franzini, J.: AIP Conf. Proc. 424, 85 (1998)
15. Klopfenstein, C., et al.: Phys. Rev. Lett. 51, 160 (1983)
16. Briere, R.A., et al.: Report CLNS 01/1742 (2001)
17. Artuso, M., et al.: Phys. Rev. Lett. 94, 032001 (2005)
18. Aad, G., et al.: Phys. Rev. Lett. 108, 152001 (2012)
19. Sirunyan, A.M., et al.: Phys. Rev. Lett. 121, 092002 (2018)
20. Aubert, B., et al.: Phys. Rev. Lett. 101, 071801 (2008)
21. Mizuk, R., et al.: Phys. Rev. Lett. 109, 23200 (2012)
22. Abashian, A., et al.: Nucl. Instrum. Methods Phys. Res. 479, 117 (2002)
23. Amsler, C., Nuclear and Particle Physics, section 20.2. IOP Publishing, Bristol (2015)
24. Cohen-Tannoudji, C., Diu, B., Laloë, F.: Quantum Mechanics, 1, p. 819. Wiley, New York (1997)
25. Eichten, E., et al.: Phys. Rev. D 21, 203 (1980)
26. Quigg, C., Rosner, J.L.: Phys. Rep. 56, 167 (1979)
27. Martin, A.: Phys. Lett. 93B, 338 (1980)
28. Pakhlova, G.V., Pakhlov, P.N., Eidelman, S.I.: Physics-Uspekhi 53, 219 (2010)
29. Brambilla, N., et al.: CERN Yellow Report CERN-2005-005 (2005). arXiv:hep-ph/0412158
30. Godfrey, S., Isgur, N.: Phys. Rev. D 32, 189 (1985)
31. Quarkonium Working Group: http://www.qwg.to.infn.it/
32. Baumgartner, B., Grosse, H., Martin, A.: Nucl. Phys. B 254, 528 (1985)
33. Buchmuller, W.: Phys. Lett. 112B, 479 (1982)
34. Cahn, R.N., Jackson, J.D.: Phys. Rev. D 68, 037502 (2003)

Chapter 10
Colour Charge

The interaction between quarks is mediated by eight gluons through the exchange of a strong interaction colour charge. Quarks carry a colour charge which is a conserved quantity. The underlying symmetry corresponding to the conservation of colour is $SU(3)_c$. The colour charge was introduced in 1965 by Han and Nambu [1] as the source of the force between quarks, although the word itself had been introduced shortly before by Greenberg [2]. The concept of coloured quarks and gluons was further developed by many other authors, among them Bogolyubov et al. [3] and Fritzsch et al. [4].

Before the introduction of colour the wavefunctions of the spin-$\frac{3}{2}$ ground state baryons with three equal flavours (Δ^{++}, Δ^- and Ω^-) appeared to violate the Pauli principle which requires the wavefunction of identical fermions to be antisymmetric under permutations. For example, three identical fermions compose the fully symmetric wavefunction

$$\Omega^- = |sss\rangle| \uparrow\uparrow\uparrow\rangle|L = 0\rangle, \tag{10.1}$$

where L denotes the overall orbital angular momentum from the three quarks. The introduction of a the new quantum number of colour solved the conundrum and forms the basis of Quantum Chromodynamics (QCD): quarks exist in three "colours", say red R, green G, and blue B for each flavour. A baryon RGB baryon is colourless (or colour neutral, the concept will be clarified below) and with the quark flavours $q_1q_2q_3$ consists of a superposition of the six permutations of colour. Antisymmetry is restored by multiplying the flavour × spin × wavefunction by the antisymmetric colour wavefunction

$$\begin{aligned}\beta(q_1q_2q_3) = \frac{1}{\sqrt{6}}|q_1(R)q_2(G)q_3(B) + q_1(B)q_2(R)q_3(G) + q_1(G)q_2(B)q_3(R) \\ -q_1(G)q_2(R)q_3(B) - q_1(B)q_2(G)q_3(R) - q_1(R)q_2(B)q_3(G)\rangle.\end{aligned} \tag{10.2}$$

© The Author(s) 2018

C. Amsler, *The Quark Structure of Hadrons*, Lecture Notes in Physics 949,
https://doi.org/10.1007/978-3-319-98527-5_10

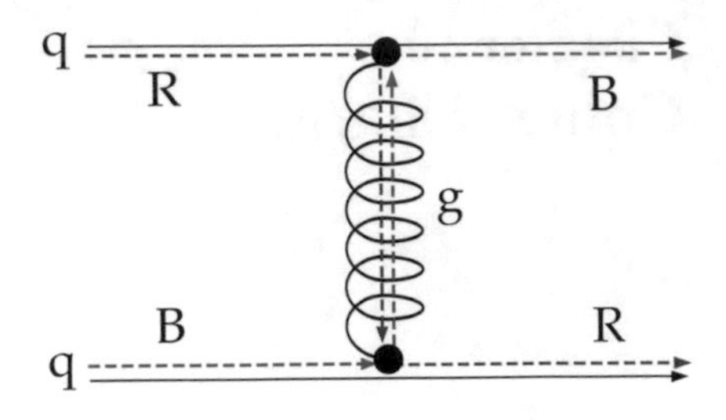

Fig. 10.1 The interaction between quarks occurs by exchanging coloured gluons. The dashed lines show the colour fluxes

For example, the colour wavefunction of the Ω^- is the antisymmetric combination

$$\beta(\Omega^-) = \frac{1}{\sqrt{6}}|s(R)s(G)s(B) + s(B)s(R)s(G) + s(G)s(B)s(R)$$
$$-s(G)s(R)s(B) - s(B)s(G)s(R) - s(R)s(B)s(G)\rangle. \quad (10.3)$$

Antiquarks also come in three "anticolours", $\overline{R}$, $\overline{G}$, $\overline{B}$ (which we shall depict below as cyan, pink and yellow). A meson is a linear superposition of quarks with colours and antiquarks with the corresponding anticolours, so that the meson system is colourless. For instance, the colour wavefunction of the π^+ is given by

$$\beta(\pi^+) = \frac{1}{\sqrt{3}}|u(R)\overline{d}(\overline{R}) + u(G)\overline{d}(\overline{G}) + u(B)\overline{d}(\overline{B})\rangle. \quad (10.4)$$

The interaction between quarks is mediated by gluons through the exchange of colour, which is conserved (Fig. 10.1). A gluon therefore consists of one colour and one anticolour. Nine gluons can be constructed from three colours and three anticolours. To identify their colour wavefunctions we can use as templates the $\mathrm{SU}(3)_f$ wavefunctions in which we replace the u, d, and s flavours with the colours R, G and B, see Table 5.1, and the isoscalar wavefunctions (7.27, 7.28). Seven $\mathrm{SU}(3)_c$ wavefunctions are readily obtained:

$$|R\overline{G}\rangle, |G\overline{R}\rangle, \frac{1}{\sqrt{2}}|G\overline{G} - R\overline{R}\rangle, |R\overline{B}\rangle, |G\overline{B}\rangle, |B\overline{G}\rangle, |B\overline{R}\rangle. \quad (10.5)$$

To the two $\mathrm{SU}(3)_f$ isoscalars correspond the octet

$$\frac{1}{\sqrt{6}}|R\overline{R} + G\overline{G} - 2B\overline{B}\rangle, \quad (10.6)$$

and the singlet

$$\frac{1}{\sqrt{3}}|R\overline{R} + G\overline{G} + B\overline{B}\rangle. \quad (10.7)$$

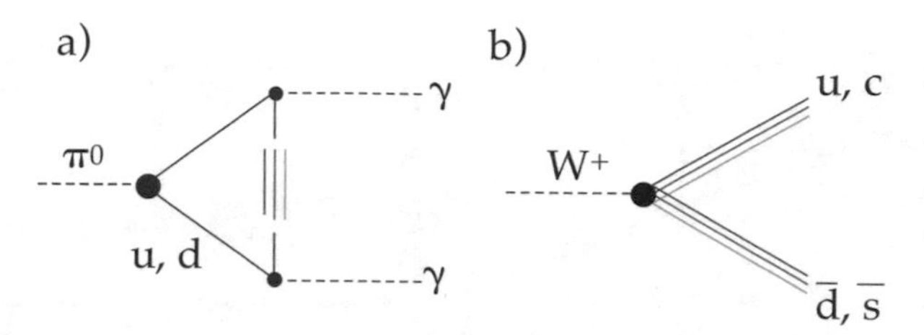

Fig. 10.2 (**a**) π^0 decay into 2γ. The quark colours increase the amplitude of the triangular diagram by a factor of 3; (**b**) the W^+ boson decays into $u\overline{d}$ and $c\overline{s}$ with a probability of 2×3 (colour) $\times$ $0.11 \simeq 66\%$. The decays into $e^+\overline{\nu}_e$, $\mu^+\overline{\nu}_\mu$ and $\tau^+\overline{\nu}_\tau$ account for the remaining 34%

Colour would increase the number of hadrons. Since there is e.g. only one proton, only colourless hadrons appear in nature. Free quarks and free gluons are not observed because quarks and gluons carry colour. However, the colourless singlet (10.7) is not observed either. Therefore only the eight gluons with colour wavefunctions (10.5) and (10.6) exist. The corresponding eight fields are the gauge fields of QCD.

The existence of colour has been demonstrated experimentally. Let us review three important experimental facts:

1. Without three colours the lifetime of the neutral pion would be a factor of 9 longer (see the decay diagram in Fig. 10.2a). The partial width for $\pi^0 - \gamma\gamma$ is given by [5]

$$\Gamma_{\gamma\gamma} = \frac{\alpha^2 m^3}{576\pi^3 (f_\pi/\sqrt{2})^2} N^2 = 7.73\,\mathrm{eV} \tag{10.8}$$

where α is the fine structure constant, m the π^0 mass, $f_\pi = 130.4\,\mathrm{MeV}$ the pion decay constant and $N = 3$ quark colours. The present experimental accuracy at the 2% level [6] ($\Gamma_{\gamma\gamma} = 7.7 \pm 0.2\,\mathrm{eV}$) determines that $N = 3$ with very high accuracy.

2. The W^+ boson decays into hadrons with a measured branching ratio of about 66%, and into $e^+\nu_e$, $\mu^+\nu_\mu$ and $\tau^+\nu_\tau$ with branching ratios of about 11% each. The hadronic decay is due to $W^+ \to u\overline{d}$ and $c\overline{s}$ (Fig. 10.2b), the decays into $u\overline{b}$, $u\overline{s}$, $c\overline{d}$ and $c\overline{b}$ being Cabibbo suppressed.
3. At sufficiently high energies quark-antiquark pairs ($u\overline{u}$, $d\overline{d}$, $s\overline{s}$, $c\overline{c}$ and $b\overline{b}$) leading to hadrons and lepton pairs (e^+e^-, $\mu^+\mu^-$ and $\tau^+\tau^-$) are produced in e^+e^- collisions via the virtual annihilation photon. Figure 10.3 (left) shows the annihilation cross section as a function of center-of-mass energy $\sqrt{s}$. The production of the vector mesons is observed above the smooth non-resonant (nr) background from $e^+e^- \to \mu^+\mu^-$. Figure 10.3 (right) shows the ratio R of the cross section for hadron production to that for non-resonant $\mu^+\mu^-$ pairs. The

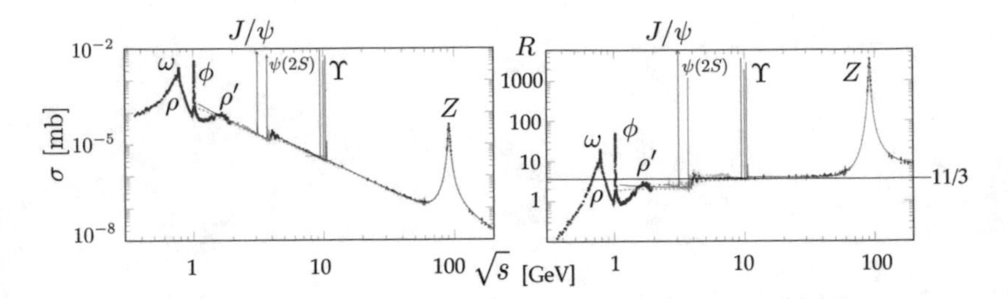

Fig. 10.3 Left: measured cross section for e^+e^- annihilation into hadrons. Right: cross section for e^+e^- annihilation into hadrons divided by the non-resonant cross section $\sigma(e^+e^- \to \gamma \to \mu^+\mu^-)_{nr} = \frac{4\pi\alpha^2(s)}{3s}$. For $N = 3$ colours $R = \frac{11}{3}$ [6]

ratio R is given by

$$R = \frac{\sigma(e^+e^- \to \text{hadrons})}{\sigma(e^+e^- \to \mu^+\mu^-)_{nr}} = \frac{\alpha^2 \sum q_i^2}{\alpha^2}\left[1 + \frac{\alpha_s(Q)}{\pi}\right] = \sum_i q_i^2 \left[1 + \frac{\alpha_s(Q)}{\pi}\right], \tag{10.9}$$

where Q is the momentum transfer between initial and final states, α the fine structure constant and α_s the strong coupling. The numerator sums over the quark charges q_i. At sufficiently high energies all masses are negligible so that the phase space factors cancel between numerator and denominator. Above the $b\bar{b}$ threshold of about 10 GeV one finds, ignoring the small term in brackets and assuming N quark colors,

$$R = N\left[\left(\frac{2}{3}\right)^2 + \left(-\frac{1}{3}\right)^2 + \left(-\frac{1}{3}\right)^2 + \left(\frac{2}{3}\right)^2 + \left(-\frac{1}{3}\right)^2\right] = \frac{11}{9}N, \tag{10.10}$$

which for $N = 3$ is in excellent agreement with measurements (Fig. 10.3, right).

Next, let us couple a coloured quark (of any flavour) to a coloured antiquark. We have already done so for $\mathrm{SU}(3)_f$ in (7.23), hence for $\mathrm{SU}(3)_c$

$$3_c \times 3_c^* = 1_c + 8_c. \tag{10.11}$$

We note the occurrence of a colour singlet meson. Coupling two quarks gives

$$\begin{array}{|c|}\hline 3 \\ \hline\end{array} \times \begin{array}{|c|}\hline 3 \\ \hline\end{array} = \begin{array}{|c|}\hline 3 \\ \hline 2 \\ \hline\end{array} + \begin{array}{|c|c|}\hline 3 & 4 \\ \hline\end{array}$$

$$= 3_c \times 3_c = 3_c^* + 6_c \tag{10.12}$$

and therefore a two-quark system is not colour neutral so that hadrons made of two quarks do not exist (the force between two quarks is repulsive). Let us now add a third quark:

$$\left(\begin{array}{|c|}\hline 3\\\hline 2\\\hline\end{array} \times \begin{array}{|c|c|}\hline 3 & 4\\\hline\end{array}\right) \times \begin{array}{|c|}\hline 3\\\hline\end{array} = \begin{array}{|c|}\hline 3\\\hline 2\\\hline 1\\\hline\end{array} + \begin{array}{|c|c|}\hline 3 & 4\\\hline 2 & \multicolumn{1}{c}{}\\\cline{1-1}\end{array} + \begin{array}{|c|c|}\hline 3 & 4\\\hline 2 & \multicolumn{1}{c}{}\\\cline{1-1}\end{array} + \begin{array}{|c|c|c|}\hline 3 & 4 & 5\\\hline\end{array}$$

$$= 3_c \times 3_c \times 3_c = 1_c + 8_c + 8_c + 10_c, \tag{10.13}$$

which includes a colour singlet baryon. It is left to the reader to check that anticolour singlet antibaryons $\overline{qqq}$ also exist.

The question now arises as to whether further quark configurations could also build colour neutral "exotic" hadrons. Let us first couple two antiquarks:

$$\begin{array}{|c|}\hline 3\\\hline 2\\\hline\end{array} \times \begin{array}{|c|}\hline 3\\\hline 2\\\hline\end{array} = \begin{array}{|c|c|}\hline 3 & 4\\\hline 2 & \multicolumn{1}{c}{}\\\cline{1-1} 1 & \multicolumn{1}{c}{}\\\cline{1-1}\end{array} + \begin{array}{|c|c|}\hline 3 & 4\\\hline 2 & 3\\\hline\end{array}$$

$$= 3_c^* \times 3_c^* = 3_c + 6_c^*. \tag{10.14}$$

A tetraquark combines two quarks and two antiquarks,

$$\begin{aligned}(q_1q_2)(\overline{q}_3\overline{q}_4) \ &: \ (3_c^* + 6_c) \times (3_c + 6_c^*)\\ &= \underbrace{3_c^* \times 3_c}_{1_c+8_c} + \underbrace{6_c \times 6_c^*}_{1_c+8_c+27_c} + \underbrace{6_c \times 3_c}_{8_c+10_c} + \underbrace{3_c^* \times 6_c^*}_{8_c+10_c^*},\end{aligned} \tag{10.15}$$

(Problem 10.1). The $3_c^* \times 3_c$ and $6_c \times 6_c^*$ decompositions both include a singlet. It is interesting to note (Fig. 10.4) that the former (T-type) couples to baryon-antibaryon pairs (such as $\overline{p}p$) but not the latter (M-type). These tetraquark mesons are predicted to exist and several candidates have been observed (Sect. 16.1).

The pentaquark ($\overline{q}qqqq$ configuration) is a baryon which includes a colour singlet. Grouping the quarks sequentially one can construct a colour singlet:

$$\begin{aligned} q_1q_2 &: 3_c \times 3_c = 3_c^* + 6_c,\\ q_3q_4 &: 3_c \times 3_c = 3_c^* + 6_c,\\ (q_1q_2)(q_3q_4) &: 3_c^* \times 3_c^* = 3_c + 6_c^*,\\ (q_1q_2q_3q_4)\overline{q} &: 3_c \times 3_c^* = 1_c + 8_c. \end{aligned} \tag{10.16}$$

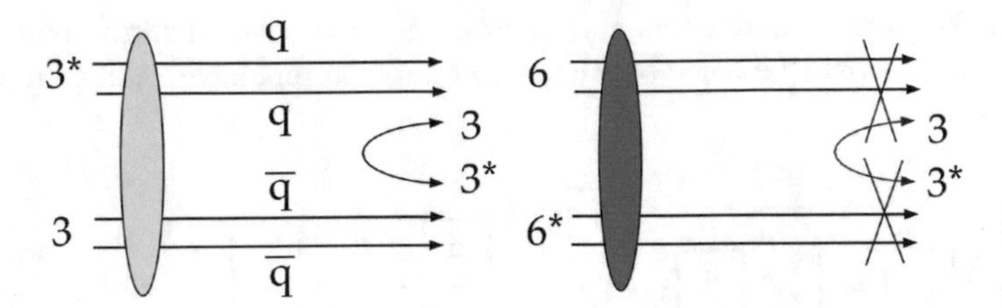

Fig. 10.4 The $3_c \times 3^*_c$ tetraquark (T-type) couples to a baryon-antibaryon pair (left) while for the M-type ($6^*_c \times 6$) the $3_c \times 6_c$ coupling does not lead to a pair of colour neutral baryon-antibaryon (right)

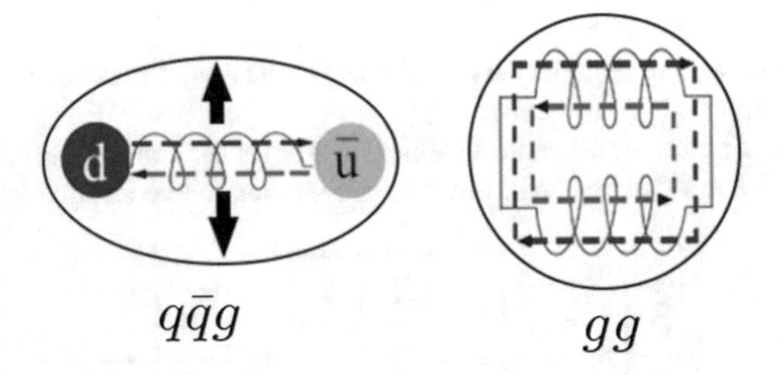

Fig. 10.5 Cartoons of a hybrid meson (left) in which the binding gluon is excited, and of a two-gluon glueball (right) which does not contain any quark

Pentaquarks made of light quarks were reported some years ago but their existence has been disputed. We shall return to pentaquarks in Sect. 16.3 and discuss the evidence for a heavy pentaquark containing a $c\bar{c}$ pair that was reported at the LHC.

More complicated quark configurations are also possible, such as $3q3\overline{q}$ (baryonium), $6q$, and $2q2q2q$ (dibaryon). Figure 10.5 (left) shows a hybrid meson ($q\overline{q}g$) in which the binding glue is excited. Candidates such as the $\pi_1(1600)$ have been reported (Sect. 11.3). Since the gluons as gauge fields of QCD carry colour charge, two or more gluons may exchange their colours to build a colour neutral hadron. The prediction of such hadrons is a remarkable feature of QCD based on its non-abelian structure, in contrast to the field quanta (the photons) of QED which do no bind. Figure 10.5b shows the cartoon of a colourless bound state of two gluons, the glueball. This hadron does not contain any quark but only the field quanta of the strong interaction. Indeed the coupling of two $SU(3)_c$ colour octets contains a colour singlet (Problem 10.2). QCD calculations on the lattice predict the lightest glueball to be a scalar meson with mass between 1500 and 1800 MeV (see Fig. 11.1 below). The $f_0(1500)$ or $f_0(1710)$, which will be discussed in more detail in the next section, have been proposed as candidates.

To complete this chapter let us illustrate the colour contents of some of these states, composed here of the three light quarks only. The pairing of two flavours $3_f \times 3_f$, say a red quark with a blue quark, decomposes into $3^* + 6$ representations

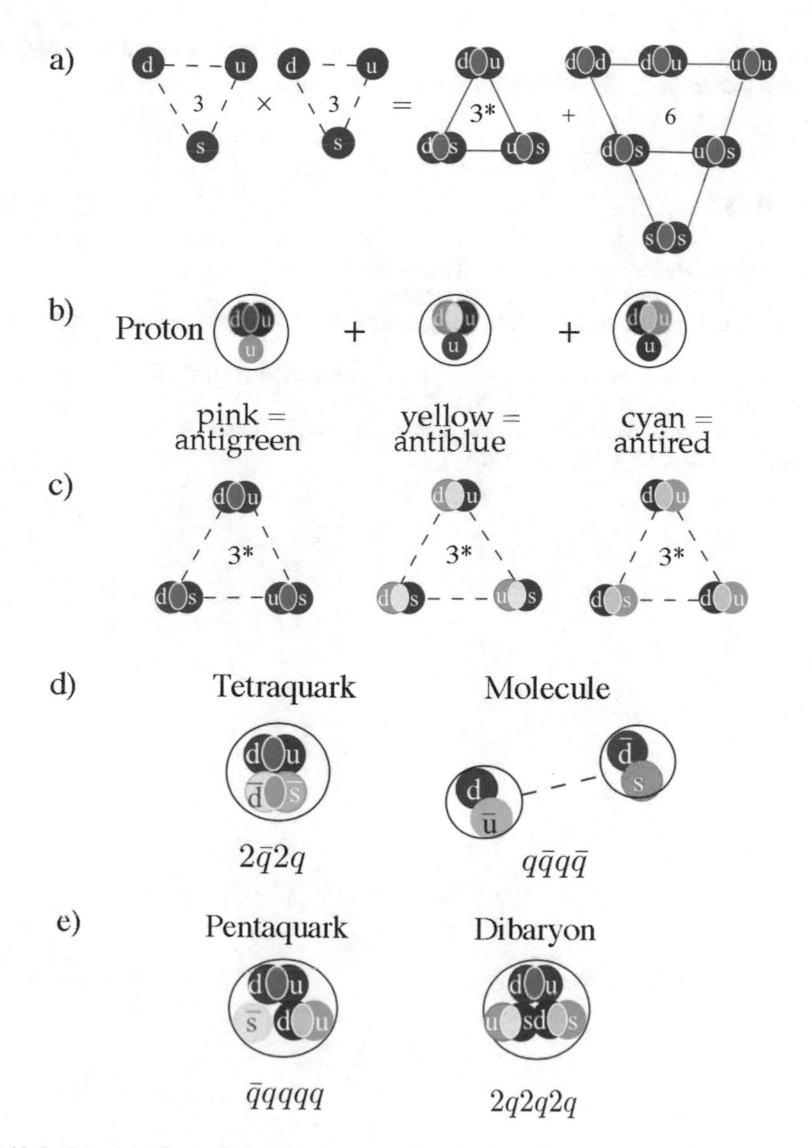

Fig. 10.6 Colour configurations of the proton and various exotic hadrons with three light quarks, see the text (adapted from [7])

of both $\mathrm{SU}(3)_f$ and $\mathrm{SU}(3)_c$ (row (a) in Fig. 10.6). For the proton in row (b), made of red, blue and green quarks, the red and blue quarks are in a 3_c^* antigreen configuration (which we depict as pink). The same reasoning applies to green-red and blue-green couplings, the anticolours yellow and cyan, respectively, row (c). The tetraquark, row (d), is either a compact colourless $2\bar{q}2q$ hadron (e.g. antigreen [pink] × green) or a $q\bar{q}q\bar{q}$ molecule made of two mesons (e.g. green × antigreen [pink] and red × antired [cyan]) loosely bound by pion exchange. Likewise, row (e) shows the pentaquark as an antigreen [pink], antired [cyan], and antiblue [yellow] state and the $2q2q2q$ dibaryon with the same anticolour configuration.

References

1. Han, M.Y.: Nambu, Y.: Phys. Rev. 139, B1006 (1965)
2. Greenberg, O.W.: Phys. Rev. Lett. 13, 598 (1964)
3. Bogolyubov, N.N., Struminsky, B.V., Tavkhelidze, A.N.: JINR Publication D-1968, Dubna (1965)
4. Fritzsch, H., Gell-Mann, M., Leutwyler, H.: Phys. Lett. B 47, 365 (1973)
5. Miskimen, R.: Annu. Rev. Nucl. Part. Sci. 61, 1 (2011)
6. Tanabashi, M., et al. (Particle Data Group): Phys. Rev. D 98, 030001 (2018)
7. Olsen, S.L.: Front. Phys. 10, 121 (2015)

Chapter 11
Glueballs and Hybrid Mesons

A brief phenomenological review on the status of glueballs and hybrid mesons will now be presented (tetraquarks will be dealt width in Chap. 16). Among the $i = 0$ scalar mesons at least two candidates, the $f_0(1500)$ and the $f_0(1710)$, dispute the status of ground state glueballs. The first excited glueball state, expected to be a tensor meson, has not been identified yet, although more tensor states have been reported than can be accommodated in the 2^{++} $q\overline{q}$ nonets. There is evidence for two isovector mesons, $\pi_1(1400)$ and $\pi_1(1600)$, with exotic quantum numbers 1^{-+} that are incompatible with $q\overline{q}$ states and which could be hybrid mesons or tetraquark states.

11.1 Glueballs

Glueballs are isoscalar states that do not fit in $q\overline{q}$ nonets, either because their decay modes are incompatible with $q\overline{q}$ states or because they are supernumerary. Their production should be enhanced in gluon-rich channels such as central production in high energy collisions (in which the incident projectiles are scattered under small angles), radiative $J/\psi(1S)$ decay into light quarks (which is mediated by gluon exchanges), and $\overline{p}p$ annihilation. Glueballs should be suppressed in $\gamma\gamma$ collisions due to the absence of γ-gluon coupling.

The ground state glueball is predicted by lattice gauge theories to be a scalar (0^{++}) and the first excited state a tensor (2^{++}). Figure 11.1 shows the predicted mass spectrum [1]. The mass of the ground state is predicted to be 1700 MeV (with an uncertainty of about 100 MeV). The first excited state has a mass of about 2400 MeV. The lightest glueball with exotic quantum numbers 2^{+-} lies around 4200 MeV.

© The Author(s) 2018

C. Amsler, *The Quark Structure of Hadrons*, Lecture Notes in Physics 949,
https://doi.org/10.1007/978-3-319-98527-5_11

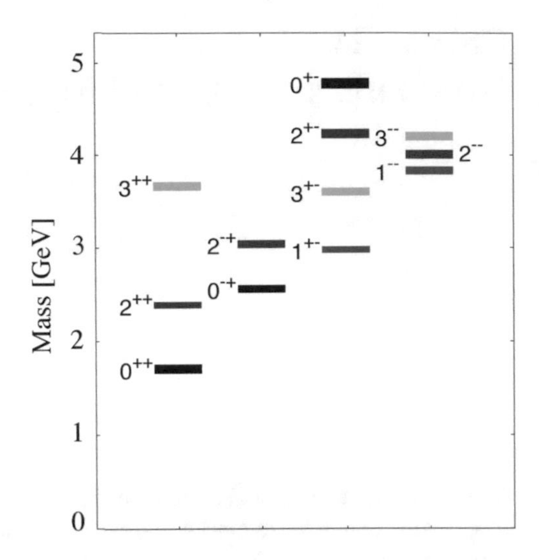

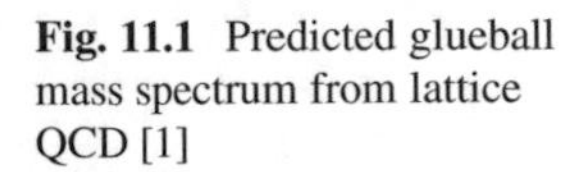
Fig. 11.1 Predicted glueball mass spectrum from lattice QCD [1]

The low mass glueballs lie in the same mass region as ordinary isoscalar $q\bar{q}$ states, that is for the ground state in the mass range of the $1^3P_0(0^{++})$ mesons and for the first excited state in the mass range of the 2^3P_2, 3^3P_2, $1^3F_2(2^{++})$ mesons. Therefore glueballs will mix with nearby $q\bar{q}$ states of the same quantum numbers. The $i = 0$ scalar mesons in the 1400–1700 MeV range will mix with the predicted pure ground state glueball to generate the observed physical states $f_0(1370)$, $f_0(1500)$, and $f_0(1710)$ [2, 3]. The lattice calculations are made in the quenched approximation that neglects virtual $q\bar{q}$ pairs, hence mixing with scalar mesons. However, first results including quark loops suggest that mass shifts are small.

The classification of scalar mesons has been the subject of vivid discussions for many years and is still controversial. However, the consensus is that there are too many scalar mesons to fit in the ground state 0^{++} $q\bar{q}$ nonet. Table 11.1 shows a plausible classification scheme, among others that have been proposed (details and references can be found in the Review of Particle Physics [4, 5]) The low mass nonet is made of four-quark states (and/or meson-meson resonances). The ground state $1^3P_0(0^{++})$ $q\overline{q}$ nonet lies in the 1400 MeV region.

We shall argue below that $f_0(1500)$ contains a large fraction of glue and that $f_0(1710)$ is dominantly $s\overline{s}$. Let us deal first with the $a_0(980)$ and $f_0(980)$ mesons. The $a_0(980)$ decays into $\eta\pi$ and $K\overline{K}$, the $f_0(980)$ into $\pi\pi$ and $K\overline{K}$. This suggests that their wavefunctions contain a significant fraction of $s\overline{s}$ pairs, which is not possible for the $q\overline{q}$ isovector $a_0(980)$, but possible for a tetraquark (e.g. $us\overline{d}\overline{s}$ for the a_0^+). The mesons below 1 GeV have been interpreted as tetraquarks [6], to which we shall return in more detail in Sect. 16.1.

Let us now discuss the isovector and isoscalar 1^3P_0 states. The $a_0(1450)$ decaying into $\eta\pi$ has been discovered by the Crystal Barrel experiment (Sect. 3.2) in $\overline{p}p$ annihilation with stopped antiprotons [7]. The $\pi^0\pi^0\eta$ Dalitz plot is shown in

Table 11.1 Tentative classification of the scalar mesons

	Γ [MeV]	i	Structure
$a_0(980)$	∼ 50	1	$K\overline{K}, qq\bar{q}\bar{q}$
$f_0(980)$	∼ 50	0	$K\overline{K}, qq\bar{q}\bar{q}$
$f_0(500)$	∼ 800	0	$\pi\pi, qq\bar{q}\bar{q}$
$\kappa(700)$	∼ 600	$\frac{1}{2}$	$K\pi, qq\bar{q}\bar{q}$
$a_0(1450)$	265	1	$u\bar{d}, d\bar{u}, d\bar{d} - u\bar{u}$
$f_0(1370)$	∼ 400	0	$d\bar{d} + u\bar{u}$
$f_0(1710)$	125	0	$s\bar{s}$
$K_0^*(1430)$	294	$\frac{1}{2}$	$u\bar{s}, d\bar{s}, s\bar{u}, s\bar{d}$

The resonances below 1 GeV would be tetraquark states which recombine at large distances to become meson-meson resonances. In the ground state $q\overline{q}$ nonet (1^3P_0, see also Fig. 5.1) there are three isoscalar states, $f_0(1370)$, $f_0(1710)$ and the additional $f_0(1500)$ (not shown). According to [2, 3], the $f_0(1370)$ and $f_0(1710)$ would be dominantly $q\overline{q}$ states mixing with glue, while the $f_0(1500)$ would be dominantly a glueball mixing with $q\overline{q}$

Fig. 11.2. For unequal final state masses it is more convenient to use as independent parameters the invariant masses squared, m_{12}^2 vs. m_{13}^2 (Appendix B). The event distribution in the Dalitz plot deviates strongly from homogeneity, indicating the presence of several interfering resonances. One observes the $a_0(980)$ and the $a_2(1320)$ decaying to $\eta\pi$ as well as the $f_0(980)$ decaying into $\pi\pi$. Due to strong interferences with these mesons the $a_0(1450)$, a 270 MeV broad state around 1450 MeV, is not directly visible, but is required by the fit. A general method to determine spins from angular distribution will be described in Chap. 18.

The $f_0(1370)$ and $f_0(1500)$ mesons were established by Crystal Barrel in their $\eta\eta$ and $\pi^0\pi^0$ decay modes [9]. Figure 11.3 shows the $\pi^0\eta\eta$ and $\pi^0\pi^0\pi^0$ Dalitz plots for $\overline{p}p$ annihilation at rest. The ∼100 MeV broad $f_0(1500)$ is clearly seen, together with other isoscalars decaying into $\pi^0\pi^0$ (recall that $\pi^0\pi^0$ has $i = 0$). For the $f_0(1370)$, which is not directly visible, the fit requires a much broader width of ∼350 MeV.

Crystal Barrel also studied $\overline{p}p$ annihilation into $K_LK_L\pi^0$ [10]. The angles of the two K_L were measured by the K_L interaction in the CsI calorimeter. The goal was to measure the decay branching ratios of the $f_0(1370)$ and $f_0(1500)$ into $K\overline{K}$. The Dalitz plot (Fig. 11.3, right), shows prominent contributions from the K^* and the ($s\overline{s}$) $f_2'(1525)$, with no sign of the $f_0(1500)$. Neither $f_0(1370)$ nor $f_0(1500)$ have a large coupling to $K\overline{K}$ [10] with the ratio of branching fractions $K\overline{K}/\pi\pi \ll 1$, indicating that neither state can have a large $s\overline{s}$ component.

The $f_0(1370)$, $f_0(1500)$ and $f_0(1710)$, decaying to $K\bar{K}$ and $\pi\pi$, were also reported by CERN's WA102 experiment in pp central production at 450 GeV [11]. The $f_0(1370)$ and $f_0(1500)$ appear to prefer $\pi\pi$ over $K\overline{K}$ decay. Hence

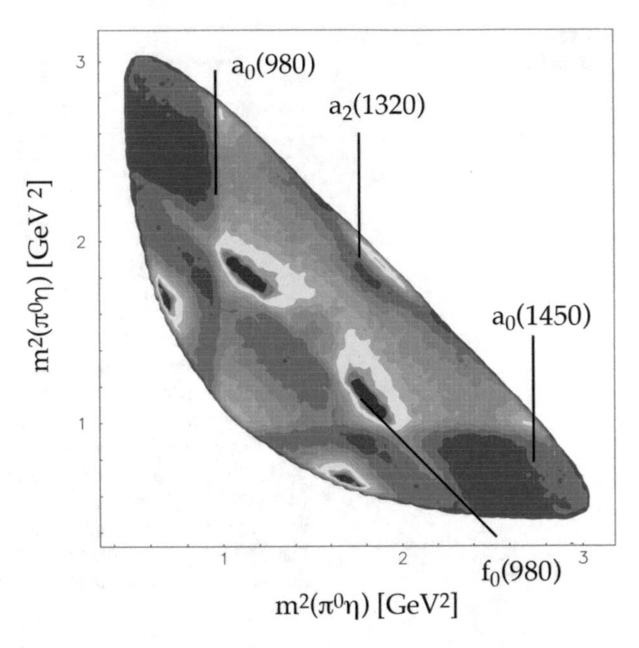

Fig. 11.2 $\pi^0\pi^0\eta$ Dalitz plot (2.8×10^5 events) in $\overline{p}p$ annihilation at rest into 6γ [7] (colour picture from [8])

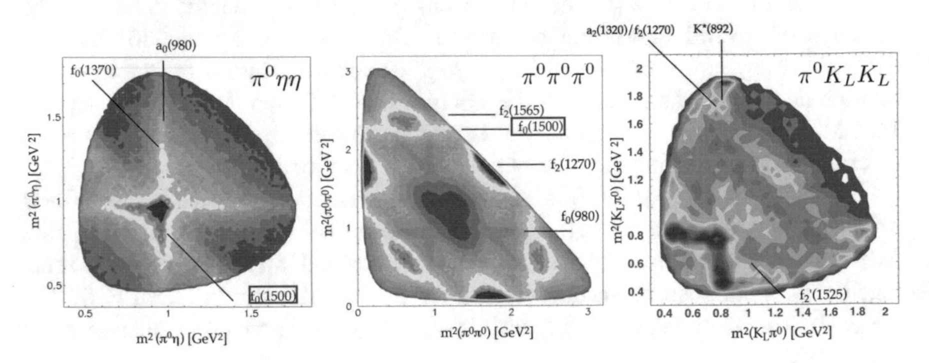

Fig. 11.3 $\pi^0\eta\eta$ (2.0×10^5 events) and $\pi^0\pi^0\pi^0$ Dalitz plots (7.1×10^5 events) in $\overline{p}p$ annihilation at rest into 6γ; $\pi^0 K_L K_L$ Dalitz plot [10] (3.7×10^4 events) (colour pictures from [8])

the $f_0(1370)$ and $f_0(1500)$ do not have large $s\bar{s}$ components, in agreement with Crystal Barrel results. As far as the $f_0(1710)$ is concerned, WA102 reports that $K\bar{K}$ decay dominates $\pi\pi$ by a factor $\sim$5 [11], thus suggesting that this state must be dominantly $s\bar{s}$.

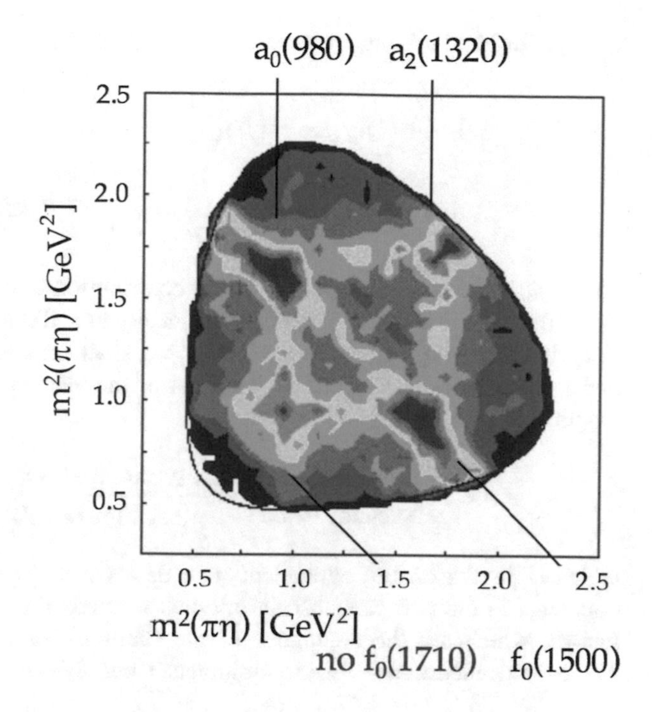

Fig. 11.4 $\pi^0\eta\eta$ Dalitz plot with 900 MeV/c antiprotons [12]

The proton having no constituent s quark, the OZI rule prevents the production of $s\overline{s}$ mesons in $\overline{p}p$ annihilation. Since there is not enough phase space to produce the $f_0(1710)$ in $\overline{p}p$ annihilation at rest, Crystal Barrel searched for the $f_0(1710)$ with higher incident antiproton momenta. Figure 11.4, shows the Dalitz plot for $\overline{p}p \to \pi^0\eta\eta$ [12]. Since the $f_0(1710) \to \eta\eta$ is not observed, while $f_0(1500)$ is clearly seen, this again suggests that the $f_0(1710)$ does not have a large $(u\overline{u} + d\overline{d})$ component.

11.1.1 SU(3) Coupling Coefficients

Before discussing the experimental evidence for the 0^{++} ground state glueball we have to digress on decay amplitudes. Let us derive the expected decay branching ratios of scalar (and tensor) mesons into two pseudoscalar mesons, $\pi\pi$, $K\overline{K}$, $\eta\eta$ and $\eta\eta'$. We shall limit ourselves to isoscalar ($i = 0$) decays which are relevant to glueballs. (The amplitudes for $i = 1$ and $i = \frac{1}{2}$ decays are listed in the appendix of Ref. [2].) We denote the mostly $s\overline{s}$ isoscalar meson with f' and the mostly $n\overline{n}$

with f. For ideal mixing we have, copying from (5.9),

$$|f'\rangle = |f_8\rangle\sqrt{\frac{2}{3}} - |f_1\rangle\sqrt{\frac{1}{3}} = -|s\bar{s}\rangle, \tag{11.1}$$

$$|f\rangle = |f_8\rangle\sqrt{\frac{1}{3}} + |f_1\rangle\sqrt{\frac{2}{3}} = \frac{1}{\sqrt{2}}|u\bar{u} + d\bar{d}\rangle| \equiv |n\bar{n}\rangle, \tag{11.2}$$

where f_8 is the octet and f_1 the singlet contribution to the two isoscalar mesons. We will first deal with the decay of these ideally mixed states and then treat the case of arbitrary mixing angles. Provisionally, we shall assume exact SU(3) symmetry, that is no penalty to produce $s\bar{s}$ pairs out of the vacuum, and set for the ratio of amplitudes

$$\rho = \frac{\gamma(\text{vacuum} \to s\bar{s})}{\gamma(\text{vacuum} \to u\bar{u})} = \frac{\gamma(\text{vacuum} \to s\bar{s})}{\gamma(\text{vacuum} \to d\bar{d})} = 1, \tag{11.3}$$

which is in reasonable agreement with data ($\rho > 0.8$) [13]. To describe the couplings to the two pseudoscalar mesons we need the so-called SU(3) isoscalar factors, which are the analogous of the Clebsch-Gordan coefficients in SU(2). The ones needed here—for the multiplets 1 and 8_1—are listed in Fig. 11.5. They

$\mathbf{1} \to \mathbf{8} \otimes \mathbf{8}$

$$(\Lambda) \to (N\overline{K}\ \Sigma\pi\ \Lambda\eta\ \Xi K) = \frac{1}{\sqrt{8}}(\boxed{2}\ \textcircled{3}\ \triangle\!\!1\ -2)^{1/2}$$

$\mathbf{8_1} \to \mathbf{8} \otimes \mathbf{8}$

$$\begin{pmatrix} N \\ \Sigma \\ \Lambda \\ \Xi \end{pmatrix} \to \begin{pmatrix} N\pi\ N\eta\ \Sigma K\ \Lambda K \\ N\overline{K}\ \Sigma\pi\ \Lambda\pi\ \Sigma\eta\ \Xi K \\ N\overline{K}\ \Sigma\pi\ \Lambda\eta\ \Xi K \\ \Sigma\overline{K}\ \Lambda\overline{K}\ \Xi\pi\ \Xi\eta \end{pmatrix} = \frac{1}{\sqrt{20}}\begin{pmatrix} 9 & -1 & -9 & -1 & \\ -6 & 0 & 4 & 4 & -6 \\ \boxed{2} & -\textcircled{12} & -\triangle\!\!4 & -2 & \\ 9 & -1 & -9 & -1 & \end{pmatrix}^{1/2}$$

$\mathbf{8_2} \to \mathbf{8} \otimes \mathbf{8}$

$$\begin{pmatrix} N \\ \Sigma \\ \Lambda \\ \Xi \end{pmatrix} \to \begin{pmatrix} N\pi\ N\eta\ \Sigma K\ \Lambda K \\ N\overline{K}\ \Sigma\pi\ \Lambda\pi\ \Sigma\eta\ \Xi K \\ N\overline{K}\ \Sigma\pi\ \Lambda\eta\ \Xi K \\ \Sigma\overline{K}\ \Lambda\overline{K}\ \Xi\pi\ \Xi\eta \end{pmatrix} = \frac{1}{\sqrt{12}}\begin{pmatrix} 3 & 3 & 3 & -3 & \\ 2 & 8 & 0 & 0 & -2 \\ 6 & 0 & 0 & 6 & \\ 3 & 3 & 3 & -3 & \end{pmatrix}^{1/2}$$

Fig. 11.5 SU(3) coupling coefficients. The symbols represent the isospin, e.g. Σ stands for $i = 1$, Λ and η for $i = 0$, Ξ, K and N for $i = \frac{1}{2}$. The 8_1 factors are used for $C = +1$, the 8_2 ones for $C = -1$ decays. $K\overline{K}$: blue squares; $\pi\pi$: red circles; two isoscalars: green triangles. A square-root is implicitly assumed over the coefficient [14]

describe—up to two unknown constants g_1 and g_8—the couplings 8×8 to singlet and octet states, which are relevant to the decay of positive C parity states such as 0^{++} and 2^{++}. For the sake of completeness we also include the coefficients 8_2 relevant to the decay of negative parity states such as 1^{--}. Isoscalar factors involving decuplets are needed for baryon decays and are listed in [14].

For example, coupling f_8 to two pions involves $8_1 \to 8\times 8$, an $i = 0$ SU(3) octet coupling to two isovectors, symbolized by a Λ coupling to $\Sigma\pi$ in Fig. 11.5. The coupling is equal to $-g_8\sqrt{\frac{3}{5}}$ (red circle in Fig. 11.5). Coupling f_1 to two kaons involves $1 \to 8\times 8$, an isospin singlet coupling to two $i = \frac{1}{2}$ states, symbolised by a Λ coupling to $N\overline{K}$. The coupling is equal to $\frac{g_1}{2}$ (blue square in Fig. 11.5).

By using (11.1) and the couplings surrounded by the red circles in Fig. 11.5, one finds the amplitude for $f' \to \pi\pi$:

$$\begin{aligned}\gamma(f' \to \pi\pi) &= \sqrt{\frac{2}{3}}\gamma(f_8 \to 8\times 8) - \sqrt{\frac{1}{3}}\gamma(f_1 \to 8\times 8)\\ &= -\sqrt{\frac{2}{3}}\sqrt{\frac{3}{5}}g_8 - \sqrt{\frac{1}{3}}\sqrt{\frac{3}{8}}g_1 = 0.\end{aligned} \tag{11.4}$$

The coupling vanishes because an $s\overline{s}$ state does not decay into two pions by virtue of the OZI rule. Therefore g_1 and g_2 are related by

$$\boxed{g_1 = -\frac{4}{\sqrt{5}}g_8}\,. \tag{11.5}$$

On the other hand, the amplitude for $f \to \pi\pi$ is given by

$$\begin{aligned}\gamma(f \to \pi\pi) &= \sqrt{\frac{1}{3}}\gamma(f_8 \to 8\times 8) + \sqrt{\frac{2}{3}}\gamma(f_1 \to 8\times 8)\\ &= -\sqrt{\frac{1}{3}}\sqrt{\frac{3}{5}}g_8 + \sqrt{\frac{2}{3}}\sqrt{\frac{3}{8}}g_1 = -g_8\frac{3}{\sqrt{5}}\end{aligned} \tag{11.6}$$

with (11.5). Similarly, one obtains the amplitudes for f' and f to $K\overline{K}$ (blue squares in Fig. 11.5):

$$\gamma(f' \to K\overline{K}) = \sqrt{\frac{2}{3}}\sqrt{\frac{1}{10}}g_8 - \sqrt{\frac{1}{3}}\cdot\frac{1}{2}g_1 = g_8\sqrt{\frac{3}{5}}, \tag{11.7}$$

$$\gamma(f \to K\overline{K}) = \sqrt{\frac{1}{3}}\sqrt{\frac{1}{10}}g_8 + \sqrt{\frac{2}{3}}\cdot\frac{1}{2}g_1 = -g_8\sqrt{\frac{3}{10}}. \tag{11.8}$$

Table 11.2 Decay amplitudes γ of the ideally mixed isoscalar scalar (or tensor) mesons (11.1) and (11.2) up to a common constant g_8, where θ is the pseudoscalar mixing angle

	γ	N
$f \to \pi\pi$	$-\frac{3}{\sqrt{5}}$	1
$f \to K\overline{K}$	$-\sqrt{\frac{3}{10}}$	2
$f \to \eta\eta$	$\frac{1}{\sqrt{15}}(\cos\theta - \sqrt{2}\sin\theta)^2$	1
$f \to \eta\eta'$	$\sqrt{\frac{2}{15}}(\cos 2\theta - \frac{\sin 2\theta}{2\sqrt{2}})$	2
$f' \to \pi\pi$	0	–
$f' \to K\overline{K}$	$\sqrt{\frac{3}{5}}$	2
$f' \to \eta\eta$	$-\sqrt{\frac{2}{15}}(\sqrt{2}\cos\theta + \sin\theta)^2$	1
$f' \to \eta\eta'$	$\frac{2}{\sqrt{15}}(\cos 2\theta - \frac{\sin 2\theta}{2\sqrt{2}})$	2

The partial widths are proportional to $N\gamma^2$, where N is a final state multiplicity factor

The amplitudes are summarised in Table 11.2. The partial width into $\pi\pi$ is proportional to γ^2 and that into $K\overline{K}$ to $(N = 2)\gamma^2$ (since there are two kaon doublets), multiplied by final state phase space and form factors, as described below.

The $\eta\eta$ and $\eta\eta'$ final states are more complicated. Let us deal with the decay $f \to \eta\eta$. According to (5.11) the two η mesons are superpositions of octet and singlet states with pseudoscalar mixing angle θ:

$$|\eta\rangle = |\eta_8\rangle \cos\theta - |\eta_1\rangle \sin\theta. \tag{11.9}$$

Coupling f_8 to $\eta\eta$ gives

$$\gamma(f_8 \to \eta\eta) = \gamma(f_8 \to 8 \times 8)\cos^2\theta - 2\underbrace{\gamma(f_8 \to 1 \times 8)}_{\equiv g_{18}}\sin\theta\cos\theta, \tag{11.10}$$

while for f_1

$$\gamma(f_1 \to \eta\eta) = \gamma(f_1 \to 8 \times 8)\cos^2\theta + \underbrace{\gamma(f_1 \to 1 \times 1)}_{g_{11}}\sin^2\theta, \tag{11.11}$$

noting that $\gamma(f_1 \to 1 \times 8) = \gamma(f_8 \to 1 \times 1) = 0$. The expressions (11.10), (11.11) involve two new constants, g_{18} and g_{11}. We will show below that

$$\boxed{g_{18} = g_{11} = \sqrt{\tfrac{2}{5}}g_8}\,. \tag{11.12}$$

Hence we get the $f \to \eta\eta$ amplitude with the superposition (11.2), by coupling an isoscalar to two isoscalar mesons (symbolised by a Λ coupling to $\Lambda\eta$, green triangles in Fig. 11.5):

$$
\begin{aligned}
\gamma(f \to \eta\eta) &= \sqrt{\frac{1}{3}}(-\sqrt{\frac{1}{5}}g_8 \cos^2\theta - 2\underbrace{g_{18}}_{\sqrt{\frac{2}{5}}g_8} \sin\theta\cos\theta) \\
&+\sqrt{\frac{2}{3}}(\underbrace{-\sqrt{\frac{1}{8}}g_1}_{\sqrt{\frac{2}{5}}g_8} \cos^2\theta + \underbrace{g_{11}}_{\sqrt{\frac{2}{5}}g_8} \sin^2\theta) \\
&= g_8\sqrt{\frac{1}{3}}\left(-\sqrt{\frac{1}{5}}\cos^2\theta - 2\sqrt{\frac{2}{5}}\sin\theta\cos\theta\right) \\
&+g_8\sqrt{\frac{2}{3}}\left(\sqrt{\frac{2}{5}}\cos^2\theta + \sqrt{\frac{2}{5}}\sin^2\theta\right) \\
&= \frac{g_8}{\sqrt{15}}(\cos^2\theta - 2\sqrt{2}\sin\theta\cos\theta + 2\sin^2\theta) \\
&= \frac{g_8}{\sqrt{15}}(\cos\theta - \sqrt{2}\sin\theta)^2.
\end{aligned}
\tag{11.13}
$$

The channels $f' \to \eta\eta$ and f or $f' \to \eta\eta'$ can be treated similarly and are left to the reader as exercises (Problem 11.1). The results are summarised in Table 11.2.

We are still left with the proof of (11.12) which follows from the OZI rule. An isovector state a cannot decay into an $s\bar{s}$ meson. With

$$
-|s\bar{s}\rangle = |8\rangle\sqrt{\frac{2}{3}} - |1\rangle\sqrt{\frac{1}{3}} \tag{11.14}
$$

the vanishing amplitude for e.g. $a \nrightarrow |s\bar{s}\rangle\pi$ reads

$$
\gamma(a \to |s\bar{s}\rangle\pi) = -\gamma(a \to 8 \times 8)\sqrt{\frac{2}{3}} + \gamma(a \to 1 \times 8)\sqrt{\frac{1}{3}} = 0. \tag{11.15}
$$

The 8×8 term describes the coupling of a pure isovector octet (symbolised by a Σ) to an isoscalar octet and isovector octet (symbolised by $\Lambda\pi$). Hence

$$
\gamma(a \to |s\bar{s}\rangle\pi) = -\sqrt{\frac{1}{5}}g_8\sqrt{\frac{2}{3}} + g_{18}\sqrt{\frac{1}{3}} = 0 \Rightarrow \boxed{g_{18} = \sqrt{\tfrac{2}{5}}g_8}\,. \tag{11.16}
$$

On the other hand, since f is pure $n\overline{n}$, $f \nrightarrow |s\overline{s}\rangle|s\overline{s}\rangle$. Using (11.2) and (11.14) one gets the amplitude

$$\gamma(f \to |s\overline{s}\rangle|s\overline{s}\rangle) = \gamma(f_8 \to 8\times 8)\sqrt{\frac{1}{3}}\cdot\frac{2}{3} - \gamma(f_8 \to 1\times 8)\,2\sqrt{\frac{1}{3}}\cdot\sqrt{\frac{1}{3}}\sqrt{\frac{2}{3}}$$
$$+\gamma(f_1 \to 8\times 8)\sqrt{\frac{2}{3}}\cdot\frac{2}{3} + \gamma(f_1 \to 1\times 1)\sqrt{\frac{2}{3}}\cdot\frac{1}{3} = 0, \tag{11.17}$$

therefore (green triangles in Fig. 11.5),

$$\gamma(f \to |s\overline{s}\rangle|s\overline{s}\rangle) = -\sqrt{\frac{1}{5}}g_8\sqrt{\frac{1}{3}}\cdot\frac{2}{3} - g_{18}\frac{2}{3}\sqrt{\frac{2}{3}}$$
$$-\sqrt{\frac{1}{8}}g_1\sqrt{\frac{2}{3}}\cdot\frac{2}{3} + g_{11}\sqrt{\frac{2}{3}}\cdot\frac{1}{3} = 0. \tag{11.18}$$

Inserting g_1 (11.5) and g_{18} (11.16) leads to the advertised result

$$\boxed{g_{11} = \sqrt{\tfrac{2}{5}}g_8}\,. \tag{11.19}$$

We have so far assumed that the decaying scalar meson is ideally mixed, either f' (11.1) or f (11.2) with mixing angle $\varphi = 35.3°$. For arbitrary mixing angle the meson becomes a linear combination of f_8 and f_1. Let us redefine the mixing angle as

$$\alpha \equiv \varphi + 54.7° \tag{11.20}$$

and write for the general case the superposition (5.11),

$$|f_\alpha\rangle = |f_8\rangle\cos\varphi - |f_1\rangle\sin\varphi = |f_8\rangle\cos(\alpha - 54.3°) - |f_1\rangle\sin(\alpha - 54.3°)$$
$$= |f_8\rangle\cos\alpha\sqrt{\frac{1}{3}} + |f_8\rangle\sin\alpha\sqrt{\frac{2}{3}} - |f_1\rangle\sin\alpha\sqrt{\frac{1}{3}} + |f_1\rangle\cos\alpha\sqrt{\frac{2}{3}}, \tag{11.21}$$

therefore

$$\boxed{|f_\alpha\rangle = |f\rangle\cos\alpha + |f'\rangle\sin\alpha}\,. \tag{11.22}$$

For $\alpha = 90°$ ($\varphi = 35.3°$) one recovers the pure $-|s\overline{s}\rangle$, and for $\alpha = 0°$ the pure $|n\overline{n}\rangle$ state. The decay couplings of the meson f_α are then obtained with the help

Table 11.3 Decay amplitudes γ_α of isoscalar scalar (or tensor) mesons up to a common arbitrary constant

	γ_α	n	γ_α ($\phi = 45°$, $\rho = 1$)
$f_\alpha \to \pi\pi$	$\cos\alpha$	$N \times 3 = 3$	$\cos\alpha$
$f_\alpha \to K\overline{K}$	$\frac{1}{2}\cos\alpha(\rho - \sqrt{2}\tan\alpha)$	$N \times 2 = 4$	$\frac{1}{2}\cos\alpha(1 - \sqrt{2}\tan\alpha)$
$f_\alpha \to \eta\eta$	$-\cos\alpha(\cos^2\phi - \rho\sqrt{2}\tan\alpha\sin^2\phi)$	1	$-\frac{1}{2}\cos\alpha(1 - \sqrt{2}\tan\alpha)$
$f_\alpha \to \eta\eta'$	$-\cos\alpha\cos\phi\sin\phi(1 + \rho\sqrt{2}\tan\alpha)$	2	$-\frac{1}{2}\cos\alpha(1 + \sqrt{2}\tan\alpha)$

The amplitudes for $\rho = 1$ and pseudoscalar mixing angle $\theta \simeq -10°$ are listed in the last column. The partial widths are proportional to $n\gamma_\alpha^2$, where n is the final state multiplicity factor. The amplitudes have been normalised to $\cos\alpha$ by multiplying with $-\sqrt{5/3}$. The multiplicity factors from Table 11.2 have been updated accordingly

of Table 11.2. However, the expressions become more transparent when rewriting Table 11.2 as a function of the pseudoscalar mixing angle also expressed as

$$\phi \equiv \theta + 54.7°. \tag{11.23}$$

Substituting θ for ϕ one finds, after a few lines of tedious algebra, the much simpler form

$$\gamma(f \to \eta\eta) = \frac{3}{\sqrt{15}}\cos^2\phi. \tag{11.24}$$

Table 11.3 lists the decay amplitudes as a function of α and ϕ. We have also taken into account the ρ factor (11.3) which has been inserted whenever an $s\overline{s}$ pair has to be created from vacuum. For $K\overline{K}$ decay ρ contributes to the first term ($n\overline{n}$). For $\eta\eta$ and $\eta\eta'$ the second terms arise from the two outgoing $s\overline{s}$ pairs, one of which is generated from vacuum. With the assumption that $\theta \sim -10°$ (Sect. 5.2)—hence for $\phi \sim 45°$—the amplitudes reduce to the simpler forms shown in the last column of Table 11.3.

Figure 11.6 illustrates how $n\gamma_\alpha^2$ varies with mixing angle α when assuming a pseudoscalar mixing angle $\theta = -17.3°$ [15] and $\rho = 1$. Ratios between partial widths are obtained by taking into account the final state phase space factor W:

$$\Gamma = \text{constant} \times n\gamma_\alpha^2 \times W, \tag{11.25}$$

with $W = pF_\ell^2(p)$ (7.56). To compare with data Ref. [2] uses the following prescription:

$$W = p^{2l+1}e^{-\frac{p^2}{8\beta^2}}, \tag{11.26}$$

with $\beta = 0.5$ GeV/c, which also leads to excellent agreement between prediction and data for tensor decays ($\ell = 2$) into pseudoscalar pairs [2]. We recall that p is the break-up momentum in the decay rest frame in GeV/c and ℓ the relative

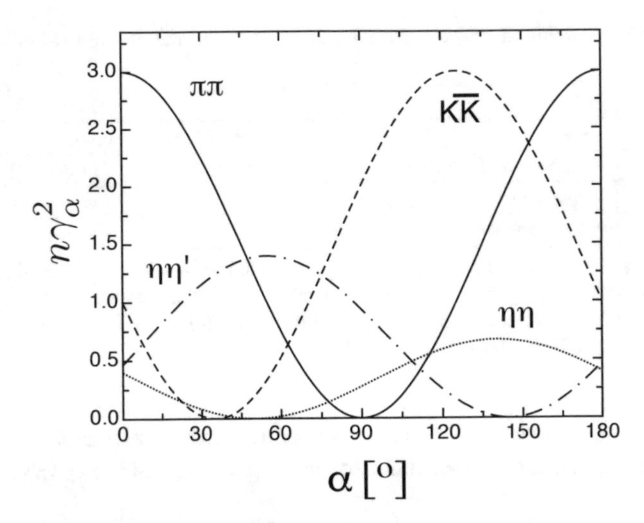

Fig. 11.6 SU(3) predictions for isoscalar $q\overline{q}$ scalar (or tensor) mesons decaying into pseudoscalar pairs, assuming $\theta = -17.3°$ and $\rho = 1$ [14]

angular momentum ($\ell = 0$ for scalar decays). The exponential form factor in (11.26) reduces the partial width for large momenta (thus counteracting the phase space) and is comparable to the damping factor in Table 7.3.

11.1.2 Two-Body Decays of Scalar Mesons

Among the three isoscalar 0^{++} states in the 1500 MeV region—the $f_0(1710)$, $f_0(1500)$ and $f_0(1370)$—the quark content of the $f_0(1370)$ is the least problematic. This very broad (∼300–600 MeV) state is required in the $\pi\pi$ scattering amplitude [16] with a small branching ratio into $K\overline{K}$. It also decays strongly into 4π [14]. When interpreted as $q\overline{q}$ the consensus for this meson is to be an $n\overline{n}$ state.

Let us now examine the decay rates of the $f_0(1500)$ and $f_0(1710)$. Figure 11.7 shows a 2-dimensional plot of the expected ratios of couplings

$$R_2 = \frac{4\gamma^2(K\overline{K})}{3\gamma^2(\pi\pi)} \text{ vs. } R_1 = \frac{\gamma^2(\eta\eta)}{3\gamma^2(\pi\pi)} \tag{11.27}$$

as a function of scalar mixing angle α, calculated from Table 11.3 with $\theta = -17.3°$ and $\rho = 1$ (Fig. 11.6). The measured values from Crystal Barrel [16] and from WA102 [18] in high energy central collisions are shown by the coloured boxes, those from Crystal Barrel in red and those from WA102 in blue. The experimental branching ratios have been corrected for phase space following the prescription (11.26). One sees that $f_0(1710)$ decays preferably into $K\overline{K}$ pairs with

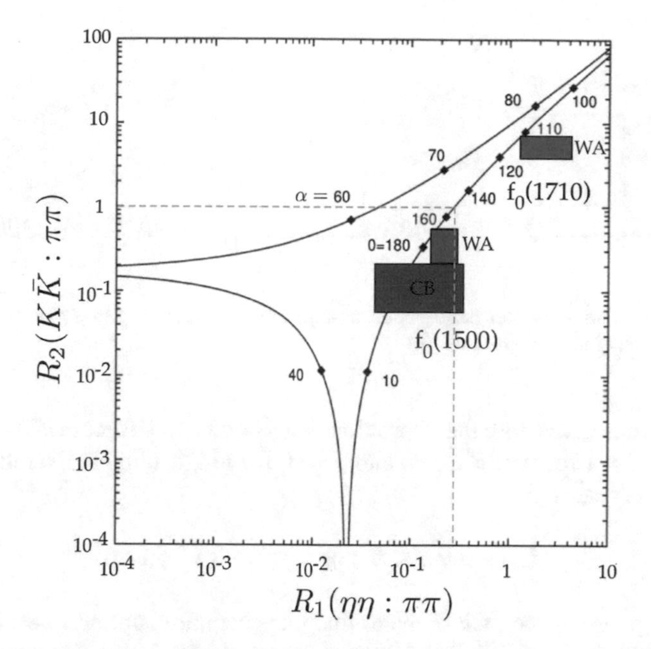

Fig. 11.7 Ratio R_2 vs. R_1 of squared couplings (11.27). The curve shows the prediction for $q\overline{q}$ mesons as a function of mixing angle α (in degrees). The boxes are data from $\overline{p}p$ annihilation at rest (Crystal Barrel, CB) and high energy proton-proton central collisions (WA). The green dashed line shows the expected glueball couplings for a pseudoscalar mixing angle of $\theta = -17.3°$ [17]

α large and not far from 90°, while $f_0(1500)$ prefers $\alpha \simeq 0$. When interpreted as $q\overline{q}$ the former would be mostly $s\overline{s}$ and the latter mostly $n\overline{n}$.

However, with the $f_0(1370)$ there is no room for the $f_0(1500)$ as $n\overline{n}$ state. Furthermore, the $f_0(1710) \to K\overline{K}$ has been observed by Belle in $\gamma\gamma$ collisions [19], while the $f_0(1500)$ was not seen, thus confirming earlier findings from LEP (for more details see [4]). A glueball should be suppressed in $\gamma\gamma$ collisions, since photons do not couple directly to gluons. These simple tentative arguments suggest that $f_0(1500)$ could be the ground state glueball.

Let us now derive the predictions for glueball decay into two pseudoscalar mesons. Pure glueballs G decaying equally into $u\overline{u}$, $d\overline{d}$ and $s\overline{s}$ are flavour singlets ($\rho = 1$). By choosing

$$\cos\alpha = \sqrt{\frac{2}{3}} \text{ and } \sin\alpha = -\sqrt{\frac{1}{3}} \tag{11.28}$$

(or $\alpha = -35.3°$) one gets from (11.22) the flavour singlet

$$|G\rangle = \frac{1}{\sqrt{3}}|u\overline{u} + d\overline{d} + s\overline{s}\rangle. \tag{11.29}$$

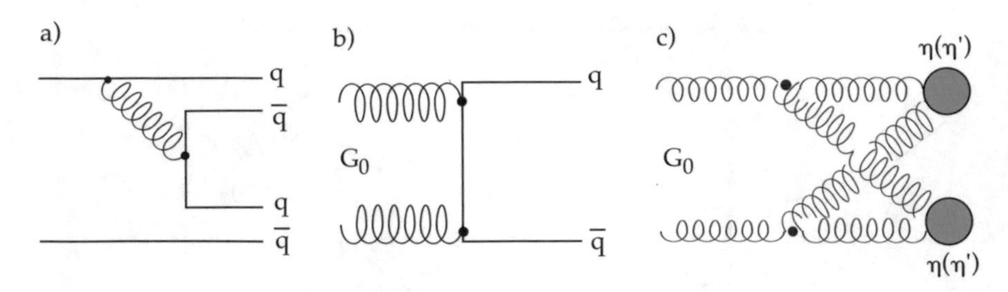

Fig. 11.8 Decay in first order perturbation of a $q\overline{q}$ meson (**a**), mixing of a glueball G_0 with $q\overline{q}$ (**b**) and G_0 decay into two glueballs (**c**)

Table 11.3 then gives with the approximation $\phi = 45°$ and exact $SU(3)_f$ ($\rho = 1$) the couplings $\gamma^2 = 1$ for $\pi\pi$, $K\overline{K}$, $\eta\eta$ and $\gamma^2 = 0$ for $\eta\eta'$, leading to the ratios between glueball decay rates

$$G \to \pi\pi : K\overline{K} : \eta\eta : \eta\eta' = 3 : 4 : 1 : 0 \tag{11.30}$$

apart from phase space factors. Note that the glueball coupling to $\eta\eta$ depends on ϕ and ρ, and the coupling to $\eta\eta'$ always vanishes for $\rho = 1$. Figure 11.7 shows the couplings of $f_0(1500)$ to $K\overline{K}$ and $\eta\eta$ relative to $\pi\pi$ (dashed lines). The ratio $\gamma^2(\eta\eta)/\gamma^2(\pi\pi)$ is consistent with that expected for a glueball, but the $K\overline{K}$ channel is suppressed by an order of magnitude.[1]

A scenario to accommodate three isoscalar 0^{++} states in the 1500 MeV region was first proposed in [2, 21]. In the flux tube model mesons are $q\overline{q}$ pairs connected by a coloured tube of gluons and glueballs are made of gluon loops (cartooned in Fig. 10.5). In first order perturbation a $q\overline{q}$ meson decays into $\pi\pi$ and $K\overline{K}$ (Fig. 11.8a), but a glueball G_0 does not decay directly into $q\overline{q}$. Instead, G_0 mixes with nearby $q\overline{q}$ mesons of the same quantum numbers, which then decay into meson pairs (Fig. 11.8b). However, G_0 may decay directly into two glueballs which couple to the glue content of two isoscalar mesons, such as the η and η' (Fig. 11.8c).[2]

In first order perturbation and under the assumption that the couplings of G_0 to $n\overline{n}$ and $s\overline{s}$ in Fig. 11.8b are equal, the pure glueball G_0 mixes with an $n\overline{n}$ state N and an $s\overline{s}$ state S to produce the observed meson G. Apart from an overall normalisation constant let us make the ansatz

$$|G\rangle = |G_0\rangle + \xi(\sqrt{2}|n\overline{n}\rangle + \omega|s\overline{s}\rangle) \tag{11.31}$$

[1] The decay $f_0(1500) \to \eta\eta'$, which occurs at the kinematical threshold, has been observed with the small branching ratio $f_{\eta\eta'} = (1.9\pm0.8)\% \ll f_{\pi\pi} = (34.9 \pm 2.3)\%$ [14, 20].

[2] The η and η' may have small gluonic components in their wavefunctions, but present data are consistent with a vanishing gluonic contribution. However, the errors are large: $< 10\%$ for the η and $29^{+18}_{-26}\%$ for the η' [22].

[2], where ξ is a mixing parameter and

$$\omega \equiv \frac{m(G_0) - m(N)}{m(G_0) - m(S)}. \tag{11.32}$$

Likewise the wavefunctions of the other two states are given by

$$|N\rangle - \xi\sqrt{2}|G_0\rangle \text{ and } |S\rangle - \xi\omega|G_0\rangle, \tag{11.33}$$

apart from normalisation constants. For the special case $m(G_0) = m(N) = m(S)$, hence $\omega = 1$, G decay becomes the flavour singlet (11.29), assuming that the G_0 component contributes negligibly. We will now show that for $\omega = -1$, that is when G_0 lies halfway between S and N, the interference between N and S decay is fully destructive and the decay into $K\overline{K}$ vanishes. We have seen that the wavefunction of a $q\overline{q}$ meson f_α is given by the superposition (11.22)

$$|f_\alpha\rangle = \cos\alpha|n\overline{n}\rangle - \sin\alpha|s\overline{s}\rangle, \tag{11.34}$$

leading with Table 11.3 to the ratio

$$\frac{\gamma(q\overline{q} \to K\overline{K})}{\gamma(q\overline{q} \to \pi\pi)} = \frac{1}{2}(1 - \sqrt{2}\tan\alpha). \tag{11.35}$$

Comparing (11.34) with the bracketed term in (11.31) we make the substitution

$$-\sin\alpha \to \omega \text{ and } \cos\alpha \to \sqrt{2} \text{ hence } \tan\alpha \to -\frac{\omega}{\sqrt{2}} \tag{11.36}$$

in (11.35) to obtain

$$\frac{\gamma(G \to K\overline{K})}{\gamma(G \to \pi\pi)} = \frac{1+\omega}{2}, \tag{11.37}$$

which proves that the decay amplitude into $K\overline{K}$ vanishes for $\omega = -1$.

The classification of the 1^3P_0 nonet shown in Table 11.1, and in particular the small $K\overline{K}$ branching ratio, can be accommodated in this scenario by Crystal Barrel and WA102, with the $f_0(1370)$ and $f_0(1710)$ being made essentially of N and S components with a small admixture of glue, while the $f_0(1500)$ is mostly glue with about ξ^2 ~20% of $q\overline{q}$ admixture [2]. The analysis was repeated later, this time by including results from central production, and led to similar conclusions [3]. Figure 11.9 shows the distribution of glue among the three pseudoscalars, derived from central collisions and J/ψ decay data, $J/\psi \to \omega K\overline{K}, \omega\pi\pi$ and $J/\psi \to \phi K\overline{K}, \phi\pi\pi$ collected at the Beijing e^+e^- collider.

Alternative schemes to the one discussed here have been proposed so that the classification presented in Table 11.1 remains controversial. In [24] and [25] the

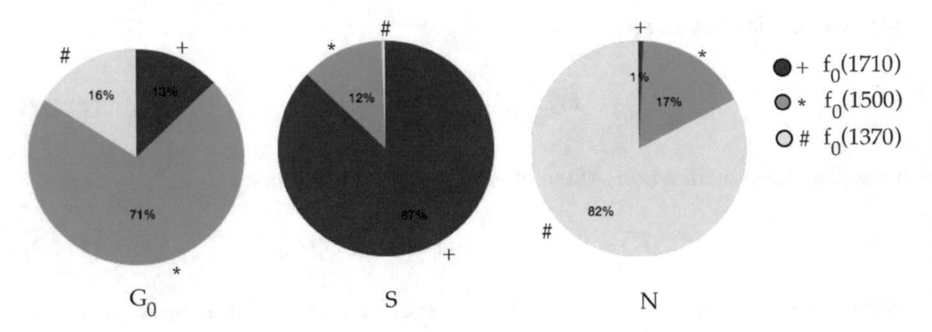

Fig. 11.9 Distribution of glue (G_0), $s\overline{s}$ (S) and $n\overline{n}$ (N) pairs in the $f_0(1710)$, $f_0(1500)$ and $f_0(1370)$ wavefunctions from central collisions and J/ψ decay (extracted from the analysis [23])

$f_0$1710) is considered a better candidate for the ground state glueball. A lighter glueball around or below 1 GeV has also been advocated [26]. Details and references can be obtained by consulting the articles on "Non-$q\overline{q}$ mesons" [4] and the "Note on scalar mesons below 2 GeV" [5] in the Review of Particle Physics. Lattice calculations may ultimately shed light on the true nature of the scalar mesons.

11.2 Tensor Mesons

The mass of the first excited glueball (2^{++}) is predicted around 2400 MeV (Fig. 11.1). Let us therefore briefly review the known isoscalar tensor sector. The ideally mixed 1^3P_2 $q\overline{q}$ mesons, $f_2(1270)$ and $f_2'(1525)$, are well known. Above 1525 MeV and below 2500 MeV eight isoscalar mesons have been reported, while six are expected to belong to the 2^3P_2, 3^3P_2, and 1^3F_2 nonets. Figure 11.10 shows that these states are broad and therefore interfere, which makes the identification of the tensor glueball rather difficult. None of these mesons can be definitively assigned to these nonets (in Fig. 5.1 we have assigned the $f_2(1640)$ and the 470 MeV broad $f_2(1950)$, which has been observed by several experiments, to the 2^3P_2 nonet).

The reaction $\overline{p}p \to \phi\phi$ is OZI suppressed but may proceed through the emission of glueballs. The cross section for $\phi\phi$ production in $\overline{p}p$ annihilation rises sharply towards very large values [27], which could signal the presence of the tensor glueball, in accord with the observation of several structures in $\pi^-p \to n\phi\phi$ reactions [28] and $pp \to pp\phi\phi$ central production [29].

The nature of the $f_2(1565)$ has remained a mystery. This meson was first reported by the ASTERIX experiment at LEAR in $\overline{p}p$ annihilation at rest into $\pi^+\pi^-\pi^0$ in gaseous hydrogen at NTP [30]. The Dalitz plot and the $\pi^+\pi^-$ invariant mass projection are displayed in Fig. 11.11. The $f_2(1565)$ state is mainly produced from the antiprotonic P states ($^3P_1$ and $^3P_2$), which are enhanced in low pressure hydrogen, as was explained in the caption of Fig. 2.7. This presumably explains

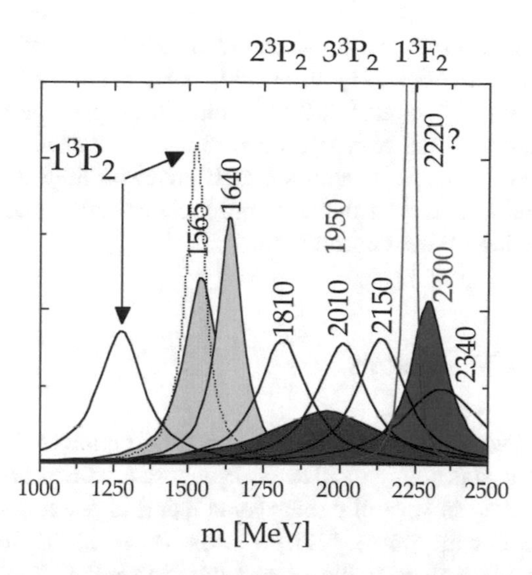

Fig. 11.10 The tensor mesons that have been observed so far, normalized to unit area. The $f_2(1270)$ and $f_2'(1525)$ (arrows) are firmly established as the ground state $n\bar{n}$ and $s\bar{s}$ tensor mesons, respectively. The quantum numbers of the narrow state at 2220 MeV are not well established

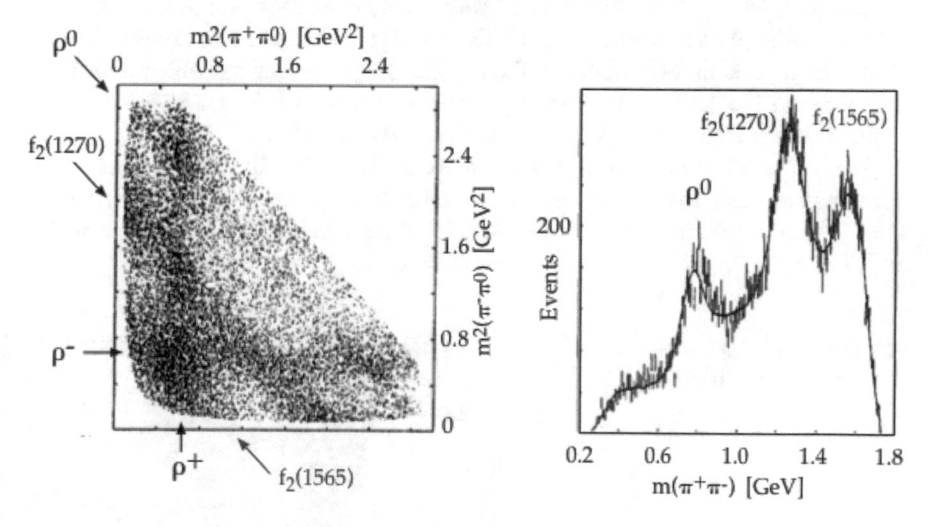

Fig. 11.11 Left: $\pi^+\pi^-\pi^0$ Dalitz plot in $\overline{p}p$ annihilation at rest in gaseous hydrogen. Right: $\pi^+\pi^-$ invariant mass projection. The curve shows the Dalitz plot fit (adapted from [31])

why it had not been spotted earlier in liquid hydrogen, e.g. in bubble chamber experiments.

The $f_2(1565)$ seems to be produced only in $\overline{p}p$ annihilation [31, 32]. In contrast to the pp interaction, repulsive at very short distances owing to the Pauli principle, the $\overline{p}p$ interaction is predicted to become attractive in some of the partial waves, leading to the existence of quasi-nuclear bound states. Most of these "baryonium" states cluster just below the $2m_p$ threshold, but a sequence of 0^{++}, 1^{--}, 2^{++} isoscalar states bound by several 100 MeV have been predicted (for a review see [33]). Experimentally, these baryonia have remained elusive, perhaps due to the onset of annihilation at short distances which increases their widths substantially. The $f_2(1565)$ could be such a nucleon-antinucleon bound state, generated from protonium by shaking off the neutral pion.

11.3 $J^{PC} = 1^{-+}$ Mesons

We have seen in Sect. 4 that mesons with quantum numbers $J^{PC} = 1^{-+}$ cannot be quark-antiquark states. A ~350 MeV broad state called $\pi_1(1400)$, decaying into $\eta\pi$, was reported in several experiments using high energy pions. Its $\ell = 1$ amplitude interferes with the $a_2(1320) \rightarrow \eta\pi$ ($\ell = 2$) amplitude, leading to a forward/backward asymmetry in the $\eta\pi$ angular distribution. The state was reported e.g. at BNL with 18 GeV pions in the reaction $\pi^- p \rightarrow \eta\pi^0 n$ [34].

The $\pi_1(1400)$ was also observed by Crystal Barrel using liquid deuterium in the annihilation $\overline{p}n \rightarrow \pi^-\pi^0\eta$ associated with a very slow spectator proton [35]. The $\pi^-\pi^0\eta$ Dalitz plot is shown in Fig. 11.12. The signal appears as an excess of events above the ρ band in the vicinity of the $a_2(1320)$ signals. The $\pi_1(1400)$ interfering with the $a_2(1320)$ and ρ is required by the fit. The mass is 1400 ± 28 MeV (slightly higher than the BNL value [34]) and the width 310 ± 70 MeV.

The $\pi_1(1600)$ is another 1^{-+} state reported in $\pi^- N$ (diffractive) interaction, decaying into $\rho\pi$, $b_1(1235)\pi$ and $f_1(1285)\pi$, but not $\eta\pi$. For example, a clear $\rho^0\pi^-$ signal is observed by the COMPASS experiment at CERN in the reaction $\pi^-\text{Pb} \rightarrow \pi^+\pi^-\pi^- X$ with 190 GeV pions. [36].

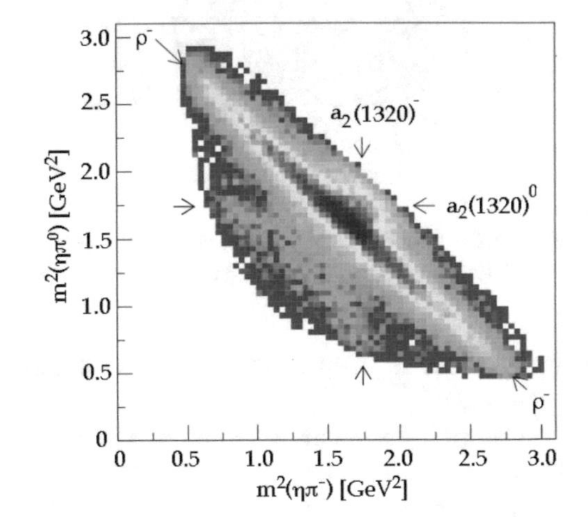

Fig. 11.12 Dalitz plot of $\overline{p}n$ annihilation into $\pi^-\pi^0\eta$ [35]

Hybrid mesons are $q\overline{q}$ mesons with a vibrating gluon flux tube (Fig. 10.5). They can be isoscalar or isovector states with exotic quantum numbers. In the flux tube model ground state hybrids with quantum numbers (0^{-+}, 1^{-+}, 1^{--}, and 2^{-+}) are expected between 1.7 and 1.9 GeV [37, 38]. Lattice calculations also predict the ground state hybrid at a mass of 1.9 GeV [39]. Most of them should be rather broad, but some can be as narrow as 100 MeV. The $\pi_1(1400)$ is much lighter, while the $\pi_1(1600)$ mass does not lie far below the predicted values. Hence the former is unlikely to be a hybrid meson, while the latter behaves like a hybrid state, decaying into a pair of P and S mesons ($b_1(1235)\pi$ and $f_1(1285)\pi$).

Clearly the search for hybrid mesons is still in its infancy. Ultimately, other members of the hybrid nonets need to be identified. Progress will be hopefully achieved with new facilities, such as GlueX at Jefferson Laboratory which will photoproduce mesons off nucleons with linearly polarized photons. Details on hybrid mesons can be found in [4] and in the comprehensive review [40].

References

1. Chen, Y., et al.: Phys. Rev. D 73, 014516 (2006)
2. Amsler, C., Close, F.E.: Phys. Rev. D 53, 295 (1996); Phys. Lett. B 483, 345 (2000)
3. Close, F.E., Kirk, A.: Eur. Phys. J . C 21, 531 (2001)
4. Amsler, C., Hanhart, C.: In: Tanabashi, M., et al. (Particle Data Group): Phys. Rev. D 98, 030001 (2018), p. 753
5. Amsler, C., et al.: In: Tanabashi, M., et al. (Particle Data Group): Phys. Rev. D 98, 030001 (2018), p. 658
6. Jaffe, R.L.: Phys. Rev. D 15, 267, 281 (1977)
7. Amsler, C., et al.: Phys. Lett. B 333, 277 (1994)
8. Amsler, C.: Rev. Mod. Phys. 70, 1293 (1998)
9. Amsler, C., et al.: Phys. Lett. B 353, 571 (1995); Amsler, C., et al.: Phys. Lett. B 342, 433 (1995)
10. Abele, A., et al.: Phys. Lett. B 385, 425 (1996)
11. Barberis, D.: Phys. Lett. B 462, 462 (1999)
12. Amsler, C., et al.: Eur. Phys. J. C 23, 29 (2002)
13. Godfrey, S., Isgur, N.: Phys. Rev. D 32 189 (1985)
14. Tanabashi, M., et al. (Particle Data Group): Phys. Rev. D 98, 030001 (2018)
15. Amsler, C., et al.: Phys. Lett. B 294, 451 (1992)
16. Amsler, C., et al.: Phys. Lett. B 425, 355 (1995)
17. Amsler, C.: Phys. Lett. B 541, 22 (2002)
18. Barberis, D., et al.: Phys. Lett. B 479, 59 (2000)
19. Uehara, S., et al.: Prog. Theor. Exp. Phys. 2013, 123C01 (2013)
20. Amsler, C., et al.: Phys. Lett. B 353, 571 (1995)
21. Amsler, C., Close, F.E.: Phys. Lett. B 353, 385 (1995)
22. Escribano, R., Nadal, J.: JHEP 05, 006 (2007); Escribano, R.: Eur. Phys. J. C 65, 467 (2010)
23. Close, F.E., Zhao, Q., Phys. Rev. D 71, 094022 (2005)
24. Brünner, F., Rebhan, A.: Phys. Rev. Lett. 115, 131601 (2015)
25. Chanowitz, M.: Phys. Rev. Lett. 95, 172001 (2005)
26. Ochs, W.: J. Phys. G 40, 043001 (2013)
27. Evangelista, C., et al.: Phys. Rev. D 57, 5370 (1998)
28. Etkin, A.: Phys. Lett. B 201, 568 (1988)

29. Barberis, D., et al.: Phys. Lett. B 432, 436 (1998)
30. May, B., et al.: Phys. Lett. B 225, 450 (1989)
31. May, B., et al.: Z. Phys. C 46, 203 (1990)
32. Bertin, A., et al.: Phys. Rev. D 57, 55 (1998)
33. Amsler, C.: Adv. Nucl. Phys. 18, 183 (1987), and references therein
34. Adams, G.S., et al.: Phys. Lett. B 657, 27 (2007)
35. Abele, A., et al.: Phys. Lett. B 423, 175 (1998)
36. Alekseev, M.G., et al.: Phys. Rev. Lett. 104, 241803 (2010)
37. Isgur, N., et al.: Phys. Rev. Lett. 54, 869 (1985)
38. Close, F.E., Page, P.R.: Nucl. Phys. B 433, 23 (1995)
39. Lacock, P., et al.: Phys. Lett. B 401, 308 (1997); Bernard, C., et al.: Phys. Rev. D 56, 7039 (1997)
40. Meyer, C.A., Swanson, E.S.: Prog. Part. Nucl. Phys. 82, 21 (2015)

Chapter 12
Resonance Analysis

The mass dependence of a resonance with nominal mass m_0 and total width Γ, decaying into the final state $|f\rangle$, for example $|\pi^+\pi^-\rangle$, is usually described by the non-relativistic Breit-Wigner amplitude

$$T(m) \propto \frac{1}{m_0 - m - i\Gamma/2}, \tag{12.1}$$

leading to the mass distribution

$$|T(m)|^2 dm \propto \frac{dm}{(m_0 - m)^2 + \Gamma^2/4}. \tag{12.2}$$

However, for broad overlapping and interfering resonances the prescription (12.1) fails. Interference effects occur when various decay modes from the initial state $|i\rangle$ lead to the same final state $|f\rangle$ (even when the decay modes involve intermediate resonances with different quantum numbers). An example of interferences between the 2^{++} a_2, the 1^{--} ρ and the 1^{-+} $\pi_1(1400)$ was mentioned earlier (Fig. 11.12). Distortions also appear in the mass spectrum due to kinematical thresholds in other channels. An example will be discussed below in the context of the $a_0(980)$ decaying into $\eta\pi$ and $K\overline{K}$.

The K-matrix formalism is better suited when dealing with resonance interference and multiple decay modes [1]. Consider for example the four scattering reactions

$$\begin{pmatrix} \pi^+\pi^- \to \pi^+\pi^- & \pi^+\pi^- \to K^+K^- \\ K^+K^- \to \pi^+\pi^- & K^+K^- \to K^+K^- \end{pmatrix}. \tag{12.3}$$

The original version of this chapter was revised. A correction to this chapter is available at https://doi.org/10.1007/978-3-319-98527-5_19

© The Author(s) 2018
C. Amsler, *The Quark Structure of Hadrons*, Lecture Notes in Physics 949,
https://doi.org/10.1007/978-3-319-98527-5_12

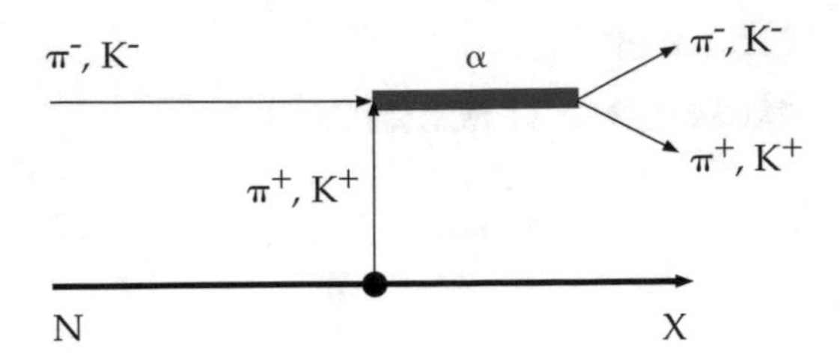

Fig. 12.1 $\pi\pi$ or $K\overline{K}$ scattering through intermediate resonances α

through one or more intermediate resonances α, for example by scattering a pion or a kaon off a target nucleon (Fig. 12.1).

The transition amplitude T for a given partial wave ℓ is described by the 2×2 T-matrix

$$\boxed{T = (1 - iK\rho)^{-1}K} \tag{12.4}$$

where K is the real and symmetric 2×2 matrix

$$K_{ij}(m) = \sum_{\alpha} \frac{\gamma_{\alpha_i}\gamma_{\alpha_j} m_\alpha \Gamma'_\alpha}{m_\alpha^2 - m^2} B_{\alpha_i}(m) B_{\alpha_j}(m), \tag{12.5}$$

(i or j = 1 for $\pi^+\pi^-$, 2 for K^+K^-). The sum runs over all resonances with K-matrix poles m_α. The 2×2 diagonal matrix $\rho(m)$ in (12.4), defined as

$$\rho_1(m) \equiv \rho_{11} = 2p_\pi/m \text{ and } \rho_2(m) \equiv \rho_{22} = 2p_K/m, \tag{12.6}$$

takes into account the two-body phase space. For masses m below $K\overline{K}$ threshold ρ_2 is imaginary and for large masses $\rho_i \simeq 1$. The factors B_{α_i} are ratios of the damping factors listed in Table 7.3:

$$B_{\alpha_i}(m) = \frac{F_\ell(p_i)}{F_\ell(p_{\alpha_i})}. \tag{12.7}$$

The decay angular momentum is ℓ, p_1 is the pion momentum, p_2 the kaon momentum, and p_{α_i} the corresponding momenta at the mass $m = m_\alpha$. The coupling constants of the resonances γ_{α_1} to $\pi\pi$ and γ_{α_2} to $K\overline{K}$ are real and fulfil the condition

$$\gamma_{\alpha_1}^2 + \gamma_{\alpha_2}^2 = 1, \tag{12.8}$$

since we have assumed that the resonances have no further decay channels. For the sake of clarity and ease of notation we shall ignore the damping factors (12.7) in the following. This is correct e.g. when dealing with the decay of scalar mesons with ℓ = 0. The K-matrix becomes simply

$$K_{ij}(m) = \sum_{\alpha} \frac{\gamma_{\alpha_i}\gamma_{\alpha_j} m_\alpha \Gamma'_\alpha}{m_\alpha^2 - m^2}. \tag{12.9}$$

The parameter Γ'_α is related to the partial width Γ_{α_i} for decay into channel i:

$$\Gamma_{\alpha_i}(m_\alpha) = \gamma^2_{\alpha_i}\Gamma'_\alpha \rho_i(m_\alpha). \tag{12.10}$$

The total resonance width is given by

$$\Gamma_\alpha = \sum_{i=1}^{2} \Gamma_{\alpha_i}. \tag{12.11}$$

Let us consider the simple case of a single resonance with mass m_0 and width Γ decaying into one channel only. Then $\Gamma'_0 = \frac{\Gamma}{\rho(m_0)}$ and

$$K = \frac{m_0\Gamma'_0}{m_0^2 - m^2}. \tag{12.12}$$

Thus (12.4) gives

$$T(m) = \frac{\frac{m_0\Gamma'_0}{m_0^2-m^2}}{1 - i\frac{m_0\Gamma'_0}{m_0^2-m^2}\rho(m)} = \frac{\frac{m_0\Gamma}{\rho(m_0)}}{m_0^2 - m^2 - im_0\Gamma(m)}, \tag{12.13}$$

the relativistic version of the Breit-Wigner amplitude with

$$\Gamma(m) \equiv \Gamma\frac{\rho(m)}{\rho(m_0)}. \tag{12.14}$$

For a narrow resonance far above decay threshold one gets with the approximation

$$m_0^2 - m^2 = (m_0 + m)(m_0 - m) \simeq 2m_0(m_0 - m) \tag{12.15}$$

and

$$p_0 \simeq p \simeq \frac{m_0}{2} \Rightarrow \rho(m) \simeq \rho(m_0) \simeq 1 \tag{12.16}$$

the familiar non-relativistic Breit-Wigner amplitude (12.1)

$$T(m) = \frac{\Gamma/2}{m_0 - m - i\Gamma/2}. \tag{12.17}$$

The mass and width of a resonance is defined as the pole of T, which lies at

$$m_P = m_0 - i\frac{\Gamma}{2} \tag{12.18}$$

in the complex plane and m_0 coincides with the pole of the K-matrix. However, for overlapping resonances the poles m_α of the K-matrix do not coincide with those of the T-matrix.

Figure 12.1 and matrix (12.4) deal with the formation of resonances in scattering experiments $ab \to \alpha \to cd$. Consider now the production of resonances in reactions associated with a recoiling system $ab \to \alpha X, \alpha \to cd$, such as $pp \to \alpha X$ or $\overline{p}p \to \alpha X$. In the isobar model the resonance is assumed not to interact with the recoiling system X. The T-matrix is replaced by the vector [2]

$$\boxed{T = (1 - iK\rho)^{-1}P}, \tag{12.19}$$

where the vector P is defined as

$$P_i(m) = \sum_\alpha \frac{\beta_\alpha \gamma_{\alpha_i} m_\alpha \Gamma'_\alpha B_{\alpha_i}(m)}{m_\alpha^2 - m^2}. \tag{12.20}$$

The coupling strength to the initial state is described by the complex number β_α while γ_{α_i} denotes as before the resonance coupling to channel i. A resonance is characterized by its complex pole position.

By comparing with (12.9) and copying from (12.13) one finds the relativistic Breit-Wigner amplitude of a single scalar resonance feeding one decay channel,

$$T(m) = \frac{\beta m_0 \Gamma / \rho(m_0)}{m_0^2 - m^2 - i m_0 \Gamma(m)} \tag{12.21}$$

with (complex) coupling strength β, and where we have again dropped the ratio of damping factors (12.7) for $\ell = 0$.

Let us now consider the production of a series of resonances with the same quantum numbers and two different decay modes, and construct the vector T. With

$$K\rho = \begin{pmatrix} K_{11} & K_{12} \\ K_{12} & K_{22} \end{pmatrix} \begin{pmatrix} \rho_1 & 0 \\ 0 & \rho_2 \end{pmatrix} = \begin{pmatrix} K_{11}\rho_1 & K_{12}\rho_2 \\ K_{12}\rho_1 & K_{22}\rho_2 \end{pmatrix} \tag{12.22}$$

and

$$(1 - iK\rho) = \begin{pmatrix} 1 - iK_{11}\rho_1 & -iK_{12}\rho_2 \\ -iK_{12}\rho_1 & 1 - iK_{22}\rho_2 \end{pmatrix}, \tag{12.23}$$

one gets

$$(1 - iK\rho)^{-1} = \begin{pmatrix} 1 - iK_{22}\rho_2 & iK_{12}\rho_2 \\ iK_{12}\rho_1 & 1 - iK_{11}\rho_2 \end{pmatrix} \frac{1}{\delta} \tag{12.24}$$

with

$$\delta \equiv (1 - iK_{11}\rho_1)(1 - iK_{22}\rho_2) + K_{12}^2\rho_1\rho_2 = 1 - \rho_1\rho_2 D - i(\rho_1 K_{11} + \rho_2 K_{22}) \tag{12.25}$$

and

$$D \equiv K_{11}K_{22} - K_{12}^2. \tag{12.26}$$

The amplitude to observe the final state $|1\rangle$ is given by (12.19):

$$T_1 = \frac{(1 - iK_{22}\rho_2)P_1 + iK_{12}\rho_2 P_2}{1 - \rho_1\rho_2 D - i(\rho_1 K_{11} + \rho_2 K_{22})}. \tag{12.27}$$

Consider for example the production of a single resonance, say $a_0(980)^0$ with mass m_0, decaying into $\eta\pi^0$ and $K\overline{K}$, hence $\ell = 0$ and $D = 0\,(\gamma_1^2\gamma_2^2 - [\gamma_1\gamma_2]^2 = 0)$. One obtains for the amplitude to decay into $\eta\pi^0$ (Problem 12.1),

$$T_1 = T(\eta\pi^0) = \frac{bg_1}{m_0^2 - m^2 - i(\rho_1 g_1^2 + \rho_2 g_2^2)}, \tag{12.28}$$

where we have defined

$$g_1 \equiv \gamma_1\sqrt{m_0\Gamma_0'}, \quad g_2 \equiv \gamma_2\sqrt{m_0\Gamma_0'}, \quad b \equiv \beta\sqrt{m_0\Gamma_0'}, \tag{12.29}$$

hence

$$g_1^2 + g_2^2 = m_0\Gamma_0'. \tag{12.30}$$

Formula (12.28) is the Flatté coupled channel amplitude [3], where g_1 and g_2 represent the coupling strengths to $\eta\pi^0$ and $K\overline{K}$, respectively, and b the $a_0(980)$ production amplitude. The amplitude T_2 to decay into $K\overline{K}$ is obtained by swapping the labels 1 and 2. The phase space factors are given by

$$\rho_1(m) = \frac{2p_\eta}{m} = \sqrt{\left[1 - \left(\frac{m_\pi + m_\eta}{m}\right)^2\right]\left[1 - \left(\frac{m_\pi - m_\eta}{m}\right)^2\right]} \tag{12.31}$$

and

$$\rho_2(m) = \frac{2p_K}{m} = \sqrt{1 - \frac{4m_K^2}{m^2}}, \tag{12.32}$$

where $p_\eta = p_\pi$ and p_K are the momenta in the $a_0(980)$ rest frame.

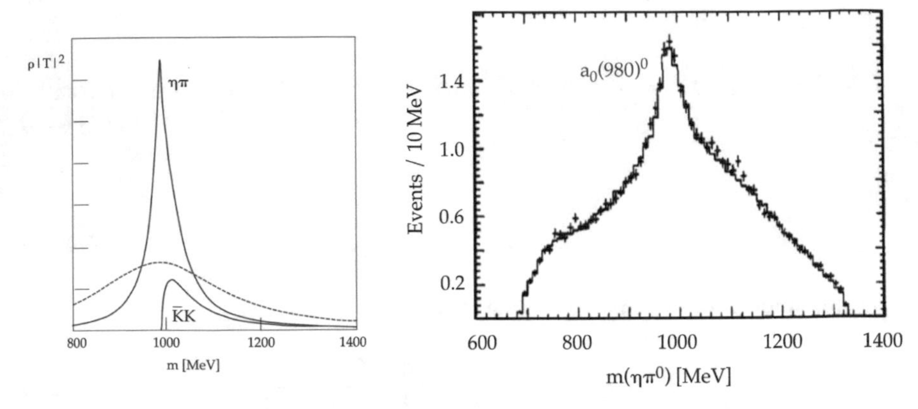

Fig. 12.2 Left: predicted $\eta\pi$ and $K\overline{K}$ mass distributions (in arbitrary units, full red curves) for $a_0(980)$ resonance production in $\overline{p}p$ annihilation into $a_0\pi$, according to the Flatté formalism with $g_1 = 324$ MeV, $g_2 = 329$ MeV ($\Gamma'_0 = 462$ MeV). The observed width is about 54 MeV. The dashed blue line shows the $\eta\pi$ mass distribution for the same width Γ'_0 in the absence of $K\overline{K}$ decay ($g_2 = 0$). The observed width increases to 300 MeV. Right: the $a_0(980) \rightarrow \eta\pi^0$ resonance in proton-antiproton annihilation at rest into $\eta\eta\pi^0$ [4]

The threshold for the running mass m is $m_{\pi^0} + m_\eta = 680$ MeV above which $\rho_1 \geq 0$. Below $K\overline{K}$ threshold ($m \leq 990$ MeV) ρ_2 is imaginary. This leads to a shift of the resonance peak and to a narrower and asymmetric distribution in the $\eta\pi$ channel, as is immediately seen from (12.28). Figure 12.2 (left) illustrates the result for realistic values of g_1 and g_2. The observed resonance width is about 54 MeV. Also shown is the expected distribution when turning the $K\overline{K}$ channel off. Figure 12.2 (right) shows the $a_0(980)$ produced in the annihilation channel $\eta\eta\pi^0$ [4].

The Dalitz plots for $\pi^0\pi^0\eta$, $\pi^0\eta\eta$ and $\pi^0\pi^0\pi^0$ in proton-antiproton annihilation have been discussed before (Figs. 11.2 and 11.3). As a further example of the K-matrix application, Fig. 12.3 shows the $\pi^0\pi^0$ and $\eta\eta$ mass distributions obtained from a coupled channel analysis using a 3×3 K-matrix for the production and decay of the resonances into $\pi^0\pi^0$, $\eta\eta$ and $K\overline{K}$, which illustrates the interference between four isoscalar 0^{++} mesons [5]. The $f_0(980) \rightarrow \pi^0\pi^0$ appears as a dip in $\pi^0\pi^0\pi^0$ and as a peak in $\pi^0\pi^0\eta$. This shows that a resonance does not necessarily appear as a peak in the invariant mass spectrum.

Conversely, not every peak can be attributed to a resonance, but could be generated by the presence of thresholds or by triangle singularities. A well-known example of threshold cusp is in the decay $K^+ \rightarrow \pi^+\pi^0\pi^0$. The $\pi^0\pi^0$ mass distribution exhibits a shoulder at $m^2 = 4m^2_{\pi^\pm} = 0.076\,\text{GeV}^2$ (Fig. 12.4a) due to the opening of the decay channel $K^+ \rightarrow \pi^+\pi^+\pi^-$. Above $2m_{\pi^\pm}$ threshold the two charged pions can rescatter into two neutral pions. This process interferes with the direct $\pi^0\pi^0$ decay and generates a distortion (cusp) in the mass spectrum.

Triangle singularities are kinematical effects associated with three collinear particles (Fig. 12.4b). Particle A decays into B and C, which in turn decays into D and E. Particle D can fly in the opposite direction to C and, if kinematics permits,

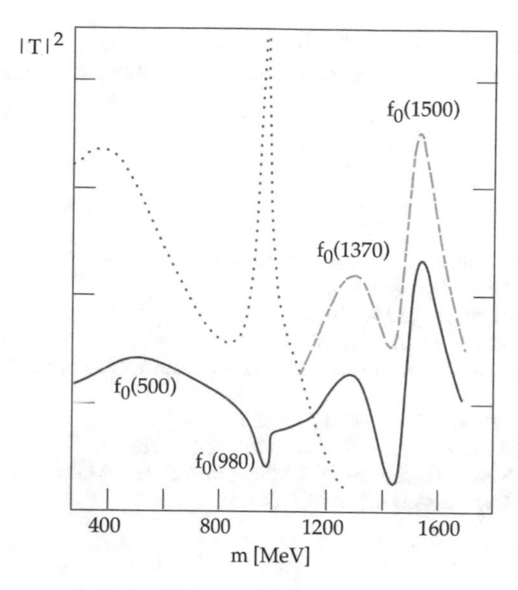

Fig. 12.3 $\pi^0\pi^0$ and $\eta\eta$ mass distributions (in arbitrary units). Full red curve: $\pi^0\pi^0\pi^0$; dotted blue curve: $\pi^0\pi^0\eta$; dashed green curve $\pi^0\eta\eta$ (compiled from [5])

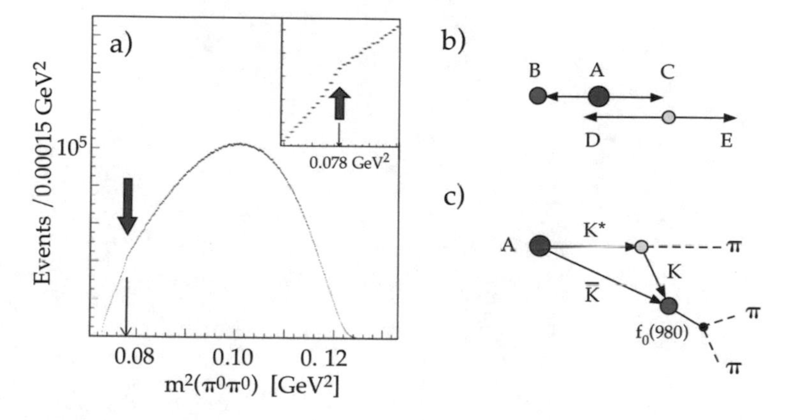

Fig. 12.4 (**a**) $m^2(\pi^0\pi^0)$ distribution in $K^+ \to \pi^+\pi^0\pi^0$ decay with a cusp (arrow) at the $\pi^+\pi^-$ threshold [6]; (**b**) triangle diagram in which particle D catches up and interacts with B; (**c**) example of a state A decaying into $K^*\bar{K}$ and 3π through triangular rescattering

eventually catch up and interact with B. This mechanism generates fake peaks in the final state and has been proposed as alternative explanations for several meson and baryon signals (see e.g. [7]). An example is shown in Fig. 12.4c: the state A decays into $\bar{K}K^* \to \bar{K}K\pi$ and by final state rescattering also into 3π. For a broad A two peaks can appear at different masses in the two final states. This mechanism has been proposed to interpret the close lying $a_1(1260) \to K^*(\to K\pi)\bar{K}$ and

$a_1(1420) \rightarrow f_0(980)(\rightarrow \pi\pi)\pi$ [8] or the $\eta(1475) \rightarrow K^*(\rightarrow K\pi)\bar{K}$ and $\eta(1410) \rightarrow a_0(980)(\rightarrow \eta\pi)\pi$ as manifestations of the same states [9].

References

1. Chung, S.U., et al.: Ann. Phys. (Leipzig) 4, 404 (1995)
2. Aitchison, I.J.R.: Nucl. Phys. A 189, 417 (1972)
3. Flatté, S.M.: Phys. Lett. B 63, 224 (1976)
4. Amsler, C., et al.: Phys. Lett. B 291, 347 (1992)
5. Spanier, S.: PhD thesis, University of Mainz (1994); Amsler, C., et al.: Phys. Lett. B 355, 425 (1995)
6. Batley, J.R., et al.: Phys. Lett. B 633, 173 (2006)
7. Liu, X.-H., Oka, M., Zhao, Q.: Phys. Lett. B 753, 297 (2016)
8. Mikhasenko, M., Ketzer, B., Sarantsev, A.: Phys. Rev. D 91, 094015 (2015)
9. Wu, X.-G., et al.: Phys. Rev. D 87, 014023 (2013)

Chapter 13
Baryons

Baryons are made of three quarks. The charge conjugated partners are the antibaryons which consist of the three corresponding antiquarks. However, candidates for pentaquark baryons made of four quarks and one antiquark have been reported (Sect. 16.3). Let us first deal with the ordinary ones made of qqq or $\overline{qqq}$. Baryons have the baryon number $B = 1$ and antibaryons $B = -1$. Those containing only light (u and d) quarks are called N for isospin $i = \frac{1}{2}$ and Δ for $i = \frac{3}{2}$ (the proton and neutron are usually abbreviated to p and n). Hyperons are baryons containing at least one s quark. Baryons with two light quarks are labelled Λ for $i = 0$ and Σ for $i = 1$. Those containing only one light quark are called Ξ and are isospin doublets. Those without any light quark are called Ω and are isospin singlets. The number of c or b quarks in a baryon is indicated by the subscripts c or b. The mass in MeV is specified in parentheses, except for the long-lived ("stable") ones which decay through the weak interaction.

For example, the $J^P = \frac{1}{2}^+$ $N(1440)$ (also known as Roper resonance) is the first radial excitation of the nucleon, decaying into $n\pi(\pi)$ or $p\pi(\pi)$. The $\frac{3}{2}^+$ $\Xi_b(5950)^0$ is a usb state decaying (strongly) into $\Xi_b^-\pi^+$, the $J^P = \frac{1}{2}^+$ Ξ_{cc}^{++} at a mass of 3621 MeV a ucc state decaying (weakly) into $\Lambda_c^+[udc]K^-\pi^+\pi^+$, the $\Omega_c(3000)^0$ an excited ssc baryon decaying (strongly) into $\Xi_c^+[usc]K^-$, see also Problem 13.1. We shall return to some of these states in Chap. 17 on heavy baryons.

By convention (Sect. 2.1) the quark has positive internal parity, hence the antiquarks negative internal parity. Following (2.5) the parity is given by

$$P(qqq) = (-1)^{\ell_\rho}(-1)^{\ell_\lambda} \tag{13.1}$$

for baryons, and

$$P(\overline{qqq}) = -(-1)^{\ell_\rho}(-1)^{\ell_\lambda} \tag{13.2}$$

© The Author(s) 2018

C. Amsler, *The Quark Structure of Hadrons*, Lecture Notes in Physics 949,
https://doi.org/10.1007/978-3-319-98527-5_13

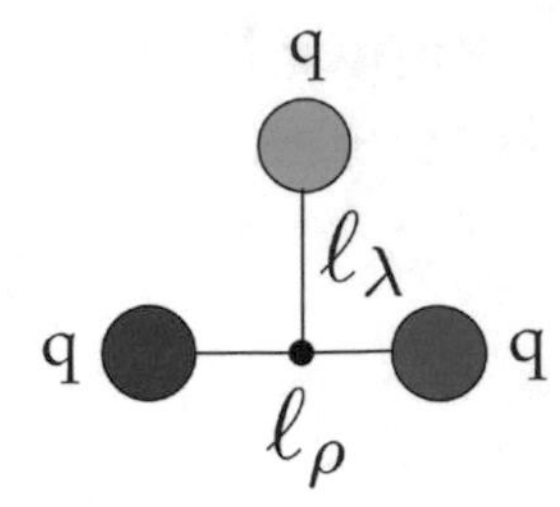

Fig. 13.1 Angular momenta ℓ_ρ and ℓ_λ in a qqq baryon

for antibaryons, where ℓ_ρ is the angular momentum of one qq (diquark) pair and ℓ_λ the angular momentum between the diquark and the third quark (Fig. 13.1). Thus ground state baryons have $\ell_\rho = \ell_\lambda = 0$, and hence the quantum numbers $J^P = \frac{1}{2}^+$ or $\frac{3}{2}^+$, while the corresponding antibaryons have the opposite parities. When adding angular momentum larger spins ($\frac{5}{2}, \frac{7}{2}, \ldots$) become possible. The C and G parities are not defined.

13.1 Ground State Light Baryons

Let us deal first with baryons made of u, d, and s quarks. We have seen in Chap. 10 on SU(3)$_c$ that coupling three coloured quarks leads to the decomposition (10.13). This can be applied to flavour symmetry SU(3)$_f$ as well:

$$3 \times 3 \times 3 = 1 + 8 + 8 + 10. \tag{13.3}$$

Thus one expects a flavour singlet, two flavour octets and one flavour decuplet. The weight diagrams of the octets and decuplets (Fig. 13.2) can be constructed by superimposing the weight diagrams of the fundamental representation of SU(3) (Fig. 7.1, left). They correspond to the 10 + 8 known ground state baryons, although with three flavours one would expect 27 states. However, we will show below that for symmetry reasons both octets are needed to describe the flavour wavefunctions of the eight spin-$\frac{1}{2}$ baryons, and that an SU(3)$_f$ ground state singlet does not exist.

The ground state "stable" (weak or electromagnetically decaying) hyperons which contain at least one s quark are listed in Table 13.1. Apart from the decuplet Ω^- they are spin-$\frac{1}{2}$ octet states decaying weakly via the strangeness changing $|\Delta S| = 1$ interaction (the Σ^0 decays electromagnetically). High energy hyperon beams can be produced thanks to their relatively long lifetimes. Section 14.2 will be devoted to measurements of their magnetic moments. Additional ground state qqq baryons beyond those predicted by SU(3)$_f$ in Fig. 13.2 have not been found. States such as Ξ^+ or Δ^{--} do not exist. Note that the charge conjugated $\overline{\Sigma^+} = |\overline{uus}\rangle$ antihyperon is negatively charged and is not identical to the $\Sigma^- = |dds\rangle$.

Figure 13.3 (left) shows for example the discovery of the Ξ^0 in a hydrogen bubble chamber, produced by the reaction $K^-p \to K^0\Xi^0$, followed by the cascade decay

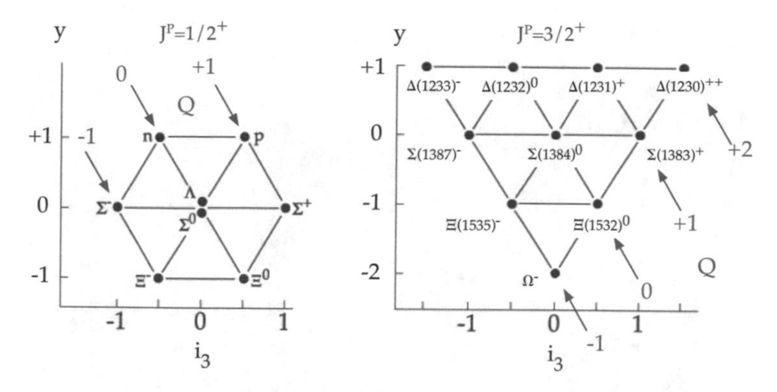

Fig. 13.2 Weight diagrams of the octet and decuplet ground state baryons. The arrows refer to the electric charges

Table 13.1 Main decay modes and mean lives of the ground state "stable" hyperons

	J^P	Mass [MeV]	Quark content	S	Main decay modes (branching ratio in %)	Mean life [$\times 10^{-10}$ s]
Λ	$\frac{1}{2}^+$	1116	*uds*	−1	$p\pi^-$ (64), $n\pi^0$ (36)	2.63
Σ^+	$\frac{1}{2}^+$	1189	*uus*	−1	$p\pi^0$ (52), $n\pi^+$ (48)	0.80
Σ^0	$\frac{1}{2}^+$	1193	*uds*	−1	$\Lambda\gamma$ ($\simeq$100)	[a]
Σ^-	$\frac{1}{2}^+$	1197	*dds*	−1	$n\pi^-$ ($\simeq$100)	1.48
Ξ^0	$\frac{1}{2}^+$	1315	*uss*	−2	$\Lambda\pi^0$ ($\simeq$100)	2.90
Ξ^-	$\frac{1}{2}^+$	1322	*dss*	−2	$\Lambda\pi^-$ ($\simeq$100)	1.64
Ω^-	$\frac{3}{2}^+$	1672	*sss*	−3	ΛK^- (68), $\Xi^0\pi^-$ (24), $\Xi^-\pi^0$ (8)	0.82

The Λ and Σ^0 have the same quark content, yet different $SU(3)_f$ wavefunctions
[a] 7.4×10^{-20} s

$\Xi^0 \rightarrow \Lambda\pi^0$, $\Lambda \rightarrow \pi^- p$ and $K^0 \rightarrow \pi^+\pi^-$, with 1.15 GeV/c kaons from the Berkeley Bevatron [1]. The line joining the Ξ^0 production point (red dot) and the Λ decay vertex (blue dot) does not lie in the plane spanned by the π^- and p tracks from Λ decay, nor is it parallel to the plane spanned by the incident kaon and the K^0 directions. This points to the emission of an additional invisible particle (the π^0). There are enough constraints to reconstruct the reaction and the Ξ^0 mass was measured to be 1326 ± 20 MeV.

Figure 13.3 (right) shows the discovery of the Ω^- hyperon at BNL [2]. The Ω^- was produced by the strong interaction $K^- p \rightarrow \Omega^- K^+ K^0$ in a 5 GeV/c kaon beam from the AGS. The bubble chamber photograph shows the decay $\Omega^- \rightarrow \Xi^0\pi^-$ followed by $\Xi^0 \rightarrow \Lambda(\rightarrow p\pi^-)\pi^0$, where the two photons from π^0 decay produce e^+e^- pairs. The Ω^- mass was calculated to be 1686 ± 12 MeV.

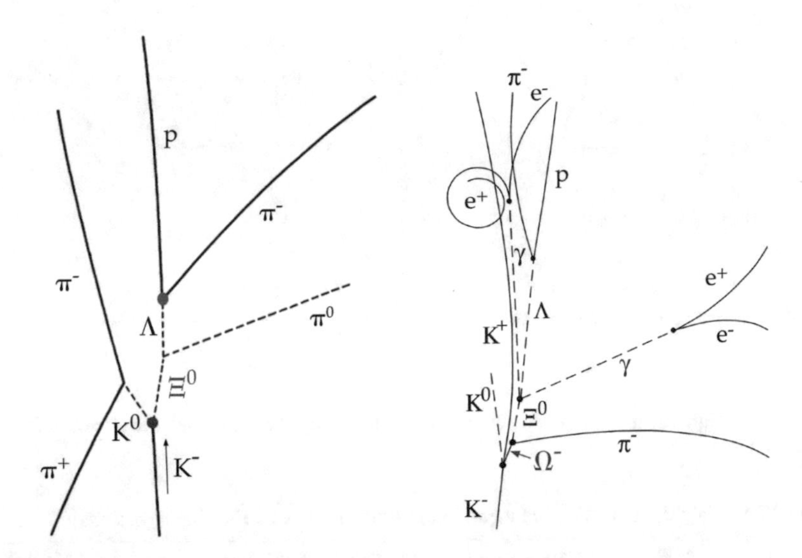

Fig. 13.3 Left: production and decay of the first Ξ^0 hyperon observed in a bubble chamber [1]. The trajectories of the invisible neutral particles are shown by dashed lines (see the text). Right: bubble chamber photograph showing the production and decay of the Ω^- [2]

The $j = \frac{3}{2}$ hyperons belong to the $\mathrm{SU}(3)_f$ decuplet. Figure 13.4 shows the $\pi^+ p$ and $\pi^- p$ total cross sections as a function of center-of-mass energy. The N resonances have isospin $\frac{1}{2}$ and therefore do not contribute to $\pi^+ p$. The low energy region is dominated by the excitation of the Δ^{++} and Δ^0 resonances in $\pi^+ p$ and $\pi^- p$, respectively.

The Δ and the strangeness $S = -1$ and -2 hyperons decay strongly (as indicated by their masses in parentheses) and have very short lifetimes. For example, the $\Xi(1532)^0$ decays into $\Xi^0 \pi^0$ and $\Xi^- \pi^+$. The Σ and Ξ have the same quark content as the $j = \frac{1}{2}$ hyperons, but different $\mathrm{SU}(3)_f$ wavefunctions, as will be discussed below.

13.2 SU(2) Wavefunctions of Three Light Quarks

In order to derive the flavour wavefunctions of 3-quark states we first need to study their spin (and isospin) structure. Let us therefore construct the spin wavefunctions of 3-quark systems. The simplest configuration is that of the Δ resonance with $j = \frac{3}{2}$ and parallel spins. The spin wavefunction is simply given by

$$|\chi^j_m\rangle \equiv |\chi^{\frac{3}{2}}_{\frac{3}{2}}\rangle = |\uparrow\uparrow\uparrow\rangle, \tag{13.4}$$

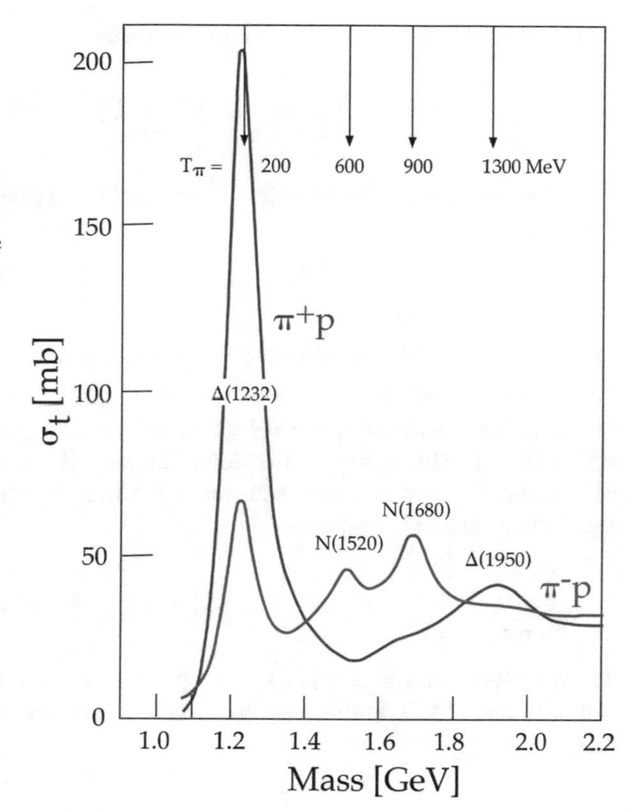

Fig. 13.4 $\pi^+ p$ and $\pi^- p$ total cross sections σ_t as a function of center-of-mass energy. The upper scale shows the corresponding laboratory kinetic energy of the incident pion. Several excited baryons contribute to the peaks above 1.4 GeV. The labels refer to the dominating ones, the spin-$\frac{3}{2}^-$ $N(1520)$, $\frac{5}{2}^+$ $N(1680)$ and $\frac{7}{2}^+$ $\Delta(1950)$

to which we apply J_- (Table 6.2):

$$J_-|\chi^{\frac{3}{2}}_{\frac{3}{2}}\rangle = \sqrt{3}|\chi^{\frac{3}{2}}_{\frac{1}{2}}\rangle. \tag{13.5}$$

Operating J_- in turn on each quark gives

$$J_-|\uparrow\uparrow\uparrow\rangle = |\downarrow\uparrow\uparrow + \uparrow\downarrow\uparrow + \uparrow\uparrow\downarrow\rangle \tag{13.6}$$

and therefore

$$|\chi^{\frac{3}{2}}_{\frac{1}{2}}\rangle = \frac{1}{\sqrt{3}}|\downarrow\uparrow\uparrow + \uparrow\downarrow\uparrow + \uparrow\uparrow\downarrow\rangle, \tag{13.7}$$

a normalized wavefunction. From now on we will apply the ladder operators on the quarks and then normalize the result. Starting from the Δ wavefunction with reversed spins

$$|\chi^{\frac{3}{2}}_{-\frac{3}{2}}\rangle = |\downarrow\downarrow\downarrow\rangle, \tag{13.8}$$

one finds by operating with J_+ and normalizing,

$$|\chi^{\frac{3}{2}}_{-\frac{1}{2}}\rangle = \frac{1}{\sqrt{3}}|\downarrow\downarrow\uparrow + \downarrow\uparrow\downarrow + \uparrow\downarrow\downarrow\rangle. \tag{13.9}$$

The four spin-$\frac{3}{2}$ wavefunctions build the quadruplet predicted by the decomposition (6.47):

$$2 \times 2 \times 2 = 2 + 2 + 4. \tag{13.10}$$

On the other hand, we also expect two doublets with spin $\frac{1}{2}$. Consider the configurations with $m = +\frac{1}{2}$ in which the spin projection of a diquark vanishes, while the third quark has spin up ($+\frac{1}{2}$). The diquark couples to $s = 0$ (antisymmetric state $|00\rangle$ in (2.30)) or to $s = 1$ (symmetric state $|10\rangle$ in (2.29)). The wavefunction of the so-called mixed symmetric case (in which the permutation of the first two quarks is symmetric) reads

$$|\chi^{\frac{1}{2}}_{\frac{1}{2}}\rangle = \frac{1}{\sqrt{2}}|\uparrow\downarrow\uparrow + \downarrow\uparrow\uparrow\rangle + \alpha|\uparrow\uparrow\downarrow\rangle. \tag{13.11}$$

The third term with $\alpha \neq 0$ is needed, otherwise the wavefunction cannot be made orthogonal to (13.7). Multiplying from the left with the bra of (13.7) and requiring that

$$2\frac{1}{\sqrt{2}}\frac{1}{\sqrt{3}} + \frac{\alpha}{\sqrt{3}} = 0, \tag{13.12}$$

leads to $\alpha = -\sqrt{2}$. The normalized wavefunction is then equal to

$$\begin{aligned} |\chi^{\frac{1}{2}}_{\frac{1}{2}}\rangle &= \frac{1}{\sqrt{6}}|\uparrow\downarrow\uparrow + \downarrow\uparrow\uparrow\rangle - \sqrt{\frac{2}{3}}|\uparrow\uparrow\downarrow\rangle \\ &= \frac{1}{\sqrt{6}}|\uparrow\downarrow\uparrow + \downarrow\uparrow\uparrow -2\uparrow\uparrow\downarrow\rangle. \end{aligned} \tag{13.13}$$

Operating with J_- on each quark gives

$$\begin{aligned} &\frac{1}{\sqrt{6}}|\downarrow\downarrow\uparrow -2\downarrow\uparrow\downarrow\rangle \text{ (1st quark)} \\ &+ \frac{1}{\sqrt{6}}|\downarrow\downarrow\uparrow -2\uparrow\downarrow\downarrow\rangle \text{ (2nd quark)} \\ &+ \sqrt{\frac{1}{6}}|\uparrow\downarrow\downarrow + \downarrow\uparrow\downarrow\rangle \text{ (3rd quark)}, \end{aligned} \tag{13.14}$$

hence

$$|\chi^{\frac{1}{2}}_{-\frac{1}{2}}\rangle = -\frac{1}{\sqrt{6}}|\uparrow\downarrow\downarrow + \downarrow\uparrow\downarrow -2\downarrow\downarrow\uparrow\rangle. \tag{13.15}$$

For the mixed antisymmetric case the wavefunction

$$|\chi^{\frac{1}{2}}_{\frac{1}{2}}\rangle' = \frac{1}{\sqrt{2}}|\uparrow\downarrow\uparrow - \downarrow\uparrow\uparrow\rangle \tag{13.16}$$

is orthogonal to all previous ones. Operating again with J_- on each quark gives

$$\begin{aligned} &\frac{1}{\sqrt{2}}|\downarrow\downarrow\uparrow\rangle \text{ (1st quark)} \\ -&\frac{1}{\sqrt{2}}|\downarrow\downarrow\uparrow\rangle \text{ (2nd quark)} \\ +&\frac{1}{\sqrt{2}}(|\uparrow\downarrow\downarrow\rangle - |\downarrow\uparrow\downarrow\rangle) \text{ (3rd quark)}, \end{aligned} \tag{13.17}$$

or

$$|\chi^{\frac{1}{2}}_{-\frac{1}{2}}\rangle' = \frac{1}{\sqrt{2}}|\uparrow\downarrow\downarrow - \downarrow\uparrow\downarrow\rangle. \tag{13.18}$$

The results are collected in Table 13.2.

The isospin wavefunctions for baryons with u, and d quarks are immediately found by formally replacing $\uparrow$ by u and $\downarrow$ by d (Table 13.3). The quadruplets are

Table 13.2 Normalized spin wavefunctions for baryons

$\lvert\chi_S\rangle =$	$\lvert\chi^{\frac{3}{2}}_{\frac{3}{2}}\rangle$	=	$\lvert\uparrow\uparrow\uparrow\rangle$	$s=\frac{3}{2}$
	$\lvert\chi^{\frac{3}{2}}_{\frac{1}{2}}\rangle$	=	$\frac{1}{\sqrt{3}}\lvert\downarrow\uparrow\uparrow + \uparrow\downarrow\uparrow + \uparrow\uparrow\downarrow\rangle$	
	$\lvert\chi^{\frac{3}{2}}_{-\frac{1}{2}}\rangle$	=	$\frac{1}{\sqrt{3}}\lvert\downarrow\downarrow\uparrow + \downarrow\uparrow\downarrow + \uparrow\downarrow\downarrow\rangle$	
	$\lvert\chi^{\frac{3}{2}}_{-\frac{3}{2}}\rangle$	=	$\lvert\downarrow\downarrow\downarrow\rangle$	
$\lvert\chi_{MS}\rangle =$	$\lvert\chi^{\frac{1}{2}}_{\frac{1}{2}}\rangle$	=	$\frac{1}{\sqrt{6}}\lvert\uparrow\downarrow\uparrow + \downarrow\uparrow\uparrow -2\uparrow\uparrow\downarrow\rangle$	$s=\frac{1}{2}$
	$\lvert\chi^{\frac{1}{2}}_{-\frac{1}{2}}\rangle$	=	$-\frac{1}{\sqrt{6}}\lvert\uparrow\downarrow\downarrow + \downarrow\uparrow\downarrow -2\downarrow\downarrow\uparrow\rangle$	
$\lvert\chi_{MA}\rangle =$	$\lvert\chi^{\frac{1}{2}}_{\frac{1}{2}}\rangle'$	=	$\frac{1}{\sqrt{2}}\lvert\uparrow\downarrow\uparrow - \downarrow\uparrow\uparrow\rangle$	$s=\frac{1}{2}$
	$\lvert\chi^{\frac{1}{2}}_{-\frac{1}{2}}\rangle'$	=	$\frac{1}{\sqrt{2}}\lvert\uparrow\downarrow\downarrow - \downarrow\uparrow\downarrow\rangle$	

The quadruplet functions $|\chi_S\rangle$ are symmetric, the doublet functions mixed symmetric ($|\chi_{MS}\rangle$) or mixed antisymmetric ($|\chi_{MA}\rangle$) under permutations of the first two quarks

Table 13.3 Normalized isospin wavefunctions for baryons made of the two lightest quarks

$\lvert\phi_S\rangle =$	$\lvert\phi^{\frac{3}{2}}_{\frac{3}{2}}\rangle$	$=$	$\lvert uuu\rangle$	Δ^{++}
	$\lvert\phi^{\frac{3}{2}}_{\frac{1}{2}}\rangle$	$=$	$\frac{1}{\sqrt{3}}\lvert duu + udu + uud\rangle$	Δ^{+}
	$\lvert\phi^{\frac{3}{2}}_{-\frac{1}{2}}\rangle$	$=$	$\frac{1}{\sqrt{3}}\lvert ddu + dud + udd\rangle$	Δ^{0}
	$\lvert\phi^{\frac{3}{2}}_{-\frac{3}{2}}\rangle$	$=$	$\lvert ddd\rangle$	Δ^{-}
$\lvert\phi_{MS}\rangle =$	$\lvert\phi^{\frac{1}{2}}_{\frac{1}{2}}\rangle$	$=$	$\frac{1}{\sqrt{6}}\lvert udu + duu - 2uud\rangle$	p
	$\lvert\phi^{\frac{1}{2}}_{-\frac{1}{2}}\rangle$	$=$	$-\frac{1}{\sqrt{6}}\lvert udd + dud - 2ddu\rangle$	n
$\lvert\phi_{MA}\rangle =$	$\lvert\phi^{\frac{1}{2}}_{\frac{1}{2}}\rangle'$	$=$	$\frac{1}{\sqrt{2}}\lvert udu - duu\rangle$	p
	$\lvert\phi^{\frac{1}{2}}_{-\frac{1}{2}}\rangle'$	$=$	$\frac{1}{\sqrt{2}}\lvert udd - dud\rangle$	n

The quadruplet functions $|\phi_S\rangle$ are symmetric, the doublet functions mixed symmetric ($|\phi_{MS}\rangle$) or mixed antisymmetric ($|\phi_{MA}\rangle$) under permutations of the first two quarks

assigned to the isospin-$\frac{3}{2}$ Δ resonance and the doublets to the isospin-$\frac{1}{2}$ nucleon. As we will show below, both doublets of spin and isospin are needed to describe the nucleon.

13.3 SU(3) Wavefunctions of Three Light Quarks

Starting from the u and d quarks we now construct the octet and decuplet wavefunctions with the ladder operators $U_\pm$ and $V_\pm$. Recall that U_- transforms a d quark into an s quark, $U_-|d\rangle = |s\rangle$, while $U_-|u\rangle = 0$. We get by starting from the proton in Fig. 13.2 (left):

$$|\Sigma^+\rangle = U_-\left[\frac{1}{\sqrt{6}}|udu + duu - 2uud\rangle\right] = \frac{1}{\sqrt{6}}|usu + suu - 2uus\rangle \tag{13.19}$$

for the MS case and

$$|\Sigma^+\rangle = U_-\left[\frac{1}{\sqrt{2}}|udu - duu\rangle\right] = \frac{1}{\sqrt{2}}|usu - suu\rangle \tag{13.20}$$

for the MA one. Successive clockwise applications of V_-, I_-, U_+, and V_+ produce the wavefunctions on the octet periphery. They are listed in Table 13.4.

The wavefunctions of the two $i = 0$ states are obtained with the same method as those for the mesons in Chap. 7. The center of the octet is reached by applying

Table 13.4 $SU(3)_f$ wavefunctions of the spin-$\frac{1}{2}$ ground state baryons made of u, d and s quarks

	ϕ_{MS}	ϕ_{MA}
p	$\frac{1}{\sqrt{6}}\lvert(ud+du)u-2uud\rangle$	$\frac{1}{\sqrt{2}}\lvert(ud-du)u\rangle$
n	$-\frac{1}{\sqrt{6}}\lvert(ud+du)d-2ddu\rangle$	$\frac{1}{\sqrt{2}}\lvert(ud-du)d\rangle$
Σ^+	$\frac{1}{\sqrt{6}}\lvert(us+su)u-2uus\rangle$	$\frac{1}{\sqrt{2}}\lvert(us-su)u\rangle$
Σ^0	$\frac{1}{2\sqrt{3}}\lvert(sd+ds)u+(su+us)d$ $-2(du+ud)s\rangle$	$\frac{1}{2}\lvert(ds-sd)u+(us-su)d\rangle$
Σ^-	$\frac{1}{\sqrt{6}}\lvert(ds+sd)d-2dds\rangle$	$\frac{1}{\sqrt{2}}\lvert(ds-sd)d\rangle$
Λ	$\frac{1}{2}\lvert(sd+ds)u-(su+us)d\rangle$	$\frac{1}{2\sqrt{3}}\lvert(sd-ds)u+(us-su)d$ $+2(ud-du)s\rangle$
Ξ^-	$-\frac{1}{\sqrt{6}}\lvert(ds+sd)s-2ssd\rangle$	$\frac{1}{\sqrt{2}}\lvert(ds-sd)s\rangle$
Ξ^0	$-\frac{1}{\sqrt{6}}\lvert(us+su)s-2ssu\rangle$	$\frac{1}{\sqrt{2}}\lvert(us-su)s\rangle$

The first two quarks are grouped in parentheses to emphasize the symmetry properties

V_- on the proton ($u \to s$). For the MS case

$$
\begin{aligned}
|\varphi\rangle &= V_-\left[\frac{1}{\sqrt{6}}|(udu+duu-2uud\rangle\right] \\
&= \frac{1}{\sqrt{6}}|sdu-2sud+dsu-2usd+dus+uds\rangle. \qquad (13.21)
\end{aligned}
$$

Applying I_- ($u \to d$) on the Σ^+ (13.19) leads to the (unnormalized) Σ^0:

$$
|\varphi'\rangle = \frac{1}{\sqrt{6}}|sdu+sud+dsu+usd-2dus-2uds\rangle. \qquad (13.22)
$$

To find the state Λ orthogonal to φ' we apply the Gram-Schmidt procedure (7.25):

$$
|\Lambda\rangle = |\varphi'\rangle - \alpha|\varphi\rangle \qquad (13.23)
$$

with

$$
\langle\Lambda|\varphi'\rangle = 0 = \langle\varphi'|\varphi'\rangle - \alpha\langle\varphi|\varphi'\rangle = 2+\alpha \Rightarrow \alpha = -2, \qquad (13.24)
$$

therefore

$$
|\Lambda\rangle = |\varphi'\rangle + 2|\varphi\rangle = \frac{3}{\sqrt{6}}|sdu-sud+dsu-usd\rangle. \qquad (13.25)
$$

After normalizing one finally obtains the wavefunction listed in Table 13.4

$$|\Lambda\rangle = \frac{1}{2}|sdu + dsu - sud - usd\rangle, \tag{13.26}$$

and with (13.22)

$$|\Sigma^0\rangle = \frac{1}{2\sqrt{3}}|sdu + dsu + sud + usd - 2dus - 2uds\rangle. \tag{13.27}$$

The wavefunctions for the MA case are derived in a similar way (Problem 13.2).

In the quark model the total wavefunction of a baryon should be symmetric under the permutation of any pair of quarks. Colour is then added to make the wavefunction antisymmetric, as required for fermions (Chap. 10). A fully symmetric wavefunction for spin-$\frac{1}{2}$ baryons is constructed by combining the spin states χ_{MS} and χ_{MA} with the flavour states ϕ_{MS} and ϕ_{MA}. It is a simple but tedious exercise to prove that the 16 ($m = \pm\frac{1}{2}$) wavefunctions

$$\boxed{|\tfrac{1}{2}^+\rangle = \tfrac{1}{\sqrt{2}}|\phi_{MS}\chi_{MS} + \phi_{MA}\chi_{MA}\rangle} \tag{13.28}$$

are fully symmetric under the permutation of any pair of quarks. In fact, using instead the combination $\frac{1}{\sqrt{2}}|\phi_{MS}\chi_{MA} - \phi_{MA}\chi_{MS}\rangle$, which turns out to be fully antisymmetric, leads to the wrong proton and neutron magnetic moments (Problem 14.1). This combination occurs for excited baryons with one unit of angular momentum, see (15.16) in Chap. 15.

Starting from the isospin-$\frac{3}{2}$ wavefunctions in Table 13.3 one can construct in a similar way the $SU(3)_f$ states ϕ_S for three quarks by applying the ladder operators $U_\pm$, $V_\pm$ and $I_\pm$. They are fully symmetric under permutations of any pair of quarks and are listed in Table 13.5. The total wavefunctions of the 40 spin-$\frac{3}{2}$ baryons are

Table 13.5 $SU(3)_f$ wavefunctions ϕ_S of the ground state spin-$\frac{3}{2}$ baryons

	ϕ_S
$\Delta^{++}(1233)$	$\|uuu\rangle$
$\Delta^{+}(1232)$	$\frac{1}{\sqrt{3}}\|uud + udu + duu\rangle$
$\Delta^{0}(1231)$	$\frac{1}{\sqrt{3}}\|udd + dud + ddu\rangle$
$\Delta^{-}(1230)$	$\|ddd\rangle$
$\Sigma^{+}(1383)$	$\frac{1}{\sqrt{3}}\|uus + usu + suu\rangle$
$\Sigma^{0}(1384)$	$\frac{1}{\sqrt{3}}\|uds + usd + sdu\rangle$
$\Sigma^{-}(1387)$	$\frac{1}{\sqrt{3}}\|dds + dsd + sdd\rangle$
$\Xi^{0}(1532)$	$\frac{1}{\sqrt{3}}\|uss + sus + ssu\rangle$
$\Xi^{-}(1535)$	$\frac{1}{\sqrt{3}}\|dss + sds + ssd\rangle$
Ω^{-}	$\|sss\rangle$

then obtained by combining the functions ϕ_S and χ_S:

$$\boxed{|\tfrac{3}{2}^+\rangle = |\phi_S \chi_S\rangle} \,. \tag{13.29}$$

Combining $SU(3)_f$ with $SU(2)_{spin}$ leads to SU(6) and it is easy to derive with the help of Young tableaux the decomposition

$$6 \times 6 \times 6 = 56 + 70 + 70 + 20. \tag{13.30}$$

Our 8×2 octet + 10×4 decuplet ground state baryons build the 56-dimensional representation. In fact, the orbital wavefunction still needs to be combined with SU(6) to build the final total wavefunction, which is required to be symmetric. We have ignored this contribution, since in the absence of angular momentum the orbital wavefunction is anyway symmetric.

The last $SU(3)_f$ orthogonal state, the singlet in (13.3), is totally antisymmetric:

$$|\phi_A\rangle = \frac{1}{\sqrt{6}}|s(du - ud) + (usd - dsu) + (du - ud)s\rangle. \tag{13.31}$$

As Table 13.2 shows, an antisymmetric spin wavefunction cannot be made with three spins and thus ϕ_A cannot be combined with any spin state to generate a totally symmetric wavefunction for ground state baryons. Therefore ϕ_A is not realized for ground state baryons but occurs for the first orbital excitations, in the 70-plet and 20-plet of SU(6), see Chap. 15.

13.4 Gell-Mann-Okubo Mass Formula

$SU(3)_f$ symmetry implies that masses within multiplets should all be equal, which is neither the case for mesons nor for baryons. The symmetry is badly broken, but many predictions can nevertheless be made with the help of the flavour wavefunctions just discussed. In the next section we will derive the magnetic moments of baryons and compare with data. Let us derive here a simple predictive mass formula for octet baryons. The weight diagram of the ground state decuplet is reproduced in Fig. 13.5. Isospin multiplets are almost degenerate. The small splittings are due to the difference between u and d masses and to electromagnetic attraction or repulsion. (We have already seen that Coulomb repulsion reduces the binding and hence increases the hadron mass, while $m_u < m_d$ acts in the opposite direction.) The equal spacing rule states that the mass difference between isospin multiplets is approximately constant with $\Delta m \sim$150 MeV. The splitting is essentially due to the heavier s quark.

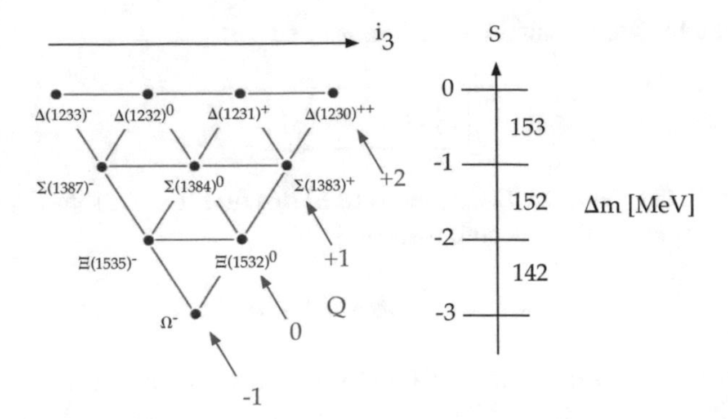

Fig. 13.5 The average isospin multiplet mass increases by $\Delta m \sim 150$ MeV with every unit $|S|$ of strangeness (equal spacing rule)

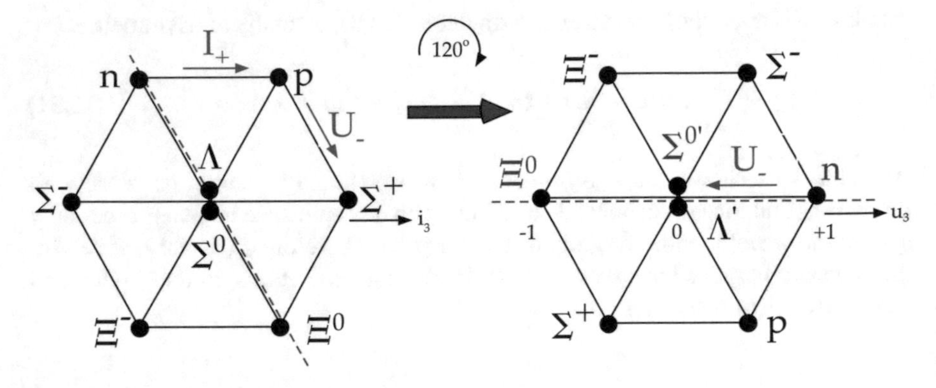

Fig. 13.6 Clockwise rotation of the baryon octet by 120°

Let us write the mass of a charged decuplet baryon as

$$m = m_0 + \Delta S \cdot \Delta m + \Delta Q \cdot \delta m \tag{13.32}$$

where m_0 is the average neutral decuplet baryon mass and δm is the electromagnetic correction. The electromagnetic interaction is u_3-independent since baryons with the same electric charge Q are connected by the ladder operator $U_\pm$ in Fig. 13.5.

Let us now rotate the weight diagram of the fundamental representation (Fig. 7.1, left) by 120° clockwise, i.e. $d \to u \to s \to d$. The u_3-axis now replaces the i_3-axis with $u_3(d) = \frac{1}{2}$, $u_3(u) = 0$ and $u_3(s) = -\frac{1}{2}$. The effect on the octet is shown in Fig. 13.6 with the neutral baryons now appearing on the abscissa. The $\Sigma^{0'}$ and the Λ' are rotated states, i.e.

$$|\Sigma^{0'}\rangle = a|\Sigma^0\rangle + b|\Lambda\rangle \tag{13.33}$$

with

$$a^2 + b^2 = 1. \tag{13.34}$$

We have proved (Problem 7.1) that $U_- I_+ = I_+ U_-$. Thus from Fig. 13.6 (left)

$$U_- I_+ |n\rangle = U_- |p\rangle = |\Sigma^+\rangle, \tag{13.35}$$

while from Fig. 13.6 (right)

$$I_+ U_- |n\rangle = I_+ \sqrt{2} |\Sigma_0'\rangle = \sqrt{2} I_+ (a|\Sigma^0\rangle + b|\Lambda\rangle) = \sqrt{2}\sqrt{2}\, a|\Sigma^+\rangle, \tag{13.36}$$

therefore $a = \frac{1}{2}$ (the factors $\sqrt{2}$ arise from the matrix elements in Table 6.2 for I-spin and U-spin triplets). Introducing a and b into (13.33) gives

$$|\Sigma_0'\rangle = \frac{1}{2}|\Sigma^0\rangle + \frac{\sqrt{3}}{2}|\Lambda\rangle. \tag{13.37}$$

Let us define $m_0 \equiv m(\Sigma^{0'})$ and apply the equal spacing rule. Then

$$m(\Xi^0) = m_0 - \Delta m \;\text{ and }\; m(n) = m_0 + \Delta m, \tag{13.38}$$

or

$$m_0 = \frac{m(\Xi^0) + m(n)}{2}. \tag{13.39}$$

One gets for (13.37) by introducing the mass Hamiltonian,

$$m(\Sigma^{0'}) = \frac{1}{4}\langle\Sigma^0|H|\Sigma^0\rangle + \frac{3}{4}\langle\Lambda|H|\Lambda\rangle = \frac{m(\Sigma^0) + 3m(\Lambda)}{4} = m_0, \tag{13.40}$$

and finally with (13.39) the Gell-Mann-Okubo mass formula

$$\boxed{\frac{m(\Sigma^0)+3m(\Lambda)}{4} = \frac{m(\Xi^0)+m(n)}{2}}. \tag{13.41}$$

The measured masses gives 1135 MeV on the left-hand side and 1127 MeV on the right-hand side, in reasonable agreement.

References

1. Alvarez, L.W., et al.: Phys. Rev. Lett. 2, 215 (1959)
2. Barnes, V., et al.: Phys. Rev. Lett. 12, 204 (1964)

Chapter 14
Magnetic Moments of Baryons

In this chapter we use the baryon SU(6) wavefunctions to predict the magnetic dipole moments of the long-lived baryons and to compare with experimental data. The magnetic moment of a particle with mass m is equal to its g-factor multiplied by $\frac{e}{2m}\vec{s}$. For a spin-$\frac{1}{2}$ baryon the g-factor departs from the value of $g = 2$ predicted by the Dirac equation, due to its internal structure. The magnetic moment of a baryon B is conventionally written relative to that of the proton and hence expressed in nuclear magnetons μ_N:

$$\vec{\mu}_B = g_B \mu_N \vec{s}, \tag{14.1}$$

where

$$\mu_N = \frac{e}{2m_p}. \tag{14.2}$$

The quark model predicts the factor g_B. The magnetic dipole moment is obtained by computing the expectation value of the operator $\vec{\mu}_i = g(\frac{q_i e}{2m_i})\vec{s}$ on quark i with charge $q_i e$, constituent mass m_i, $g = 2$ for point-like fermions, and by adding the contributions of the three quarks. We assume that the constituent masses of the u and d quarks are equal, hence that $m_u = m_d \equiv m$, while the s quark has the mass m_s. The magnetic moments of the quarks are then given by

$$\mu_u = \frac{2}{3}\left(\frac{e}{2m}\right), \quad \mu_d = -\frac{1}{3}\left(\frac{e}{2m}\right), \quad \mu_s = -\frac{1}{3}\left(\frac{e}{2m_s}\right). \tag{14.3}$$

© The Author(s) 2018

C. Amsler, *The Quark Structure of Hadrons*, Lecture Notes in Physics 949,
https://doi.org/10.1007/978-3-319-98527-5_14

14.1 Magnetic Dipole Moment of the Nucleon

The magnetic dipole moment of the proton, obtained by adding the quark contributions, is equal to

$$\mu_p = \sum_{i=1}^{3} \langle p \uparrow | 2\mu_i s_{zi} | p \uparrow \rangle = \sum_{i=1}^{3} \langle p \uparrow | \mu_i \sigma_{zi} | p \uparrow \rangle, \tag{14.4}$$

where $|p \uparrow\rangle$ is the wavefunction for a proton with spin along the z-axis and $\vec{s} \equiv \frac{1}{2}\vec{\sigma}$. The proton wavefunction is (Tables 13.2 and 13.4)

$$\begin{aligned} |p \uparrow\rangle &= \frac{1}{\sqrt{2}}\frac{1}{\sqrt{6}}\frac{1}{\sqrt{6}}|udu + duu - 2uud\rangle|\uparrow\downarrow\uparrow + \downarrow\uparrow\uparrow -2\uparrow\uparrow\downarrow\rangle \\ &+\frac{1}{\sqrt{2}}\frac{1}{\sqrt{2}}\frac{1}{\sqrt{2}}|udu - duu\rangle|\uparrow\downarrow\uparrow - \downarrow\uparrow\uparrow\rangle \\ &= \frac{1}{6}\frac{1}{\sqrt{2}}[|udu + duu - 2uud\rangle|\uparrow\downarrow\uparrow + \downarrow\uparrow\uparrow -2\uparrow\uparrow\downarrow\rangle \\ &+|3udu - 3duu\rangle|\uparrow\downarrow\uparrow - \downarrow\uparrow\uparrow\rangle. \end{aligned} \tag{14.5}$$

Grouping the terms in udu and duu gives

$$\begin{aligned} |p \uparrow\rangle = \frac{1}{6}\frac{1}{\sqrt{2}} \\ &\times 4|udu \uparrow\downarrow\uparrow\rangle - 2|udu \downarrow\uparrow\uparrow\rangle - 2|udu \uparrow\uparrow\downarrow\rangle \\ &-2|duu \uparrow\downarrow\uparrow\rangle + 4|duu \downarrow\uparrow\uparrow\rangle - 2|duu \uparrow\uparrow\downarrow\rangle \\ &-2|uud \uparrow\downarrow\uparrow\rangle - 2|uud \downarrow\uparrow\uparrow\rangle + 4|uud \uparrow\uparrow\downarrow\rangle. \end{aligned} \tag{14.6}$$

The eigenvalues of $\sum \mu_i \sigma_{zi}$ are

$$\begin{aligned} &\overbrace{\mu_u - \mu_d + \mu_u}^{2\mu_u - \mu_d}, \; \cancel{-\mu_u} + \mu_d + \cancel{\mu_u}, \; \cancel{\mu_u} + \mu_d - \cancel{\mu_u}, \\ &\mu_d - \cancel{\mu_u} + \cancel{\mu_u}, \; \overbrace{-\mu_d + \mu_u + \mu_u}^{2\mu_u - \mu_d}, \; \mu_d + \cancel{\mu_u} - \cancel{\mu_u}, \\ &\cancel{\mu_u} - \cancel{\mu_u} + \mu_d, \; -\cancel{\mu_u} + \cancel{\mu_u} + \mu_d, \; \overbrace{\mu_u + \mu_u - \mu_d}^{2\mu_u - \mu_d}, \end{aligned} \tag{14.7}$$

operating on each of the nine kets in (14.6), respectively. Taking into account the orthonormality of the kets, one obtains by left-multiplying with $\langle p \uparrow |$

$$\mu_p = \left[\frac{1}{6}\frac{1}{\sqrt{2}}\right]^2 [3 \times (2\mu_u - \mu_d) \times 16 + 6 \times \mu_d \times 4]$$

$$= \left[\frac{1}{6}\frac{1}{\sqrt{2}}\right]^2 (96\mu_u - 24\mu_d) = \frac{4}{3}\mu_u - \frac{1}{3}\mu_d, \tag{14.8}$$

or with (14.3)

$$\boxed{\mu_p = \frac{e}{2m}}\,. \tag{14.9}$$

Let us repeat the calculation for the magnetic moment of the neutron. The wavefunction is given by

$$|n\uparrow\rangle = -\frac{1}{\sqrt{2}}\frac{1}{\sqrt{6}}\frac{1}{\sqrt{6}}|udd + dud - 2ddu\rangle|\uparrow\downarrow\uparrow + \downarrow\uparrow\uparrow -2\uparrow\uparrow\downarrow\rangle$$

$$+\frac{1}{\sqrt{2}}\frac{1}{\sqrt{2}}\frac{1}{\sqrt{2}}|udd - dud\rangle|\uparrow\downarrow\uparrow - \downarrow\uparrow\uparrow\rangle$$

$$= \frac{1}{6}\frac{1}{\sqrt{2}}[-|udd + dud - 2ddu\rangle|\uparrow\downarrow\uparrow + \downarrow\uparrow\uparrow -2\uparrow\uparrow\downarrow\rangle$$

$$+|3udd - 3dud\rangle|\uparrow\downarrow\uparrow - \downarrow\uparrow\uparrow\rangle. \tag{14.10}$$

Grouping the terms in udd and dud gives

$$|n\uparrow\rangle = \frac{1}{6}\frac{1}{\sqrt{2}}$$

$$\times 2|udd\uparrow\downarrow\uparrow\rangle - 4|udd\downarrow\uparrow\uparrow\rangle + 2|udd\uparrow\uparrow\downarrow\rangle$$

$$-4|dud\uparrow\downarrow\uparrow\rangle + 2|dud\downarrow\uparrow\uparrow\rangle + 2|dud\uparrow\uparrow\downarrow\rangle$$

$$2|ddu\uparrow\downarrow\uparrow\rangle + 2|ddu\downarrow\uparrow\uparrow\rangle - 4|ddu\uparrow\uparrow\downarrow\rangle. \tag{14.11}$$

The eigenvalues of $\sum \mu_i\sigma_{zi}$ are

$$\mu_u - \cancel{\mu_d} + \cancel{\mu_d},\ \overbrace{-\mu_u + \mu_d + \mu_d}^{2\mu_d-\mu_u},\ \mu_u + \cancel{\mu_d} - \cancel{\mu_d},$$

$$\overbrace{\mu_d - \mu_u + \mu_d}^{2\mu_d-\mu_u},\ -\cancel{\mu_d} + \mu_u + \cancel{\mu_d},\ \cancel{\mu_d} + \mu_u - \cancel{\mu_d},$$

$$\cancel{\mu_d} - \cancel{\mu_d} + \mu_u,\ -\cancel{\mu_d} + \cancel{\mu_d} + \mu_u,\ \overbrace{\mu_d + \mu_d - \mu_u}^{2\mu_d-\mu_u}, \tag{14.12}$$

operating on each of the nine kets in (14.11), respectively. By left-multiplying with $\langle n \uparrow |$ one gets

$$\mu_n = \left[\frac{1}{6}\frac{1}{\sqrt{2}}\right]^2 [6 \times \mu_u \times 4 + 3 \times (2\mu_d - \mu_u) \times 16]$$
$$= \left[\frac{1}{6}\frac{1}{\sqrt{2}}\right]^2 (-24\mu_u + 96\mu_d) = -\frac{1}{3}\mu_u + \frac{4}{3}\mu_d, \quad (14.13)$$

or with (14.3)

$$\boxed{\mu_n = -\tfrac{2}{3}\left(\tfrac{e}{2m}\right)}. \quad (14.14)$$

Hence with (14.9) one arrives at the prediction for the ratio of magnetic dipole moments

$$\boxed{\frac{\mu_n}{\mu_p} = \frac{g_n}{g_p} = -\frac{2}{3}}. \quad (14.15)$$

(see Problem 14.1).

The magnetic dipole moment of the proton was measured for the first time with a molecular H_2 beam by the much celebrated experiment of Frisch and Stern [1]. In ortho-hydrogen the two protons form a spin triplet, in para-hydrogen a spin singlet. At room temperature the ratio of ortho- to para-hydrogen is 3:1, while at low temperature the more strongly bound para-hydrogen dominates. In a Stern-Gerlach experiment the force exerted by the inhomogeneous magnetic field on the magnetic dipole moment operates only on the spin-1 state. (The magnetic moments of the electrons cancel.) However, the two-proton wavefunction must be antisymmetric and therefore the relative angular momentum is odd, at least equal to one. The rotation of the molecule leads to a contribution to the magnetic dipole moment which can be measured with para-hydrogen in a low temperature beam, before being subtracted from the measurement with ortho-hydrogen. The result of the experiment (g_p between 4 and 6) led to the stunning conclusion that the g-factor was much larger than the value of 2 expected from the Dirac theory.

Modern measurements are performed in electromagnetic (Penning) traps. The spin-flip frequency is measured with an alternating B-field perpendicular to the main magnetic field [2]). The magnetic moment of the proton ($g_p \simeq 5.58$) is now known to a precision of 0.3×10^{-9} [3].

The magnetic moment of the neutron ($g_n \simeq -3.82$) was measured in a cold neutron beam at the ILL nuclear reactor in Grenoble and compared to that of the proton using flowing water [4]. The neutrons, polarized by a magnetic mirror, passed through an homogeneous magnetic field spectrometer equipped with spin flipping RF cavities, before being detected by a ^{6}Li loaded glass scintillator. The protons from the water were polarized by a strong magnetic field and detected by NMR methods. The result was confirmed more recently by comparing to the well-known magnetic moment of ^{199}Hg , using ultra-cold neutrons stored in a magnetic bottle [5]. The magnetic moment of the neutron is known to a precision of 0.2×10^{-6} [6]. The value for the ratio $\frac{g_n}{g_p}$ is –0.6849793 ± 0.0000002 [6]) for which the naive prediction (14.15) is an impressive approximation.

The mass m of the u and d quarks can be estimated from (14.1) and (14.9):

$$\mu_p = \frac{g_p}{2}\mu_N = 2.79\left(\frac{e}{2m_p}\right) = \frac{e}{2m} \Rightarrow m = \frac{m_p}{2.79} = 336\,\text{MeV}. \tag{14.16}$$

14.2 Magnetic Dipole Moments of Hyperons

The magnetic moment of the lightest hyperon, the Λ (uds), is easy to predict: the isospin-$\frac{1}{2}$ u and d quarks couple to $i = 0$, since $i(s) = 0$ and $i(\Lambda) = 0$, and are therefore in an antisymmetric state. Hence their spins also couple to zero and the magnetic moment of the Λ stems from the s quark. One expects that

$$\mu_\Lambda = \mu_s = \frac{e}{2m_s}\left(-\frac{1}{3}\right). \tag{14.17}$$

14.2.1 Λ Decay

Before discussing a typical measurement of μ_Λ we need to derive the angular distribution of the daughter products in $\Lambda \to \pi^- p$ decay. A heuristic approach will suffice here while a more precise treatment will be described in Sect. 18.1. The Λ being a spin-$\frac{1}{2}$ particle, the relative angular momentum between the pion and the proton is $\ell = 0$ or 1, hence parity is not conserved. Indeed, the decay is mediated by the strangeness changing weak interaction and is therefore parity violating.

Let us define as z-axis the flight direction of the proton in the Λ rest frame and as z_1 the direction of the hyperon spin. Let us assume for the moment that the proton flies along $z = z_1$ ($\theta = 0$, Fig. 14.1) . Since the projection of the angular momentum is equal to zero along z, the projection of the proton spin is along the Λ spin, $m_s = +\frac{1}{2}$. The 2-dimensional final state spinor is given by the sum of the $\ell =$ 0 and 1 contributions

$$|\pi^- p\rangle = \alpha_0\left\langle\frac{1}{2}\frac{1}{2}\middle|0\frac{1}{2}0+\frac{1}{2}\right\rangle + \alpha_1\left\langle\frac{1}{2}\frac{1}{2}\middle|1\frac{1}{2}0+\frac{1}{2}\right\rangle = \begin{pmatrix}\alpha_0 - \frac{1}{\sqrt{3}}\alpha_1 \\ 0\end{pmatrix}, \tag{14.18}$$

using the Clebsch-Gordan coefficients in Fig. 2.8. The unknown constants α_0 and α_1 are complex. Assume now that the proton flies along $z = z_2$ in the direction

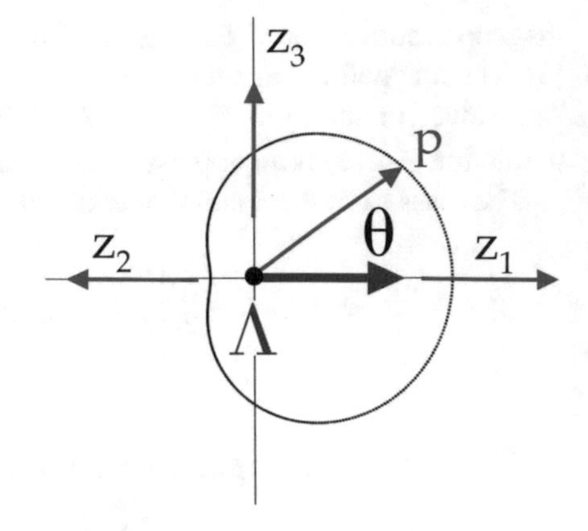

Fig. 14.1 $\Lambda \to \pi^- p$: angular distribution of the proton in the rest frame of the Λ

opposite to the Λ spin (θ = 180°). The final state spinor is with $m_s = -\frac{1}{2}$

$$|\pi^- p\rangle = \alpha_0 \left\langle \frac{1}{2} - \frac{1}{2} \middle| 0 \frac{1}{2} 0 - \frac{1}{2} \right\rangle + \alpha_1 \left\langle \frac{1}{2} - \frac{1}{2} \middle| 1 \frac{1}{2} 0 - \frac{1}{2} \right\rangle = \begin{pmatrix} 0 \\ \alpha_0 + \frac{1}{\sqrt{3}}\alpha_1 \end{pmatrix}. \quad (14.19)$$

Finally, let us assume that the proton flies along z_3, in the direction perpendicular to the hyperon spin, that is with θ = 90° in the Λ rest frame. The spinor is an equal superposition of (14.18) and (14.19). This can be achieved by multiplying the spinors with $\cos\frac{\theta}{2}$ and $\sin\frac{\theta}{2}$, respectively. The decay intensity for any proton angle θ is then proportional to

$$w(\theta) = \langle \pi^- p | \pi^- p \rangle = \cos^2\frac{\theta}{2} \left| \alpha_0 - \frac{\alpha_1}{\sqrt{3}} \right|^2 + \sin^2\frac{\theta}{2} \left| \alpha_0 + \frac{\alpha_1}{\sqrt{3}} \right|^2, \quad (14.20)$$

which describes the angular distribution of the proton in the rest frame of a 100% polarized Λ along z_1. Defining $S \equiv \alpha_0$ and $P \equiv -\frac{\alpha_1}{\sqrt{3}}$, and expressing w as a function of $\cos\theta$ gives

$$\begin{aligned} w(\theta) &= |\alpha_0|^2 + \frac{|\alpha_1|^2}{3} - 2\mathrm{Re}\frac{\alpha_0^*\alpha_1}{\sqrt{3}} \left(\cos^2\frac{\theta}{2} - \sin^2\frac{\theta}{2} \right) \\ &= (|S|^2 + |P|^2)\{1 + \underbrace{\frac{2\mathrm{Re}[S^* P]}{|S|^2 + |P|^2}}_{\equiv \alpha_\Lambda} \cos\theta\}. \end{aligned} \quad (14.21)$$

Normalizing over 4π leads to the angular distribution of the proton in the rest frame of the 100% polarized Λ:

$$w(\theta) = \frac{1}{4\pi}(1 + \alpha_\Lambda \cos\theta). \quad (14.22)$$

The parity violating forward-backward asymmetry is due to the interference of the S ($\ell = 0$) and P ($\ell = 1$) waves.

For hyperons with partial polarization P_Λ one needs to add the fractional contributions from decays with Λ spins along the z_1 and the z_2 directions:

$$w(\theta) = \frac{1}{4\pi}\left[\frac{1+P_\Lambda}{2}\right](1+\alpha_\Lambda\cos\theta) + \frac{1}{4\pi}\left[\frac{1-P_\Lambda}{2}\right](1-\alpha_\Lambda\cos\theta),$$
$$= \frac{1}{4\pi}(1+\alpha_\Lambda P_\Lambda\cos\theta). \quad (14.23)$$

For an unpolarized hyperon it is easy to see that the proton is longitudinally polarized with polarization $P_p = \alpha_\Lambda$: let us divide the number of unpolarized hyperons in 50% with polarizations $P_\Lambda = +1$ and 50% $P_\Lambda = -1$ along the z-axis, taken as the direction of the proton. The number of protons flying in $+z$ direction is proportional to $1+\alpha_\Lambda$ and $1-\alpha_\Lambda$, respectively. The former are polarized in the forward direction, the latter in the backward direction. The proton polarization is therefore given by

$$P_p = \frac{1+\alpha_\Lambda-(1-\alpha_\Lambda)}{1+\alpha_\Lambda+(1-\alpha_\Lambda)} = \alpha_\Lambda. \quad (14.24)$$

This property has been used to measure α_Λ [7]. The Λ hyperons were produced from the reaction $\pi^- p \to K^0\Lambda$ at the BNL Cosmotron with 1 GeV/c pions impinging on a polyethylene target (Fig. 14.2). A veto counter suppressed reactions producing charged particles and the decays $\Lambda \to \pi^- p$ and/or $K^0 \to \pi^+\pi^-$ were photographed with a spark chamber array. The proton track was identified by kinematics constraints and through its heavy ionization of the chamber gas.

The hyperons produced in $\pi^- p \to K^0\Lambda$ happen to be polarized. The polarization is orthogonal to the scattering plane, due to parity conservation in the strong production process [8]. However, the Λ is on average unpolarized when the orientation of the scattering plane is ignored in the subsequent analysis. The proton is then polarized along its flight direction in the Λ rest frame. In the non-relativistic limit the spin direction does not change (in contrast to the momentum vector) when boosting the proton momentum from the Λ rest frame into the laboratory. This is a good approximation in the present case. In the laboratory the proton polarization therefore acquires a component $P_\perp$ perpendicular to the proton momentum (see Fig. 14.2), which can be obtained by measuring the asymmetry a in proton-carbon scattering:

$$a(\theta_p) = \frac{n_L - n_R}{n_L - n_R} = P_\perp A_C(\theta_p). \quad (14.25)$$

The scattering rates to the left and to the right in the plane perpendicular to the proton polarization are denoted by n_L and n_R, respectively. A_C is the known analyzing power of carbon which depends on the proton scattering angle θ_p and on

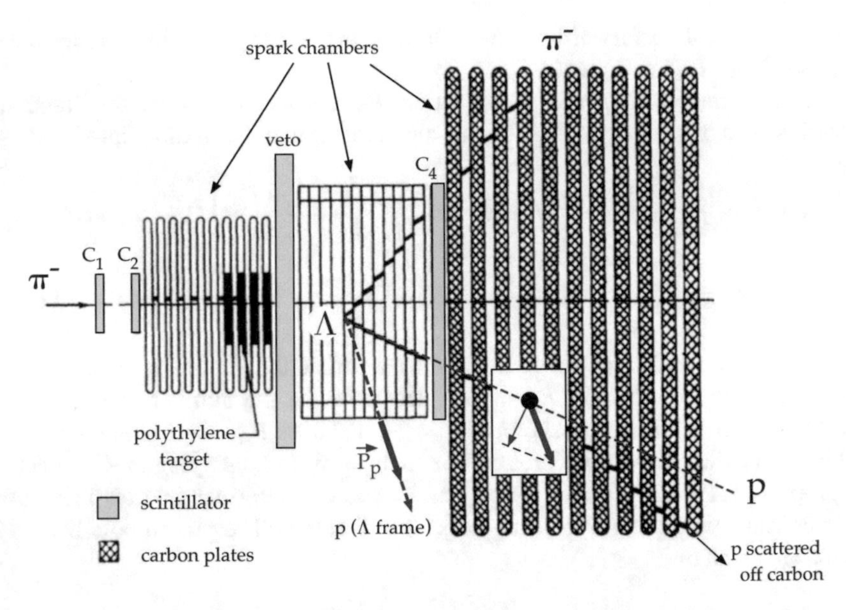

Fig. 14.2 Measurement of the asymmetry parameter α_Λ in Λ decay [7]. The polarization of the decay proton is analyzed by measuring the left-right asymmetry in proton-carbon scattering. The blue arrows show the momentum and polarization of the proton in the Λ rest frame for an unpolarized hyperon. The angle between the proton momentum and the polarization vector in the laboratory (red arrow) can be calculated with the usual Lorentz transformations

its energy. The result $\alpha_\Lambda = P_p = +0.63 \pm 0.08$ leads to a large forward-backward asymmetry in Fig. 14.1. The current best value[1] is $\alpha_\Lambda = 0.642 \pm 0.013$.[2] From the angular distribution (14.23) one then finds that the Λ hyperons are produced with the polarization $P_\Lambda \simeq 50\%$.

14.2.2 Magnetic Moment of the Λ

The magnetic moment of the Λ was measured at Fermilab by using the property that spin-$\frac{1}{2}$ ground state hyperons produced inclusively in high energy proton-nucleon interactions ($pN \rightarrow$ hyperon + X) are polarized, with typical polarizations in the range of 10–30%, depending on hyperon momentum (Fig. 14.3). Figure 14.4 shows

[1]Since CP conservation is a very good approximation, one also expects for the decay of the antihyperon $\overline{\Lambda} \rightarrow \overline{p}\pi^+$ an angular distribution of the form $1 + \overline{\alpha}_\Lambda \cos\theta$ where $\overline{\alpha}_\Lambda = -\alpha_\Lambda$. This is because parity reverses the (polar) momentum vectors but not the (axial) spin vectors. For $\overline{\Lambda}$ decay the tabulated value is $\overline{\alpha}_\Lambda = -0.71 \pm 0.08$ [6] which is within large errors in agreement with CP invariance.

[2]BESIII reports the larger value $\alpha_\Lambda = 0.75 \pm 0.01$ from J/ψ decay into $\Lambda\overline{\Lambda}$ [9].

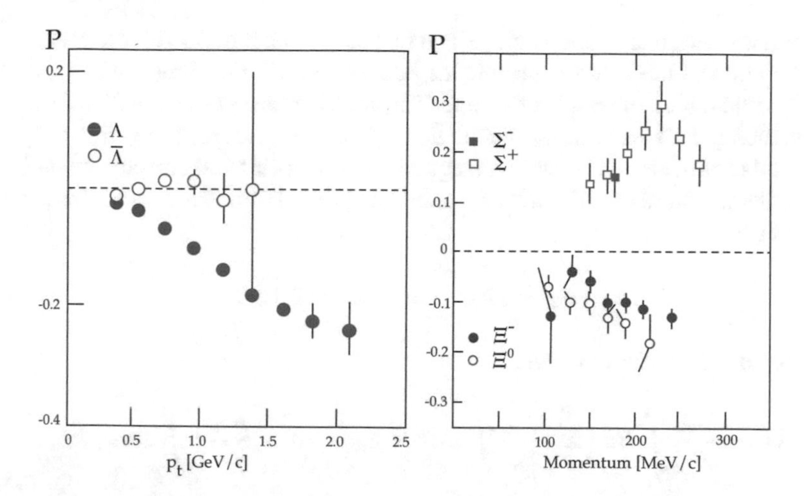

Fig. 14.3 Left: polarizations of the Λ and $\overline{\Lambda}$ produced at 400 GeV, as a function of transverse momentum. Right: polarizations of Σ and Ξ produced at small angles (5 to 8 mrad), as a function of hyperon momentum (adapted from [10])

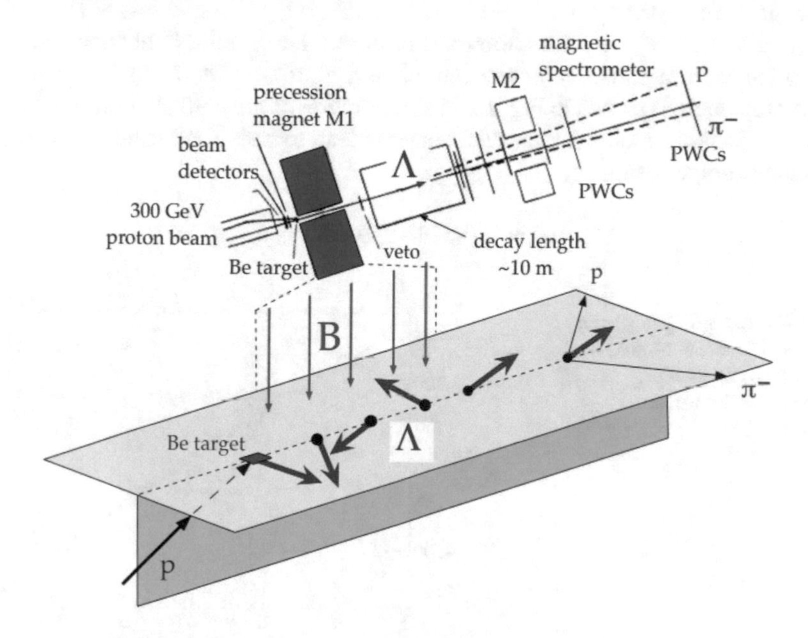

Fig. 14.4 Sketch of the apparatus to measure the magnetic moment of the Λ hyperon [11]

a sketch of the experimental setup using 300 GeV protons impinging on a beryllium target [11].

The hyperon beam is defined by a collimator and charged particles are swept away by a vertical magnetic field B. Owing to parity conservation, the Λ hyperons are polarized in the direction perpendicular to the plane spanned by the incident proton and outgoing Λ momenta. Since this plane is not defined for a Λ emitted in the direction of the incident proton, the beam is directed towards the target under a small vertical incident angle of 7 mrad. The polarization is typically 8%. The spin of the Λ then precesses in the magnetic field before the Λ decays, e.g. into $\pi^- p$. The pion and proton are analyzed in a magnetic spectrometer and detected by multiwire proportional chambers. The average decay length is 10 m. The Larmor frequency is given by

$$\omega_L = (\mu_\Lambda \uparrow - \mu_\Lambda \downarrow) B = 2\mu_\Lambda B, \tag{14.26}$$

leading to the precession angle

$$\Phi = \frac{2\mu_\Lambda}{\beta} \int B d\ell \simeq \frac{2\mu_\Lambda}{\mu_N} \int B d\ell \times \mu_N = 18.3^\circ \frac{\mu_\Lambda}{\mu_N} \left(\int B d\ell \, [\mathrm{Tm}] \right), \tag{14.27}$$

for $\beta \simeq 1$. We have introduced the numerical value 3.15×10^{-14} MeV/T for the nuclear magneton (14.2), applied the transformation of units (1.7) and expressed the angle Φ in degrees.

The direction of the polarization vector at the decay point is obtained by measuring the asymmetric angular distribution of the proton (Fig. 14.1). The measured precession angle is shown in Fig. 14.5a as a function of the field integral along the Λ path. The measurements for various momenta lead to consistent results (Fig. 14.5b) with the average value

$$\mu_\Lambda = -0.6138 \pm 0.0047 \, \mu_N. \tag{14.28}$$

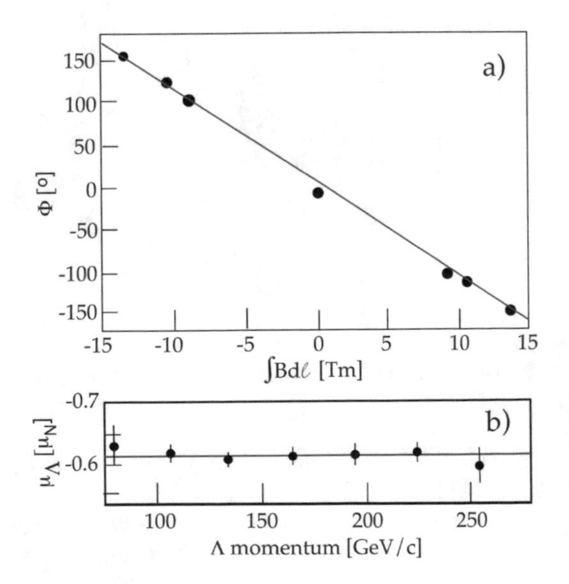

Fig. 14.5 (**a**) Spin precession angle as a function of field integral; (**b**) magnetic moment μ_Λ for various momenta [11]

The mass of the s quark can be estimated with (14.9) and (14.17):

$$\frac{\mu_\Lambda}{\mu_p} = -\frac{0.614}{2.79} = -\frac{m}{3m_s} \Rightarrow m_s = 1.51\,m = 509\,\text{MeV}. \tag{14.29}$$

The magnetic moment of the Σ^+ (uus) is easily predicted from that of the proton by replacing the d quark in (14.8) by an s quark:

$$\mu_{\Sigma^+} = \frac{4}{3}\mu_u - \frac{1}{3}\mu_s = \frac{8}{9}\overbrace{\left(\frac{e}{2m}\right)}^{2.79\mu_N} + \frac{1}{9}\overbrace{\left(\frac{e}{2m_s}\right)}^{-3\mu_\Lambda} = 2.68\,\mu_N \tag{14.30}$$

with (14.17) and the numerical value (14.28). The magnetic moment of the Σ^+ has also been measured with the precession method using its decay into $p\pi^0$. The angular dependence of the decay nucleon is of the form (14.22) with $\alpha_{\Sigma^+} \simeq -0.98$ [6]. Since the hyperon trajectories are curved in the magnetic field, one measures the spin precession angle with respect to the direction of the momentum. This angle Φ varies as a function of time according to

$$\Phi(t) = \theta_L(t) - \theta_C(t) = \frac{eBt}{m_{\Sigma^+}}\left(\frac{g-2}{2}\right), \tag{14.31}$$

where θ_L is the Larmor rotation angle and θ_C the cyclotron angle. With $g(\frac{e}{2m_{\Sigma^+}}) = g_{\Sigma^+}(\frac{e}{2m_p})$ one gets the precession angle

$$\Phi = \frac{e}{m_{\Sigma^+}}\left(\frac{g_{\Sigma^+}m_{\Sigma^+}}{2m_p} - 1\right)\int B d\ell \tag{14.32}$$

for hyperons with $\beta \simeq 1$. The result is

$$\mu_{\Sigma^+} = \frac{g_{\Sigma^+}}{2}\mu_N = 2.4613 \pm 0.0052\,\mu_N, \tag{14.33}$$

consistent with that of the $\overline{\Sigma^+}$ (–2.428 $\pm$ 0.037 μ_N) [12]. The magnetic moment of the $\Sigma^- \to n\pi^-$ (Problem 14.2) can be measured with the same method, or with stopped hyperons that are captured in the Coulomb shells of the target atoms. One then detects the transition X-rays in the Σ^- atoms and measures the fine-structure splittings which depend on g_{Σ^-} [13].

The magnetic moments of the Ξ^0, Ξ^- and Ω^- are obtained from the polarization of the $\Lambda(\to \pi^- p)$ in the decays $\Lambda\pi^0$, $\Lambda\pi^-$ and ΛK^-, respectively. The magnetic moment of the (spin-$\frac{3}{2}$) Ω^- is particularly difficult to measure, because the Ω^- is unpolarized at high energies, in contrast to the other hyperons. Polarized Ω^- hyperons have been obtained by first producing polarized Λ and Ξ^0 hyperons with 800 GeV protons, and then by transferring the polarization to the Ω^- with the reaction (Λ, Ξ^0) Be $\rightarrow \Omega^- X$ on a beryllium target. The magnetic moment measured from the decay $\Omega^- \rightarrow \Lambda K^-$ is $\mu_{\Omega^-} = -2.024 \pm 0.056 \mu_N$ [14]. Let us compare with the quark model prediction, assuming m_s= 509 MeV (14.29). The wavefunction of the Ω^- is given by the product (13.29) of the symmetric wavefunctions ϕ_S and χ_S. For the spin projection $+\frac{3}{2}$ along the z-axis the wavefunction is then simply

$$|\Omega^- \uparrow\rangle = |sss \uparrow\uparrow\uparrow\rangle. \tag{14.34}$$

The magnetic moment is predicted by summing over the three s quarks:

$$\mu_{\Omega^-} = \sum_{i=1}^{3} \langle \Omega^- \uparrow |\mu_i \sigma_{zi}| \Omega^- \uparrow\rangle = 3\mu_s. \tag{14.35}$$

Since $\mu_s = \mu_\Lambda$ (14.17) we predict with the measured Λ magnetic moment (14.28) that

$$\mu_{\Omega^-} = 3\mu_\Lambda = -1.84 \pm 0.01 \; \mu_N, \tag{14.36}$$

which deviates from the measured value [14] by three standard deviations.

Table 14.1 lists the current experimental values for the magnetic moments of the ground state baryons, together with the predictions from the quark model. There are clearly significant discrepancies, but given its crudeness, our quark model performs surprisingly well (Problem 14.3).

Table 14.1 Comparison of quark model predictions and measured magnetic dipole moments of the ground state baryons; $\kappa \equiv \frac{e}{2m} = 2.79 \; \mu_N$ and $\kappa_s \equiv \frac{e}{2m_s} = -3\mu_\Lambda = 1.84 \; \mu_N$

Baryon	Quark model			Experimental value
			$[\mu_N]$	$[\mu_N]$ [6]
p	κ		input	2.793
n	$-\frac{2}{3}\kappa$	=	−1.86	−1.913
Λ	$-\frac{1}{3}\kappa_s$		input	−0.6138 ± 0.0047
Σ^+	$\frac{8}{9}\kappa + \frac{1}{9}\kappa_s$	=	2.68	2.458 ± 0.010
Σ^0	$\frac{2}{9}\kappa + \frac{1}{9}\kappa_s$	=	0.82	–
Σ^{0a}	$-\frac{1}{\sqrt{3}}\kappa$	=	−1.61	−1.61 ± 0.08
Σ^-	$-\frac{4}{9}\kappa + \frac{1}{9}\kappa_s$	=	−1.04	−1.160 ± 0.025
Ξ^0	$-\frac{2}{9}\kappa - \frac{4}{9}\kappa_s$	=	−1.44	−1.250 ± 0.014
Ξ^-	$\frac{1}{9}\kappa - \frac{4}{9}\kappa_s$	=	−0.51	−0.6507 ± 0.0025
Ω^-	$-\kappa_s$	=	−1.84	−2.024 ± 0.056

[a]This line refers to the $\Sigma^0 \rightarrow \Lambda$ transition magnetic moment since the mean life of the Σ^0 is too short for precession experiments

References

1. Frisch, R., Stern, O.: Z. Phys. 85, 4 (1933)
2. DiSciacca, J., Gabrielse, G.: Phys. Rev. Lett. 108, 153001 (2012)
3. Schneider, G., et al.: Science 358, 1081 (2017)
4. Greene, G.L., et al.: Phys. Rev. D 20, 2139 (1979)
5. Afach, S., et al.: Phys. Lett. B 739, 128 (2014)
6. Tanabashi, M., et al. (Particle Data Group): Phys. Rev. D 98, 030001 (2018)
7. Cronin, J.W., Overseth, O.E.: Phys. Rev. 129, 1795 (1963)
8. Amsler, C.: Nuclear and Particle Physics, section 11.3. IOP Publishing, Bristol (2015)
9. Ablikim, M., et al.: (2018). arXiv: 1808.08917
10. Lach, J.: Nucl. Phys. B (Proc. Suppl.) 50, 216 (1996)
11. Schachinger, L., et al.: Phys. Rev. Lett. 41, 1348 (1978)
12. Morelos, A., et al.: Phys. Rev. Lett. 71, 3417 (1993)
13. Hertzog, D.W., et al.: Phys. Rev. D 37, 1142 (1988)
14. Wallace, N.B., et al.: Phys. Rev. Lett. 74, 3732 (1995)

Chapter 15
Light Baryon Excitations

15.1 Harmonic Oscillator

We have dealt so far with the well established $J^P = \frac{1}{2}^+$ and $\frac{3}{2}^+$ ground state baryons. Let us now turn on angular momenta $\ell_\rho > 0$ within one of the qq pairs and/or $\ell_\lambda > 0$ between the diquark and the third quark, as shown in Fig. 13.1. The three-body system involving two-body forces can be treated analytically with the harmonic oscillator model. To simplify, we shall assume that the three quarks have the same mass m (for the treatment with unequal masses see [1]). Let us introduce the (Jacobi) parameters

$$\vec{\rho} = \frac{1}{\sqrt{2}}(\vec{x}_1 - \vec{x}_2), \tag{15.1}$$

$$\vec{\lambda} = \frac{1}{\sqrt{6}}(\vec{x}_1 + \vec{x}_2 - 2\vec{x}_3), \tag{15.2}$$

where $\vec{x}_{1,2,3}$ are the spatial coordinates of the quarks. The distance between quark 1 and 2 is $\sqrt{2}\rho$, while the third quark lies at the distance $\sqrt{\frac{3}{2}}\lambda$ (Fig. 15.1):

$$\vec{x}_1 - \vec{x}_2 = \sqrt{2}\vec{\rho}, \tag{15.3}$$

$$\vec{x}_0 - \vec{x}_3 = \vec{x}_2 + \frac{\vec{\rho}}{\sqrt{2}} - \vec{x}_3 = \vec{x}_2 + \frac{(\vec{x}_1 - \vec{x}_2)}{2} - \vec{x}_3$$
$$= \frac{\vec{x}_1 + \vec{x}_2 - 2\vec{x}_3}{2} = \sqrt{\frac{3}{2}}\vec{\lambda}. \tag{15.4}$$

© The Author(s) 2018

C. Amsler, *The Quark Structure of Hadrons*, Lecture Notes in Physics 949,
https://doi.org/10.1007/978-3-319-98527-5_15

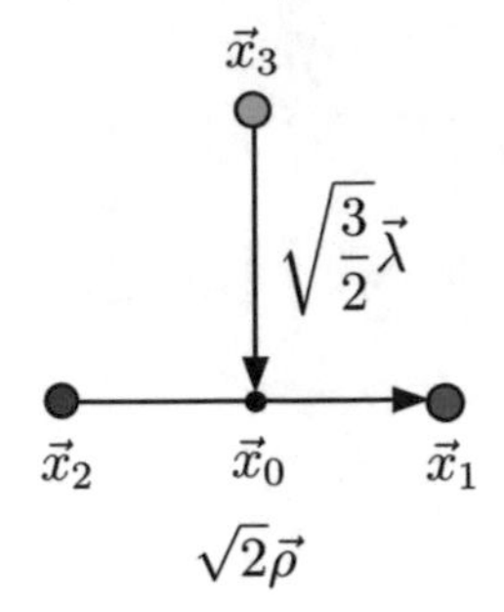

Fig. 15.1 The baryon excitation can be decomposed into two harmonic oscillations with amplitude vectors $\vec{\rho}$ and $\vec{\lambda}$

The momenta associated with $\vec{\rho}$ and $\vec{\lambda}$ are

$$\vec{p}_\rho = \frac{1}{\sqrt{2}}(\vec{p}_1 - \vec{p}_2), \tag{15.5}$$

$$\vec{p}_\lambda = \frac{1}{\sqrt{6}}(\vec{p}_1 + \vec{p}_2 - 2\vec{p}_3). \tag{15.6}$$

It is straightforward to show that the Hamiltonian

$$H = \sum_{i=1}^{3} \frac{\vec{p}_i^{\,2}}{2m} + \frac{1}{2}K\sum_{i<j}(\vec{x}_i - \vec{x}_j)^2 \tag{15.7}$$

of the (non-relativistic) harmonic oscillator can be recast as

$$H = \frac{\vec{P}^2}{2 \times 3m} + \left[\frac{\vec{p}_\rho^{\,2}}{2m} + \frac{3}{2}K\vec{\rho}^{\,2}\right] + \left[\frac{\vec{p}_\lambda^{\,2}}{2m} + \frac{3}{2}K\vec{\lambda}^2\right], \tag{15.8}$$

where the first term describes the kinetic energy of the center-of-mass system with momentum $\vec{P} = \sum \vec{p}_i$, which vanishes in the rest frame of the baryon. The problem thus reduces to two decoupled 3-dimensional harmonic oscillators along $\vec{\rho}$ and $\vec{\lambda}$, both with frequency $\omega = \sqrt{\frac{3K}{m}}$.

The level energies are given by $\omega(2k + \ell + \frac{3}{2}) = \omega(n + \frac{3}{2})$ (Fig. 15.2), with $\hbar = 1$, $k = 0, 1, 2\ldots$ and $\ell = 0, 1, 2\ldots$ for each oscillator [2], hence the baryon mass is given by

$$\begin{aligned} M &= 3m + \text{constant} + \omega(2k_\rho + 2k_\lambda + \ell_\rho + \ell_\lambda + 3) \\ &= 3m + \text{constant} + \omega(n_\rho + n_\lambda + 3) \\ &= 3m + \text{constant} + \omega(N + 3), \end{aligned} \tag{15.9}$$

where $N = n_\rho + n_\lambda = 2k_\rho + 2k_\lambda + \ell_\rho + \ell_\lambda$ is the band number.

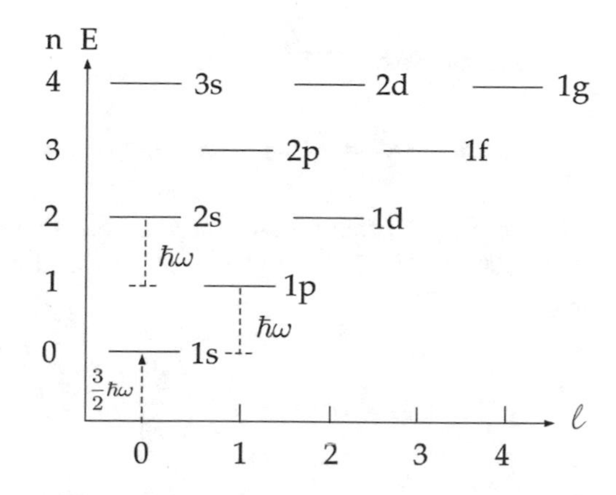

Fig. 15.2 Energy levels of the 3-dimensional harmonic oscillator

The baryon angular momentum is $\vec{L} = \vec{\ell}_\rho + \vec{\ell}_\lambda$, to which the spin is added, $\vec{J} = \vec{L} + \vec{s}$ (with $s = \frac{1}{2}$ and $\frac{3}{2}$). The spatial wavefunction being the product of the two oscillator wavefunctions, the parity is given by $(-1)^{\ell_\rho + \ell_\lambda}$. The mass degeneracy of the levels is given by

$$d_N = \frac{(N+1)(N+2)(N+3)(N+4)(N+5)}{5!}. \tag{15.10}$$

Note that $k_\rho = k_\lambda = 0$ for $N \leq 1$. For example, for the ground state $N = 0$ ($\ell_\rho = \ell_\lambda = 0$) one expects with $d_0 = 1$ the spin-$\frac{1}{2}$ and $\frac{3}{2}$ baryons to have the same masses. For $N = 1$, $(\ell_\rho, \ell_\lambda) = (1, 0)$ or $(\ell_\rho, \ell_\lambda) = (0, 1)$, hence $d_1 = 6$ (including the $[2\ell_\rho + 1]$ m_ρ and $[2\ell_\lambda + 1]$ m_λ projections). For $N = 2$ we have 2 states with $k_\rho = 1$ or $k_\lambda = 1$ and $\ell_\rho = \ell_\lambda = 0$, and 3 states with $(\ell_\rho, \ell_\lambda) = (1, 1)$, $(\ell_\rho, \ell_\lambda) = (2, 0)$, $(\ell_\rho, \ell_\lambda) = (0, 2)$ with $k_\rho = k_\lambda = 0$, giving 19 states, hence $d_2 = 21$. The degeneracy increases very rapidly with N. This simple model, which does not include spin-spin and spin-orbit interactions, cannot reproduce the mass spectrum, but is still useful to predict the number of baryon excitations and their quantum numbers, as we now demonstrate.

We have seen in Sect. 13.3 that the total wavefunction, including the spatial part, must be fully symmetric under the exchange of any pair of quarks. Hence let us combine the spin $\times$ SU(3)$_f$ wavefunctions with those of the harmonic oscillator and retain the symmetric ones. The ground state solution for one oscillator is given by

$$\psi_{n=0}(r) = \left(\frac{\beta^2}{\pi}\right)^{\frac{3}{4}} \mathrm{e}^{-\frac{\beta^2 r^2}{2}} \tag{15.11}$$

with $\beta \equiv \sqrt{\frac{m\omega}{2}}$ [2]. The baryon orbital wavefunction is given by the product of the two oscillations in ρ and λ. The ground state wavefunction is proportional to the fully symmetric $e^{-\frac{\beta^2(\rho^2+\lambda^2)}{2}}$, since

$$\rho^2 + \lambda^2 = \frac{1}{3}[(\vec{x}_1 - \vec{x}_2)^2 + (\vec{x}_1 - \vec{x}_3)^2 + (\vec{x}_2 - \vec{x}_3)^2]. \tag{15.12}$$

The 56 SU(6) ground state baryons are therefore described by the symmetric combinations (13.28) and (13.29). The wavefunction of the first ($1p$) excited state is given by [2]

$$\psi_{n=1,m}(r) = \sqrt{\frac{8}{3}} \left(\frac{\beta^{\frac{3}{2}}}{\pi^{\frac{1}{4}}} \right) \beta r Y_1^m(\theta, \phi) e^{-\frac{\beta^2 r^2}{2}}. \tag{15.13}$$

The orbital wavefunctions of the first excited state ($L = 1$) are, expressing $\vec{\rho}$ and $\vec{\lambda}$ in spherical coordinates,

$$\psi_{n=1,m,\ell_\rho=1,\ell_\lambda=0}(\rho) \propto \rho Y_1^m(\theta_\rho, \phi_\rho)\, e^{-\frac{\beta^2(\rho^2+\lambda^2)}{2}}, \tag{15.14}$$

$$\psi_{n=1,m,\ell_\rho=0,\ell_\lambda=1}(\lambda) \propto \lambda Y_1^m(\theta_\lambda, \phi_\lambda)\, e^{-\frac{\beta^2(\rho^2+\lambda^2)}{2}}. \tag{15.15}$$

Now, $\vec{\rho}$ is antisymmetric and $\vec{\lambda}$ symmetric under permutations of the two first quarks, and so are (15.14) and (15.15). These two orbital wavefunctions have to be combined with a mixed antisymmetric and a mixed symmetric spin × flavour wavefunction, respectively, to obtain totally symmetric wavefunctions. Table 15.1 shows the symmetry properties of the SU(2)×SU(3)$_f$ functions.

There are four ways to obtain SU(6) mixed symmetries M, leading to 70 negative parity states. The 70-plet decomposes into one spin-$\frac{3}{2}$ octet and three spin-$\frac{1}{2}$ multiplets, a decuplet, an octet and a singlet (Table 15.2). The 19 baryons with $s = \frac{1}{2}$ can have the quantum numbers $J^P = \frac{1}{2}^-$ and $\frac{3}{2}^-$ and the 8 baryons with $s = \frac{3}{2}$ in addition $J^P = \frac{5}{2}^-$. The total wavefunctions are obtained by combining the MS functions with (15.15) and adding the MA ones combined with (15.14), for details see [3].

The complexity increases with N. For $N = 2$ one obtains two 56-plets, two 70-plets, and one 20-plet with the totally antisymmetric configurations composed of

Table 15.1 SU(6) spin × flavour symmetry properties

	SU(3)$_f$ ϕ →	S	M	A
SU(2) χ ↓		10	8 + 8	1
S	4	S	M	A
M	2 + 2	M	SAM	M

Table 15.2 Symmetry properties of the SU(6) wavefunctions for the $L = N = 1$ baryons (see the text)

$\phi \times \chi$	MS	MA	s	$SU(3)_f$	Nr. of states
$S \times M$	$\phi_S \chi_{MS}$	$\phi_S \chi_{MA}$	$\frac{1}{2}$	10	20
$M \times S$	$\phi_{MS} \chi_S$	$\phi_{MA} \chi_S$	$\frac{3}{2}$	8	32
$M \times M$ ($\frac{1}{\sqrt{2}} \times$)	$-\phi_{MS}\chi_{MS} + \phi_{MA}\chi_{MA}$	$\phi_{MS}\chi_{MA} + \phi_{MA}\chi_{MS}$	$\frac{1}{2}$	8	16
$A \times M$	$\phi_A \chi_{MA}$	$\phi_A \chi_{MS}$	$\frac{1}{2}$	1	2

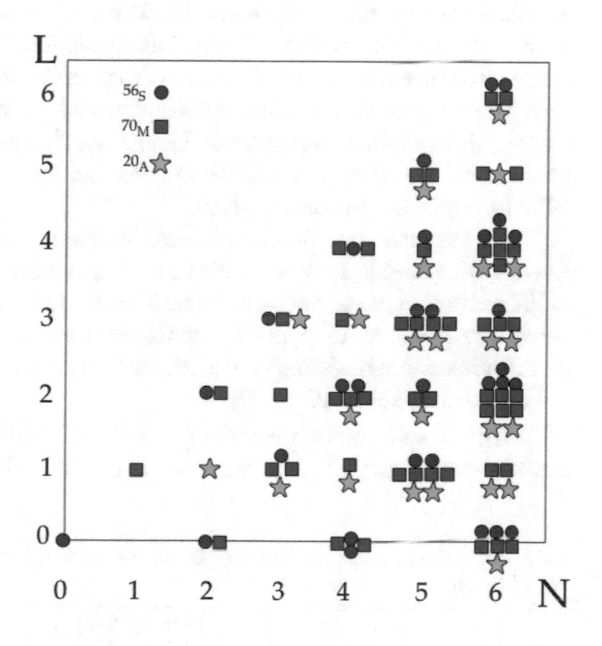

Fig. 15.3 Baryon multiplets as a function of total angular momentum L and band number N in the harmonic oscillator model [5]

one spin-$\frac{3}{2}$ singlet and one spin-$\frac{1}{2}$ octet [4, 5]:

$$\phi_A \chi_S \ \left(s = \frac{3}{2} \text{ singlet}\right) \text{ and } \frac{1}{\sqrt{2}}[\phi_{MS}\chi_{MA} - \phi_{MA}\chi_{MS}] \ \left(s = \frac{1}{2} \text{ octet}\right). \tag{15.16}$$

Figure 15.3 shows the distributions of the SU(6) supermultiplets in the harmonic oscillator model, as a function of L and N. To summarize, the SU(6) supermultiplets decompose into the following SU(3) multiplets:

$$\begin{aligned} 56 &: s = \frac{1}{2} \text{ octet}, s = \frac{3}{2} \text{ decuplet}, \\ 70 &: s = \frac{1}{2} \text{ decuplet}, s = \frac{3}{2} \text{ octet}, s = \frac{1}{2} \text{ octet}, s = \frac{1}{2} \text{ singlet}, \\ 20 &: s = \frac{3}{2} \text{ singlet}, s = \frac{1}{2} \text{ octet}. \end{aligned} \tag{15.17}$$

15.2 Experimental Status

Baryon excitations have been studied using various reactions such as $\pi^{\pm}p$ and $K^{-}p$ elastic and inelastic scattering (e.g. $\pi^{-}p \to n\pi^{0}$, $n\eta$, $n2\pi^{0}$, $K^{-}p \to \Lambda\pi^{0}$, $\Lambda\eta$, $\Sigma^{0}\pi^{0}$) and photoproduction (such as $\gamma p \to p\pi\pi$, $K^{+}\Lambda$, $K^{0}\Sigma^{+}$). For experimental and theoretical details we refer to the comprehensive review [6]. Table 15.3 lists the firmly established octet and singlet baryons up to the $N = 2$ band, Table 15.4 the decuplet ones [7]. Further states that need to be confirmed are not included here but are listed in [8]. Most states up to the $N = 1$ band have been observed. There is so far no established candidate in the 20-plet. The number of predicted baryon excitations exceeds by far the number of states observed. This could be due to the fact that many resonances do not couple strongly to pion-nucleon or kaon-nucleon. They are therefore only weakly produced with pion and kaon beams and, with increasing masses, decay into a multitude of kinematically open final states. As is the case for mesons, the density of overlapping excitations increases with mass and width, which complicates the data analysis.

There are also mass ordering issues. For example, the $N(1440)$ (also known as Roper resonance) is an $N = 2$ state in the harmonic oscillator model and is the first radial excitation of the nucleon. According to (15.9) this state should lie above, not below, the $N = 1$ $N(1535)$ state. The Roper resonance could be more complex than qqq, for instance consisting of a quark core surrounded by a meson cloud that would reduce its mass substantially [9].

The mass splitting between the $\frac{1}{2}^{-}$ $N(1535)$ and the $\frac{3}{2}^{-}$ $N(1520)$, presumably due spin-orbit forces, is tiny (and has the opposite sign) compared to that between

Table 15.3 Quark model assignments of the established baryon octet and singlet states in the $N \leq 2$ bands

					Octet [MeV]				Singlet [MeV]
J^P	SU(6)	L	N	s	N	Λ	Σ	Ξ	Λ
$\frac{1}{2}^{+}$	56	0	0	$\frac{1}{2}$	939	1116	1193	1318	–
$\frac{1}{2}^{-}$	70	1	1	$\frac{1}{2}$	1535	1670	1620[a]		1405
$\frac{3}{2}^{-}$	70	1	1	$\frac{1}{2}$	1520	1690	1670	1820	1520
$\frac{1}{2}^{-}$	70	1	1	$\frac{3}{2}$	1650	1800	1750[a]		–
$\frac{3}{2}^{-}$	70	1	1	$\frac{3}{2}$	1700		1940[a]		–
$\frac{5}{2}^{-}$	70	1	1	$\frac{3}{2}$	1675	1830	1775	1950[a]	–
$\frac{1}{2}^{+}$	56	0	2	$\frac{1}{2}$	1440	1600	1660	1690[a]	–
$\frac{3}{2}^{+}$	56	2	2	$\frac{1}{2}$	1720	1890			–
$\frac{5}{2}^{+}$	56	2	2	$\frac{1}{2}$	1680	1820	1915	2030	–
$\frac{1}{2}^{+}$	70	0	2	$\frac{1}{2}$	1710	1810	1880		1810[a]

Empty slots: not observed or not established yet
[a]For alternative assignments see [7]

Table 15.4 Quark model assignments of the established baryon decuplets

J^P	SU(6)	L	N	s	Δ	Σ	Ξ	Ω
$\frac{3}{2}^+$	56	0	0	$\frac{3}{2}$	1232	1385	1530	1672
$\frac{1}{2}^-$	70	1	1	$\frac{1}{2}$	1620	1750[a]		
$\frac{3}{2}^-$	70	1	1	$\frac{1}{2}$	1700			
$\frac{3}{2}^+$	56	0	2	$\frac{3}{2}$	1600	1690[a]		
$\frac{5}{2}^+$	56	2	2	$\frac{3}{2}$	1905			
$\frac{7}{2}^+$	56	2	2	$\frac{1}{2}$	1950	2030		

[a]Alternative assignments have been proposed (see [7])

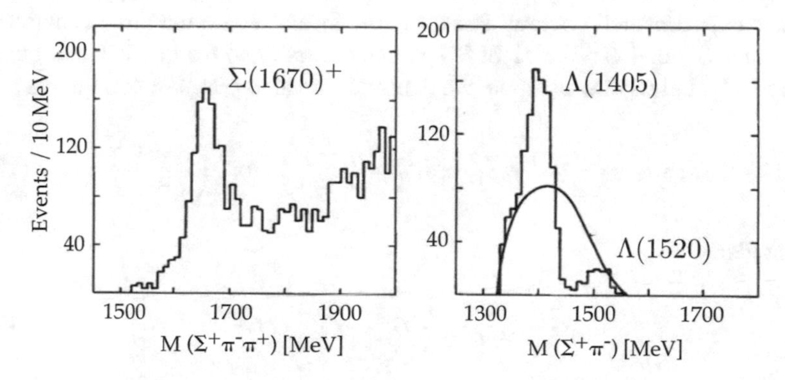

Fig. 15.4 Left: invariant $\Sigma^+\pi^-\pi^+$ mass distribution in the reaction $K^-p \rightarrow \Sigma^+\pi^-\pi^+\pi^-$, showing the production of the $\Sigma(1670)^+$. Right: $\Sigma^+\pi^-$ mass distribution after having applied a cut around the $\Sigma(1670)^+$ peak. The curve describes the $\Sigma\pi\pi$ phase space distribution in $\Sigma(1670)$ decay. The small peak around 1500 MeV is due to the contribution from $\Sigma(1750) \rightarrow \Lambda(1520)\pi$ [10]

the $\Lambda(1520)$ and the $\Lambda(1405)$, which have the same configurations, but a light quark substituted by an s quark.

The $\frac{1}{2}^-$ $\Lambda(1405)$ is much lighter than its $N(1535)$ partner, notwithstanding its s quark content. The threshold for coupling to K^-p lies at 1432 MeV, hence K^-p decay is kinematically forbidden and $\Lambda(1405)$ decays exclusively into $\Sigma\pi$. Figure 15.4 shows invariant mass plots from the process $K^-p \rightarrow \Sigma^+\pi^-\pi^+\pi^-$ measured in a bubble chamber with 4.2 GeV/c kaons [10]. The reaction proceeds through the production of the $\Sigma(1670)^+$, which in turn decays into $\Lambda(1405)\pi^+$. The asymmetric $\Lambda(1405) \rightarrow \Sigma^+\pi^-$ distribution in Fig. 15.4 (right) is due to the opening of the K^-p decay channel at 1432 MeV. The nearby K^-p threshold leads to significant mass shifts compared to the quark model prediction (akin to the $a_0(980)$ meson discussed in Fig. 12.2). The nature of the $\Lambda(1405)$ has been controversial for many years and is still not settled. It could consist of a bound K^-p (or $qqqq\overline{q}$) state mixed with a genuine qqq baryon, see e.g. [11].

15.3 Colour Hyperfine Splitting

Before closing this chapter let us examine how colour hyperfine splitting affects the baryon masses, in particular those of the ground states. Let us write the hyperfine interaction mediated by gluon exchange between two quarks in a baryon (or between the quark and antiquark in a meson) as

$$H = -c\,(\vec{G}^1 \cdot \vec{G}^2)(\vec{s}_1 \cdot \vec{s}_2). \tag{15.18}$$

This ansatz is inspired by the spin-spin interaction (9.18) in QED and will be justified experimentally below. The operators $\vec{s}_1$ and $\vec{s}_2$ are the three generators of SU(2) and $\vec{G}^1$ and $\vec{G}^2$ the eight SU(3) generators (7.4) for quark 1 and quark 2, respectively. Let us now compute the expectation value $\langle H\rangle$. For two quarks

$$\langle \vec{s}_1 \cdot \vec{s}_2\rangle = \frac{1}{2}\langle \vec{s}^{\,2} - \vec{s}_1^{\,2} - \vec{s}_1^{\,2}\rangle = \begin{pmatrix} -\frac{3}{4} \\ \frac{1}{4} \end{pmatrix} \text{ for } \begin{pmatrix} s=0 \\ s=1 \end{pmatrix}, \tag{15.19}$$

and similarly

$$\langle \vec{G}^1 \cdot \vec{G}^2\rangle = \frac{1}{2}\langle \vec{G}^{\,2} - \vec{G^1}^2 - \vec{G^2}^2\rangle. \tag{15.20}$$

The expectation values $\langle \vec{G}^{\,2}\rangle$ are computed in Appendix E. Let us deal first with mesons by using $SU(3)_c$. Since $q\overline{q}$ states are colour singlets and quarks or antiquarks colour triplets we have for the expectation values (see Table E.1 in Appendix E)

$$\langle \vec{G}^{\,2}\rangle = 0, \quad \langle \vec{G^1}^2\rangle = \langle \vec{G^2}^2\rangle = \frac{4}{3}, \tag{15.21}$$

and thus

$$\langle \vec{G}^1 \cdot \vec{G}^2\rangle = -\frac{4}{3}. \tag{15.22}$$

The energy levels between $s = 0$ an 1 are split by $\frac{4c}{3}$:

$$\langle H\rangle = -c\begin{pmatrix} -\frac{3}{4} \\ \frac{1}{4} \end{pmatrix}\left[-\frac{4}{3}\right] = \begin{pmatrix} -c \\ \frac{c}{3} \end{pmatrix} \text{ for } \begin{pmatrix} s=0 \\ s=1 \end{pmatrix}. \tag{15.23}$$

Taking for the ground states the masses of the π and the ρ mesons one finds that c is positive and roughly equal to 480 MeV.

We now repeat the calculation for the ground state baryons. From (15.18)

$$\langle H\rangle = -c(\langle \vec{G}^1 \cdot \vec{G}^2\rangle\langle \vec{s}_1 \cdot \vec{s}_2\rangle + \langle \vec{G}^1 \cdot \vec{G}^3\rangle\langle \vec{s}_1 \cdot \vec{s}_3\rangle + \langle \vec{G}^2 \cdot \vec{G}^3\rangle\langle \vec{s}_2 \cdot \vec{s}_3\rangle). \tag{15.24}$$

We use (15.20) to evaluate the pairs $\langle\vec{G}^i \cdot \vec{G}^j\rangle$. However, according to (10.12, 10.13) each pair must couple to 3* in order to obtain with the third quark a colour singlet qqq state. Hence $\langle\vec{G}^2\rangle = \frac{4}{3}$ from Table E.1 and we obtain

$$\langle\vec{G}^i \cdot \vec{G}^j\rangle = \frac{1}{2}\left(-\frac{4}{3}\right) = -\frac{2}{3}. \tag{15.25}$$

The spin contribution gives

$$\langle\vec{s}_1 \cdot \vec{s}_2 + \vec{s}_1 \cdot \vec{s}_3 + \vec{s}_2 \cdot \vec{s}_3\rangle = \frac{1}{2}\langle\vec{s}^{\,2} - \vec{s}_1^{\,2} - \vec{s}_1^{\,2} - \vec{s}_3^{\,2}\rangle = \frac{1}{2}\begin{pmatrix} \frac{1}{2}\cdot\frac{3}{2} - 3\cdot\frac{1}{2}\cdot\frac{3}{2} \\ \frac{3}{2}\cdot\frac{5}{2} - 3\cdot\frac{1}{2}\cdot\frac{3}{2} \end{pmatrix}$$

$$= \begin{pmatrix} -\frac{3}{4} \\ \frac{3}{4} \end{pmatrix} \text{ for } \begin{pmatrix} s = \frac{1}{2} \\ s = \frac{3}{2} \end{pmatrix}. \tag{15.26}$$

Finally one gets with (15.25) the splitting

$$\langle H\rangle = c\left(\frac{2}{3}\right)\begin{pmatrix} -\frac{3}{4} \\ \frac{3}{4} \end{pmatrix} = c\begin{pmatrix} -\frac{1}{2} \\ \frac{1}{2} \end{pmatrix} \text{ for } \begin{pmatrix} s = \frac{1}{2} \\ s = \frac{3}{2} \end{pmatrix}. \tag{15.27}$$

One predicts an average splitting $c \sim 480\,\text{MeV}$ between the spin-$\frac{3}{2}$ and $\frac{1}{2}$ ground state baryons, less than that between the $s = 0$ and 1 mesons. Note that without $\langle\vec{G}^i \cdot \vec{G}^j\rangle$ colour interaction the spin-$\frac{3}{2}$ baryons would be lighter than the spin-$\frac{1}{2}$ ones!

The physics of excited baryons is still an area of work for the future. Paraphrasing Ref. [6] one can state that "the underlying mechanisms leading to baryon excitations are not (fully) understood".

References

1. Kalman, C.S., Tran, B.: Nuov Cimento A 102, 835 (1989)
2. Cohen-Tannoudji, C., Diu, B., Laloë, F.: Quantum Mechanics, 1, p. 819. Wiley, New York (1997)
3. Capstick, S., Roberts, W.: Prog. Part. Nucl. Phys. 45, S241 (2000)
4. Close, F.E.: An Introduction to Quarks and Partons, p. 61. Academic Press, London (1979)
5. Flamm, D., Schöberl, F.: Introduction to the Quark Model of Elementary Particles, p. 247. Gordon and Breach, New York (1982)
6. Klempt, E., Richard, J.M.: Rev. Mod. Phys. 82, 1095 (2010)
7. Amsler, C., DeGrand, T., Krusche, B.: In: Tanabashi, M., et al. (Particle Data Group): Phys. Rev. D 98, 030001 (2018), p. 287
8. Tanabashi, M., et al. (Particle Data Group): Phys. Rev. D 98, 030001 (2018)

9. Burkert, V.D., Roberts, C.D.: (2017). arXiv:1710.02549
10. Hemingway, R.J.: Nucl. Phys. B 253, 742 (1985)
11. Mai, M., Meissner, U.-G.: Nucl. Phys. A 900, 51 (2013); Meissner, U.-G., Hyodo, T.: In: Tanabashi, M., et al. (Particle Data Group): Phys. Rev. D 98, 030001 (2018), p. 766

Chapter 16
Multiquark States

A new hadron spectroscopy has emerged, in particular in the heavy quark sector, with the experimental evidence for "exotic" states that cannot (or cannot easily) be accommodated as $q\overline{q}$ mesons or qqq baryons. This chapter deals with some of these hadrons that have been reported recently by several experiments, the nature of which is highly controversial and currently under vivid discussion (some of them might not withstand the test of time). The next two sections deal with tetraquarks and pentaquarks. Some of these states may be molecular structures made of pairs of mesons such as D, D_s and D^*, D_s^* excitations, or their B and B^* counterparts. They could also be mimicked by kinematical effects. For recent reviews and references see [1, 2, 3, 4].

16.1 Light Tetraquark

A tetraquark is a compact colour neutral object made of a diquark-antidiquark pair $q_1q_2\overline{q}_3\overline{q}_4$. Let us deal first with the light ones containing only u, d and s quarks. By copying the SU(3)$_c$ colour decomposition (10.15) one obtains the SU(3)$_f$ multiplets

$$3 \times 3 \times 3^* \times 3^* = \underbrace{3 \times 3^*}_{9} + \underbrace{6 \times 6^*}_{36} + \underbrace{3 \times 6}_{18} + \underbrace{3^* \times 6^*}_{18^*} . \tag{16.1}$$

With the constituent mass $m_u = m_d \simeq 350\,\text{MeV}$ the lightest tetraquark would lie in the 1400 MeV mass region, and the ones with one or more s quarks would be heavier. However, one needs to take into account the colour hyperfine splitting induced by gluon exchange that was introduced in Sect. 15.3. Recall the colour interaction between two quarks given by (15.18):

$$H = -c\,(\vec{G}^1 \cdot \vec{G}^2)(\vec{s}_1 \cdot \vec{s}_2), \tag{16.2}$$

© The Author(s) 2018

C. Amsler, *The Quark Structure of Hadrons*, Lecture Notes in Physics 949,
https://doi.org/10.1007/978-3-319-98527-5_16

with $c \simeq 480\,\text{MeV}$ and $\langle \vec{s}_1 \cdot \vec{s}_2 \rangle = -\frac{3}{4}\left(\frac{1}{4}\right)$ for $s = 0$ (1). On the other hand,

$$\langle \vec{G}^1 \cdot \vec{G}^2 \rangle = \frac{1}{2} \langle \vec{G}^2 - \vec{G}^{1^2} - \vec{G}^{2^2} \rangle$$
$$= \frac{1}{2}\left(\frac{4}{3} - 2 \times \frac{4}{3}\right) = -\frac{2}{3} \text{ for } 3 \text{ and } 3^* \qquad (16.3)$$

see (15.25), and

$$\langle \vec{G}^1 \cdot \vec{G}^2 \rangle = \frac{1}{2}\left(\frac{10}{3} - 2 \times \frac{4}{3}\right) = +\frac{1}{3} \text{ for } 6 \text{ and } 6^*, \qquad (16.4)$$

with Table E.1 in Appendix E. Let us discuss the ground states in which all angular momenta vanish. The wavefunction of the diquark needs to be antisymmetric when taking into account the spin, flavour and colour components. For $s = 0$ the spin function is antisymmetric (A) and for $s = 1$ symmetric (S). With three quarks there are 6 symmetric (S) flavour wavefunctions and 3 antisymmetric (A) ones, which correspond to the 6 and 3^* representations of $\text{SU}(3)_f$:

$$S\ (6):\ \frac{1}{\sqrt{2}}(ud + du), \frac{1}{\sqrt{2}}(us + su), \frac{1}{\sqrt{2}}(ds + sd), uu, dd, ss, \qquad (16.5)$$

$$A\ (3^*):\ \frac{1}{\sqrt{2}}(ud - du), \frac{1}{\sqrt{2}}(us - su), \frac{1}{\sqrt{2}}(ds - sd). \qquad (16.6)$$

The same reasoning applies to $\text{SU}(3)_c$ (replace u, d, s by R, G, B) and to the antidiquark (6^* and 3 representations). The corresponding $\text{SU}(3)_f$ weight diagrams of 3^* and 3 are depicted in Fig. 16.1a, those for 6 and 6^* in Fig. 16.2.

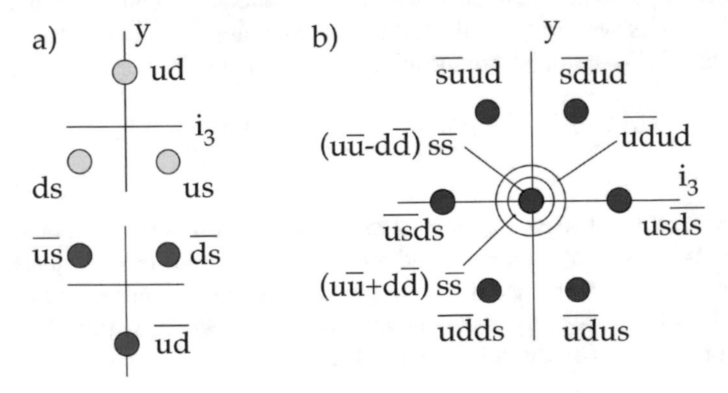

Fig. 16.1 (**a**) Weight diagrams of the 3^* and 3 representations of $\text{SU}(3)_f$ for diquarks and antidiquarks, respectively; (**b**) scalar nonet of the lightest tetraquark configuration (see the text)

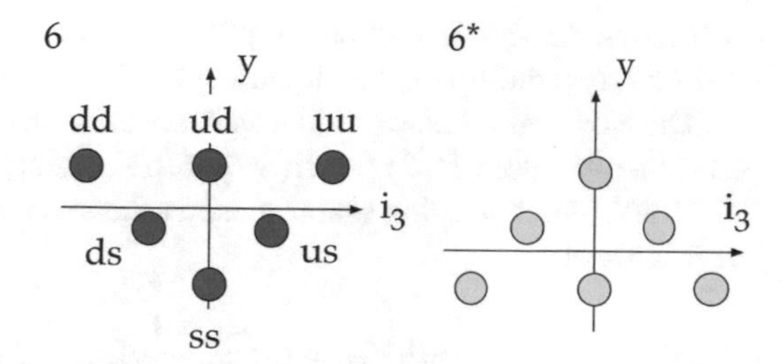

Fig. 16.2 Weight diagrams of the 6 and 6* representations of $SU(3)_f$ for diquarks and antidiquarks, respectively

Table 16.1 Antisymmetric combinations of spin, flavour and colour of the qq and $\overline{qq}$ pairs

Symmetry		qq		$\overline{qq}$		
Spin, flavour, colour	s	Flavour	Colour	Flavour	Colour	$\langle H \rangle$
ASS	0	6	6	6*	6*	$\frac{c}{4}$
SAS	1	3*	6	3	6*	$-\frac{c}{12}$
SSA	1	6	3*	6*	3	$\frac{c}{6}$
AAA	0	3*	3*	3	3	$-\frac{c}{2}$

Table 16.2 Ground state ($L = 0$) tetraquark flavour multiplets and colour hyperfine splitting (HfS)

$(qq)(\overline{qq})$	J^P	Multiplet	HfS
$(ASS)(ASS)$	0^+	36	$\frac{c}{2}$
$(ASS)(SAS)$	1^+	18	$\frac{c}{6}$
$(SAS)(ASS)$	1^+	18*	$\frac{c}{6}$
$(SAS)(SAS)$	$0, 1, 2^+$	9	$-\frac{c}{6}$
$(SSA)(SSA)$	$0, 1, 2^+$	36	$\frac{c}{3}$
$(SSA)(AAA)$	1^+	18	$-\frac{c}{3}$
$(AAA)(SSA)$	1^+	18*	$-\frac{c}{3}$
$(AAA)(AAA)$	0^+	9	$-c$

The M-type states are listed in the top half, the T-type ones in the bottom half

The four possible diquark and antidiquark fully antisymmetric configurations are listed in Table 16.1. The expectation values of the operator (16.2) are listed in the last column. For example, for the ASS configuration

$$\langle \vec{s}_1 \cdot \vec{s}_2 \rangle = -\frac{3}{4}, \quad \langle \vec{G}^1 \cdot \vec{G}^2 \rangle = \frac{1}{3} \tag{16.7}$$

with (16.4), and therefore

$$\langle H \rangle = \frac{c}{4}. \tag{16.8}$$

Finally, the total tetraquark wavefunction is a colour singlet, hence only the $3_c^* \times 3_c$ (T-type, see Fig. 10.4) and $6_c \times 6_c^*$ (M-type) configurations are kept. Table 16.2 lists the possible combinations, the multiplet dimensions and the hyperfine splitting

$\langle H(qq)\rangle + \langle H(\overline{qq})\rangle$. Without angular momentum the tetraquarks can have spin $j = 0, 1$ or 2, depending on the diquark spins. The parity is positive.

The most remarkable prediction is the existence of a light tetraquark scalar nonet (last line in Table 16.2) which would lie roughly in the mass region $4m_u - c \simeq 920\,\text{MeV}$. The low-lying scalar mesons shown in Table 11.1 could be the members of this nonet:

$$
\begin{aligned}
a_0(980) &= (us\bar{d}\bar{s}, \frac{1}{\sqrt{2}}[u\bar{u} - d\bar{d}]s\bar{s}, \overline{us}ds),\\
f_0(980) &= \frac{1}{\sqrt{2}}[u\bar{u} + d\bar{d}]s\bar{s},\\
f_0(500) &= \bar{u}\bar{d}ud,\\
K_0^*(700) &= (\bar{s}\bar{d}ud, \overline{su}ud) \text{ and } (\bar{u}\bar{d}us, \bar{u}\bar{d}ds),
\end{aligned}
\tag{16.9}
$$

with weight diagram depicted in Fig. 16.1b. These assignments have been proposed a long time ago [5]. The $s\bar{s}$ component in the wavefunctions of the $a_0(980)$ and $f_0(980)$ must be large since they strongly couple to the $K\bar{K}$ channel, yet lying below $K\bar{K}$ threshold. The latter strongly modifies the resonance parameters which are observed in the $\eta\pi$ and $\pi\pi$ mass distributions, respectively, shifting the masses and reducing the widths (Chap. 12). Kaon-antikaon molecular states have also been suggested [6] as well as tetraquark cores surrounded by virtual $K\overline{K}$ clouds [7]. The branching ratios for the radiative decays $\phi \to \gamma f_0(980) \to \gamma\pi^0\pi^0$ and $\phi \to \gamma a_0(980) \to \gamma\pi^0\eta$ also argue in favour of four-quark structures for the f_0 and a_0 [8].

Doubly charged tetraquarks are also predicted, such as $uu\overline{dd}$ which occurs in the 36-dimensional representations, see (16.5). The large number of higher lying tetraquark states predicted in Table 16.2 are expected to be broad [5], which might explain why no obvious candidate has been reported so far.

16.2 Heavy Tetraquark

Some 20 mesons in the charmonium spectrum, which are hard (or impossible) to interpret as $c\bar{c}$ states, have been observed at e^+e^- and high energy hadronic colliders [9], several of them needing experimental confirmation. Figure 16.3 shows the established ones together with the known $c\bar{c}$ mesons.

The most prominent candidate is the $X(3872)$ established by several experiments at e^+e^-, $\overline{p}p$ and pp colliders. This state had in fact been spotted long before in high energy πN data [10]. Figure 16.4a shows the $J/\psi\,\pi^+\pi^-$ invariant mass from Belle in $B^+ \to K^+J/\psi\,\pi^+\pi^-$, where the J/ψ decays into e^+e^- or $\mu^+\mu^-$ pairs. The B mesons were produced in e^+e^- collisions at the $\Upsilon(4S)$ [11]. Figure 16.4b shows the signal observed by CDF at the Fermilab Tevatron running at the $\overline{p}p$ collision

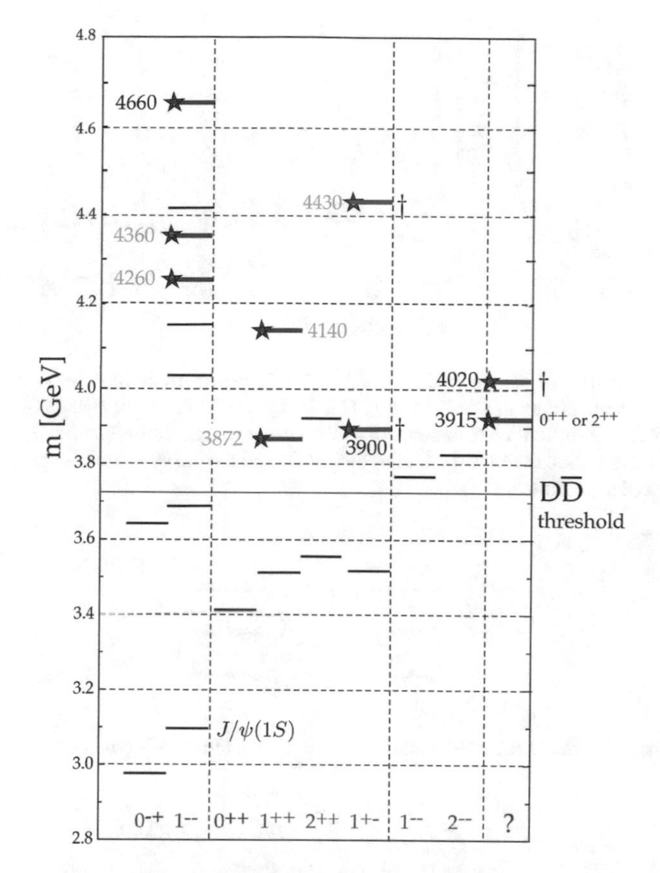

Fig. 16.3 States populating the charmonium spectrum that have been established so far [9]. The $c\bar{c}$ states are shown in black (see also Fig. 9.4), the exotic ones are tagged by stars. The quantum numbers of the two states on the right are not firmly established. The $X(3872)$ [now called $\chi_{c1}(3872)$ according to the nomenclature in Table 4.1], $X(4140)$ [or $\chi_{c1}(4140)$], $X(4260, 4360)$ [or $\psi(4260, 4360)$], $X(3900)^{\pm}$ [or $Z_{c1}(3900)^{\pm}$] and $X(4430)^{\pm}$ [or $Z_{c1}(4430)^{\pm}$] are discussed in the text; † $i = 1$ meson, the C parity is not defined but attributed from that of the neutral partner

energy of 1.96 TeV, where the J/ψ decays into $\mu^+\mu^-$ pairs. Figure 16.4c shows the signal also observed by CMS in pp collisions at 7 TeV.

The mass of the $X(3872)$ is 3871.7 ± 0.2 MeV and its width is less than 1.2 MeV. Its quantum numbers have been determined by CDF to be either $J^{PC} = 1^{++}$ or 2^{-+}, from the correlation between the planes spanned by the $\pi^+\pi^-$ and $\mu^+\mu^-$ pairs and their angular distributions [14]. The 1^{++} assignment was then firmly established by LHCb [15]. Since the $X(3872) \equiv \chi_{c1}(3872)$ decays into J/ψ it should contain $c\bar{c}$ pairs and could be either a $c\bar{c}$ state or a tetraquark (Fig. 16.5).

The $X(3872)$ could be the missing 2^3P_1 charmonium state $\chi_{c1}(2P)$ (Fig. 9.4). However, the latter is expected to lie about 100 MeV higher in mass [16]. On the other hand, the $X(3872)$ lies at the $D^0\overline{D}^{*0}$ threshold (3871.7 MeV), which suggests

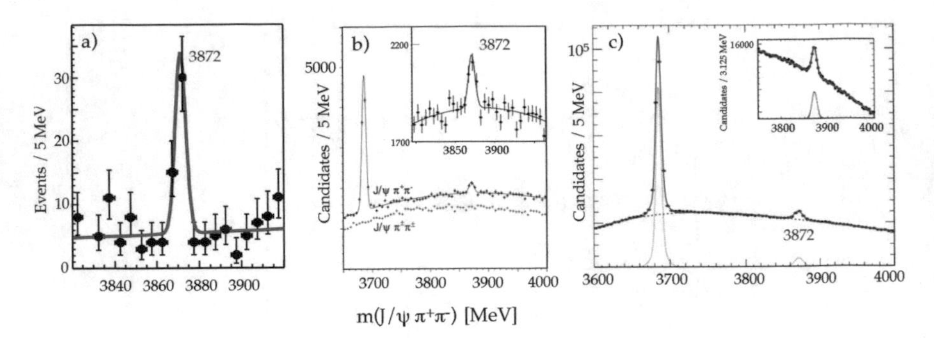

Fig. 16.4 Observation of the $X(3872) \to J/\psi\,\pi^+\pi^-$; (**a**) by Belle in $B^+ \to K^+ J/\psi\,\pi^+\pi^-$ (including the charge conjugated B^- decay) [11]; (**b**) by CDF, the strong peak at 3686 MeV stems from the $\psi(2S)$, the bottom distribution shows the wrong charge combinations $J/\psi\,\pi^\pm\pi^\pm$ [12]; (**c**) by CMS, the top blue curve is the fit, the bottom dotted red curve is the background subtracted signal (enhanced in the inset) [13]

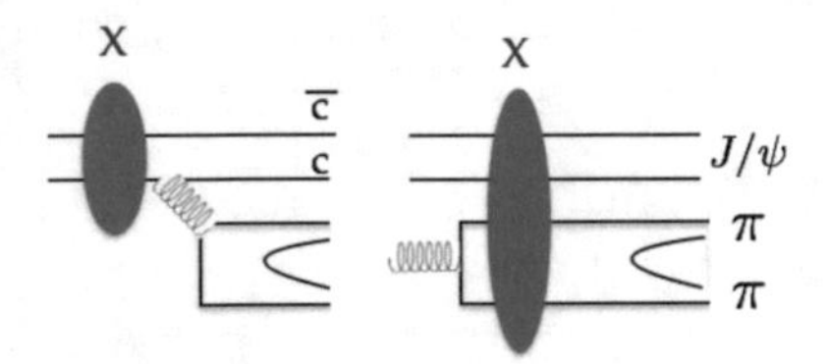

Fig. 16.5 The $X(3872)$ as a charmonium meson (left) or a tetraquark (right)

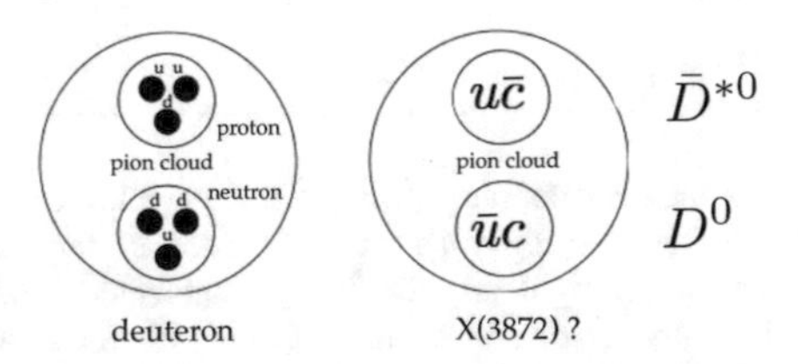

Fig. 16.6 The $X(3872)$ as a loosely bound $D\overline{D}^*$ system similar to the deuteron

a weakly bound $D^0\overline{D}^{*0}$ "molecule" bound by one pion exchange (Fig. 16.6). The existence of such a state was predicted many years ago in analogy to the weakly bound deuteron [17]. Indeed, a $D\overline{D}^*$ mass enhancement is observed at threshold by Belle in B decays [18]. One also expects the $D^+\overline{D}^{*-}$ decay (threshold at 3879.8 MeV), which with $D^0\overline{D}^{*0}$ leads to $i = 0$ and $i = 1$ isospin mixing (analogous to the $K\overline{K}$ or $K\overline{K}^*$ systems, see (6.45)), leading to signals in $J/\psi\,\rho$ and $J/\psi\,\omega$, which have both been observed [9]. Note that a molecular structure is of the type $(c\overline{q})(\overline{c}q)$, in contrast to the $cq\overline{cq}$ tetraquark which is a more compact system. The latter is perhaps unlikely for the $X(3872)$, since the charged partner (e.g. $cu\overline{cd}$) has so far not been observed.

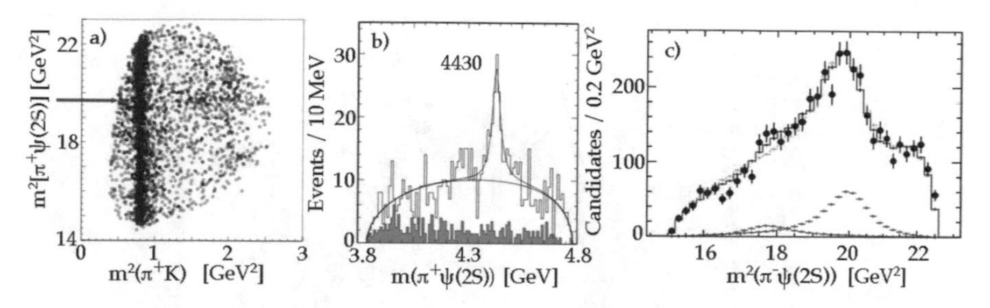

Fig. 16.7 Observation of the $Z_{c1}(4430)^{\pm}$; **(a)** $\pi^+\psi(2S)$ vs. π^+K Dalitz plot in $B \to K\pi^+\psi(2S)$ from Belle [19]; **(b)** $\pi^+\psi(2S)$ mass projection with fit excluding the K^* band. The dark (blue) area shows the excluded K^* contribution; **(c)** $m^2(\pi^-\psi(2S))$ projection in $B^0 \to K^+\pi^-\psi(2S)$ from LHCb. The strong peak around 20 GeV^2 (with contribution shown by the blue bottom distribution) is due to the $Z_{c1}^-(4430)$. A possible additional resonance around 4200 MeV is also suggested [20]

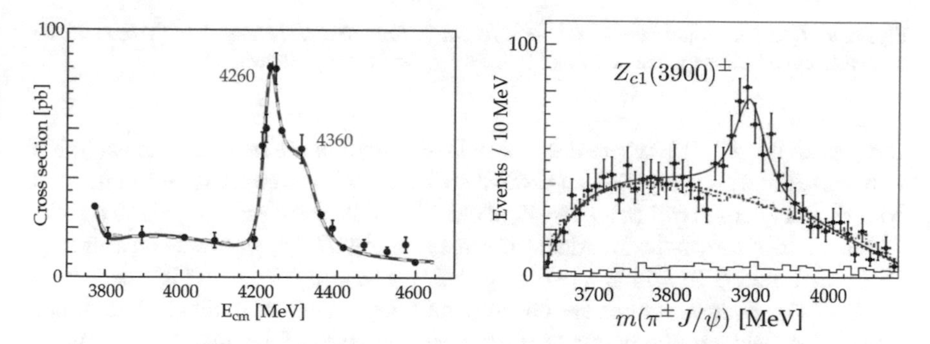

Fig. 16.8 Left: cross section for $e^+e^- \to J/\psi\,\pi^+\pi^-$ measured by BESIII as a function of energy with fits (for details see [21]). Right: decay of the $X(4260)$ into $Z_{c1}(3900)^{\pm}\pi^{\mp}$ with fits. The histogram shows the background contribution estimated from the J/ψ sidebands [22]

States decaying into charmonium and a charged pion cannot be pure $c\bar{c}$ states, but must contain both $c\bar{c}$ and charged $q\bar{q}'$ pairs. Figure 16.7a shows the first observation of a charged state by Belle, containing both $c\bar{c}$ and charged $q\bar{q}'$ pairs, and decaying into $\pi^{\pm}\psi(2S)$. The state is produced in B decay into $K\pi^{\pm}\psi(2S)$ with $\psi(2S) \to J/\psi\,\pi^+\pi^-$ or $\psi(2S) \to$ lepton-antilepton pairs [19]. The strong vertical band stems from the K^* background. The horizontal red arrow shows a faint signal which is enhanced in the $\pi^{\pm}\psi(2S)$ projection when applying a K^* cut (Fig. 16.7b). The LHCb collaboration has confirmed this charged state in the channel $B^0 \to K^+\pi^-\psi(2S)$ (Fig. 16.7c) and has determined its spin and parity to be 1^+ [20]. The $\pi^0\psi(2S)$ decay mode is also expected, hence $C = -1$ is assumed.

Figure 16.8 (left) shows the cross section for $e^+e^- \to J/\psi\,\pi^+\pi^-$ measured with the BESIII detector (see Fig. 9.5) as a function of collision energy. Two states are observed at 4260 and 4360 MeV, assumed to have the quantum numbers 1^{--} (since they are produced in e^+e^- collisions), in a region where no charmonium state is expected. These mesons have been seen earlier in $e^+e^- \to \gamma J/\psi\,\pi^+\pi^-$

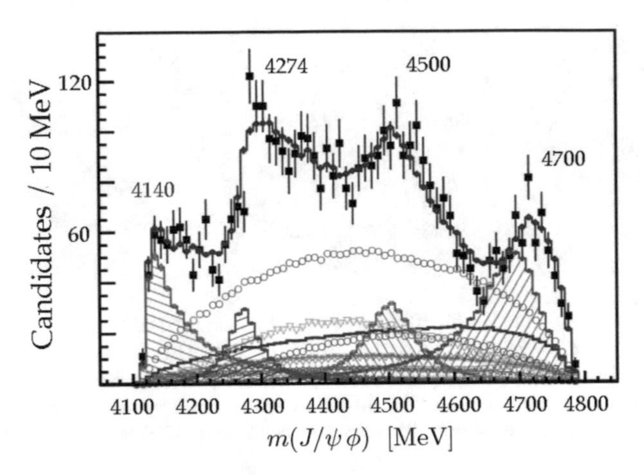

Fig. 16.9 $J/\psi\,\phi$ invariant mass in $B^+ \to J/\psi\,\phi K^+$. The top (red) histogram is the total fit. The individual contributions to the peaks and from background are also shown [24]

and $\gamma\psi(2S)\pi^+\pi^-$ [9], where the γ is emitted by one of the incident leptons (initial state radiation, ISR) [23]. The $X(4260)$ then decays into a charged meson containing both $c\bar{c}$ and charged $q\bar{q}'$ pairs, the $Z_{c1}(3900)^{\pm}$ ($\to J/\psi\pi^{\pm}$) $\pi^{\mp}$ (Fig. 16.8 right).

Figure 16.9 shows the invariant mass distribution of $J/\psi\,\phi$ events measured by LHCb at 7 and 8 TeV in $B^+ \to J/\psi\,\phi K^+$, where $J/\psi \to \mu^+\mu^-$ and $\phi \to K^+K^-$ [24]. Four structures are observed and their quantum numbers determined. The $J/\psi\,\phi$ decay mode points to a structure containing four quarks $c, \bar{c}, s$ and $\bar{s}$. The two lower states have $J^{PC} = 1^{++}$ and the two upper ones $J^{PC} = 0^{++}$. The lowest one at 4140 MeV had been reported before at the Tevatron and by CMS, who may have also observed the one at 4274 MeV [25]. So far, only the 4140 MeV state is considered to be well established with a mass of 4146.8 ± 2.4 MeV and a rather narrow width of 22 ± 8 MeV [9].

States have also been reported that do not fit in the traditional bottomonium spectrum shown in Fig. 9.6. Two (10–20 MeV) narrow $J^P = 1^+$ charged mesons, $Z_{b1}(10610)^{\pm}$ and $Z_{b1}(10650)^{\pm}$ have been observed by Belle in the decay of the radial excitation $\Upsilon(5S)$:

$$\begin{aligned}
&\Upsilon(5S) \to Z_{b1}(10610)^{\pm}\pi^{\mp},\ Z_{b1}(10650)^{\pm}\pi^{\mp},\\
\text{with }\ &Z_{b1}(10610)^{\pm} \to [\Upsilon(1S), \Upsilon(2S), \Upsilon(3S)]\pi^{\pm},\\
\text{and }\ &Z_{b1}(10650)^{\pm} \to [\Upsilon(1S), \Upsilon(2S), \Upsilon(3S)]\pi^{\pm},
\end{aligned} \tag{16.10}$$

where $\Upsilon(1S, 2S, 3S) \to \mu^+\mu^-$ [26]. The decays into $h_b(1P, 2P)\pi^{\pm}$ have also been observed. The $\Upsilon(nS)\pi^{\pm}$ mass distributions are shown in Fig. 16.10. The angular distributions of the charged pions favour the quantum numbers $J^P = 1^+$. The $Z^0_{b1}\pi^0$ decay modes are also expected, hence $C = -1$ is assumed for the Z_{b1}. Again, the charged mesons cannot be charmonium states. Since they both lie very

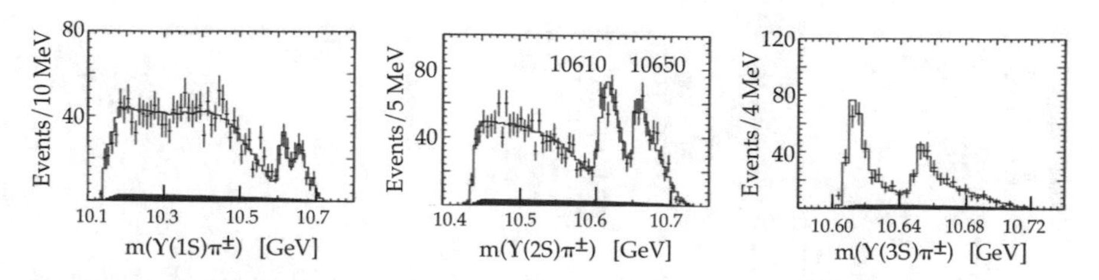

Fig. 16.10 $\Upsilon(nS)\pi^{\pm}$ invariant mass spectrum in $\Upsilon(5S) \to \Upsilon(nS)\pi^+\pi^-$ decay; from left to right $n = 1, 2$ and 3 [26]

close to the $B^*\overline{B}$ and $B^*\overline{B}^*$ thresholds (10,604.6 and 10,650.2 MeV), they could be two-meson molecular states.

16.3 Pentaquark

The pentaquark is a conjectural baryon made of four quarks and one antiquark which form a colour singlet, see (10.16). These states have been sought for many years by looking for hyperons with the wrong strangeness sign $S = +1$, e.g. decaying into nK^+ or pK_S associated with the production of an $S = -1$ hyperon. Several candidates with masses between 1780 and 2500 MeV had been reported in the seventies from analyses of the K^+p cross section, which were later not confirmed (for a review see [27]). Pentaquark states were again reported at the turn of the century. A baryon with putative $uudd\overline{c}$ quark content, the $\Theta_c(3100)$, was observed to decay into $D^{*-}p$. Another candidate, the doubly charged $\Phi(1680)$ was reported to decay into $\Xi^-\pi^-(ssdd\overline{u})$.

The most prominent one was the $\Theta(1540)^+$ with quark content $uudd\overline{s}$, emitting an $S = +1$ kaon. Figure 16.11a, b show for example the K^-n and K^+n invariant mass spectra obtained in photoproduction at the SPring-8 facility in Japan, $\gamma n \to K^+K^-n$ [28]. The incident photon (of typically 2 GeV) was produced by backward scattering laser photons off 8 GeV electrons. A CH scintillator was used as the neutron target. The outgoing neutron was not observed, but the contribution from $\gamma p \to K^+K^-p$ was vetoed by detecting the emitted proton. While there is no prominent structure in Fig. 16.11a, a narrow peak is observed in the "wrong" strangeness mass distribution 16.11b. The significance of the signal was estimated to be about 4.6σ (standard deviations, close to the 5σ required to claim a discovery by today's standards). An even stronger signal was reported in $K^+n \to K^0p$ from a xenon bubble chamber, published only recently [30]. Several collaborations had also claimed evidence for the $\Theta(1540)^+$, to be rebutted by many others [31]. For example, Fig. 16.11c shows the negative result of the high statistics CLAS experiment at Jefferson Lab in $\gamma d \to pK^-K^+n$.

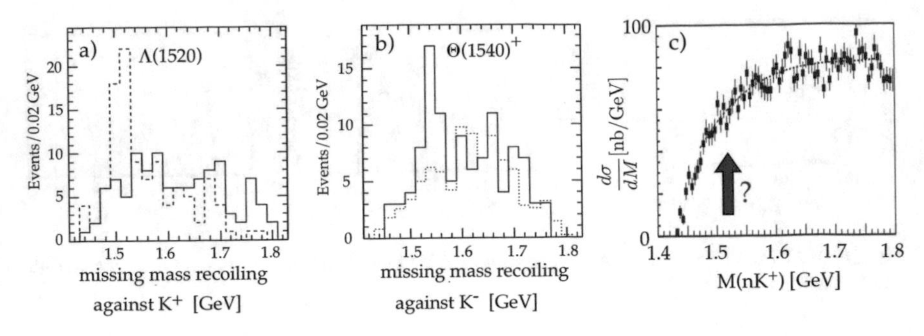

Fig. 16.11 (**a**) K^-n mass distribution in $\gamma n \to K^+K^-n$ (full histogram). The dotted histogram shows the mass distribution when using hydrogen as a target. The peak is due to the reaction $\gamma p \to K^+K^-p$, where the intermediate $\Lambda(1520)$ decays into K^-p [28]; (**b**) K^+n mass showing the positive strangeness baryon candidate $\Theta(1540)^+$ [28]; (**c**) the K^+n mass distribution in $\gamma d \to pK^-K^+n$ does not show any signal [29]

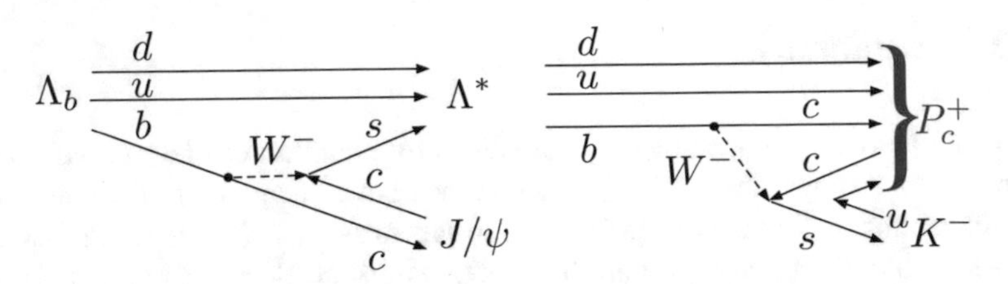

Fig. 16.12 Left: Λ_b^0 decay into Λ^* excitations. Right: production of the pentaquark baryon P_c^+

None of these pentaquark states seem to have survived more detailed investigations. However, evidence for two heavy pentaquark states in Λ_b^0 decay have been presented by the LHCb collaboration in pp collisions at 7 and 8 TeV [32]. The branching fraction for $\Lambda_b^0 \to K^-J/\psi\, p$ is only 3×10^{-4}. However, sufficient and clean data can be collected by triggering on $\mu^+\mu^-$ pairs from J/ψ decay and by reconstructing the Λ_b^0 decay vertex, which is displaced from the pp collision point by several millimeters. The Feynman diagrams contributing to $\Lambda_b^0 \to K^-J/\psi\, p$ are displayed in Fig. 16.12. The weak process $b \to c$ leads to several intermediate Λ^* resonances decaying into K^-p (left), but could also contribute to the production of pentaquarks P_c^+ with quark content $ducu\bar{c}$, decaying into $J/\psi\, p$ (right).

Figure 16.13 (left) shows the Dalitz plot of the decay $\Lambda_b^0 \to K^-J/\psi\, p$ (including the charge conjugate $\overline{\Lambda}_b^0 \to K^+J/\psi\,\overline{p}$) in which a band is visible around the $(J/\psi)p$ squared mass of 20 GeV^2. The signal is even more striking in the $(J/\psi)p$ mass projection, Fig. 16.13 (right). An amplitude analysis was performed, leading to the presence of two resonances [32], the narrow $P_c^+(4450)$ with width $\simeq 40$ MeV and the broad $P_c^+(4380)$ with width $\simeq 200$ MeV. The quantum numbers could not be assigned unambiguously: $j = \frac{1}{2}$ for one of the two states and $\frac{3}{2}$ with opposite parity for the other.

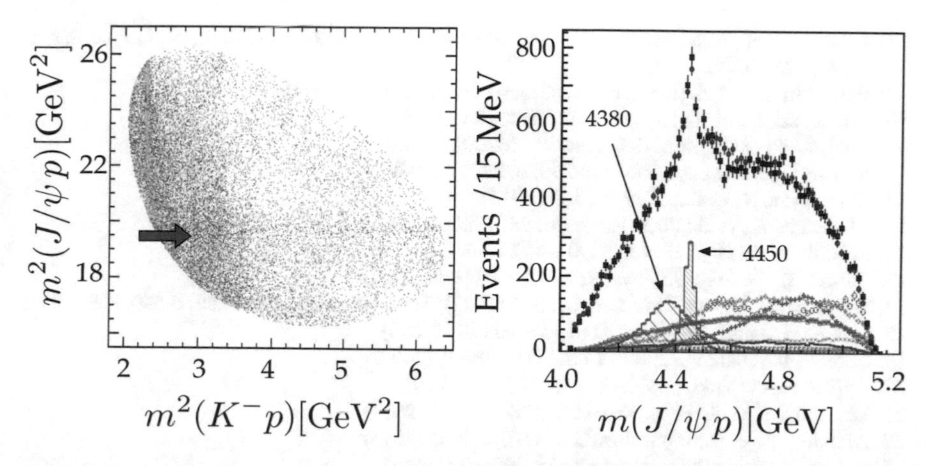

Fig. 16.13 Dalitz plot (left) and $J/\psi\, p$ mass projection (right) of the decay $\Lambda_b^0 \to K^- J/\psi\, p$ from LHCb. The red arrow on the left shows the enhancement at a mass of about 4.5 GeV. The black squares on the right show the data, the red dots the fit. The bottom distributions represent the fitted contribution from the many Λ^* decay channels. The contributions from the two pentaquark candidates at 4380 and 4450 MeV are shown by the hatched areas [32]

The nature of these states—pentaquarks, molecules, or less mundane kinematic effects—remains to be clarified (for a review see [33]). Worth noting are several nearby thresholds such as $\Sigma_c^{*+}\overline{D}^0$ (4382 MeV) and $\Lambda_c^{*+}\overline{D}^0$ (4457 MeV). These pentaquarks qualify as molecular states made of heavy baryons, weakly bound to heavy mesons. A triangle singularity involving the nearby $p\chi_{c1}$ threshold (4449 MeV) was also invoked: $\Lambda_b^0 \to \Lambda^*(\to K^-p)\chi_{c1}$ followed by rescattering, $p\chi_{c1} \to P_c^+ \to pJ/\psi$ [34].

These pentaquarks have been seen so far by one experiment in one decay channel only, and therefore need confirmation. They could be produced directly as resonances in $\pi^-p \to J/\psi\, n$ [35] or in photoproduction with the inverse reaction $(\gamma \Rightarrow J/\psi)p \to P_c^+$ [36], where the J/ψ is one of the hadronic components of the incident photon ($c\bar{c}$ in addition to $u\bar{u}$, $d\bar{d}$ and $s\bar{s}$, see Fig. 7.7).

References

1. Amsler, C., Hanhart, C.: In: Tanabashi, M., et al. (Particle Data Group): Phys. Rev. D 98, 030001 (2018), p. 753
2. Godfrey, A., Olsen, S.L.: Annu. Rev. Nucl. Part. Sci. 58, 51 (2008)
3. Olsen, S.L.: Front. Phys. 10, 121 (2015)
4. Amsler, C., Törnqvist, N.A.: Phys. Rep. 389, 61 (2004)
5. Jaffe, R.L.: Phys. Rev. D 15 267, 281 (1977)
6. Achasov, N.N., et al.: Phys. Rev. D 56 203 (1997)
7. Close, F.E., Törnqvist, N.: J. Phys. G 28 R249 (2002)

8. Achasov, N.N., Ivanchenko, V.N.: Nucl. Phys. B 315, 465 (1989); Achasov, N.N.: Nucl. Phys. A 728, 425 (2003)
9. Tanabashi, M., et al. (Particle Data Group): Phys. Rev. D 98, 030001 (2018)
10. Antoniazzi, L., et al.: Phys. Rev. D 50, 4258 (1994)
11. Choi, S.-K., et al.: Phys. Rev. Lett. 91, 262001 (2003)
12. Acosta, D., et al.: Phys. Rev. Lett. 93, 072001 (2004)
13. Chatrchyan, S., et al.: JHEP 04, 154 (2013)
14. Abulencia, A., et al.: Phys. Rev. Lett. 98, 132002 (2007)
15. Aaij, R., et al.: Phys. Rev. D 92, 011102 (2015)
16. Barnes, T., Godfrey, S.: Phys. Rev. D 69, 054008 (2004)
17. Törnqvist, N.A.: Phys. Rev. Lett. 67, 556 (1992); Törnqvist, N.A.: Phys. Lett. B 590, 209 (2004)
18. Aushev, T., et al.: Phys. Rev. D 81, 031103 (R) (2010)
19. Choi, S.-K., et al.: Phys. Rev. Lett. 100, 142001 (2008)
20. Aaij, R., et al.: Phys. Rev. Lett. 112, 222002 (2014)
21. Ablikim, M., et al.: Phys. Rev. Lett. 118, 092001 (2017)
22. Ablikim, M., et al.: Phys. Rev. Lett. 110, 252001 (2013)
23. Aubert, B., et al.: Phys. Rev. Lett. 95, 142001 (2005)
24. Aaij, R., et al.: Phys. Rev. Lett. 118, 022003 (2017)
25. Chatrchyan, S., et al.: Phys. Lett. B 734, (2014) 261 (2014)
26. Bondar, A., et al.: Phys. Rev. Lett. 108, 122001 (2012)
27. Trippe, T.G., et al.: Rev. Mod. Phys. 48 S1 (1976)
28. Nakano, T., et al.: Phys. Rev. Lett. 91 012002 (2003)
29. McKinnon, B., et al.: Phys. Rev. Lett. 96, 212001 (2006)
30. Barmin, V.V., et al.: Phys. Rev. C 89 045204 (2014)
31. Wohl, C.G.: Phys. Lett. B 667, 1124 (2008)
32. Aaij, R., et al.: Phys. Rev. Lett. 115 072001 (2015)
33. Karliner, M., Swarnicki, T.: In: Tanabashi, M., et al. (Particle Data Group): Phys. Rev. D 98, 030001 (2018), p. 772
34. Guo, F.-K., et al.: Eur. Phys. J. A 52, 318 (2016); Phys. Rev. D 92, 071502 (R) (2015)
35. Lü, Q.-F., et al.: Phys. Rev. D 93, 034009 (2016)
36. Kubarovsky, V., Voloshin, M.B.: Phys. Rev. D 92, 031502 (R) (2015)

Chapter 17
Heavy Baryons

We have introduced the nomenclature in Chap. 13: the Λ is an isoscalar and the Σ and isovector baryon with one heavy (s, c or b) quark. The Ξ contains two and the Ω three heavy quarks. The number of c or b quarks is indicated by the subscripts c or b. Hyperons are baryons with at least one s quark. Many baryons containing c or b quarks have been identified (Table 1.2).

17.1 Charmed Baryons

With the additive charm quantum number C baryons can be classified in a 3-dimensional representation in terms of the three coordinates (i_3, S, C), or alternatively (i_3, y, C). Although flavour symmetry is badly broken with the much heavier c quark (and even more so with the b quark), the $SU(4)_f$ representation is still useful for classification/inventory purposes. With four quarks the 64 possible configurations decompose into

$$4 \times 4 \times 4 = 4^* + 20 + 20 + 20 \tag{17.1}$$

(Problem 17.1). Figure 17.1 shows the weight diagrams. There is one flavour symmetric 20-plet associated with spin-$\frac{3}{2}$ baryons, which contains the charmless $SU(3)_f$ decuplet at its bottom. The two other 20-plets correspond to the mixed symmetric and mixed antisymmetric flavour wavefunctions of spin-$\frac{1}{2}$ baryons, in which one recognizes the charmless octet baryons at the bottom (Fig. 17.1, right). Note that there are two dsc and two usc spin-$\frac{1}{2}$ states, labelled Ξ_c^0, $\Xi_c'^{\,0}$ and Ξ_c^+, $\Xi_c'^{+}$, respectively. This is because one of the qq pairs can have spin 1 (symmetric) or spin 0 (antisymmetric), leading both to $j = \frac{1}{2}$ with the third quark. The $C = 1$ level forms with two among the u, d, s quarks a sextet 6 for $s = 1$ and a triplet 3^* of $SU(3)_f$ for $s = 0$. These are the colour configurations 3^* SSA and AAA of

© The Author(s) 2018

C. Amsler, *The Quark Structure of Hadrons*, Lecture Notes in Physics 949,
https://doi.org/10.1007/978-3-319-98527-5_17

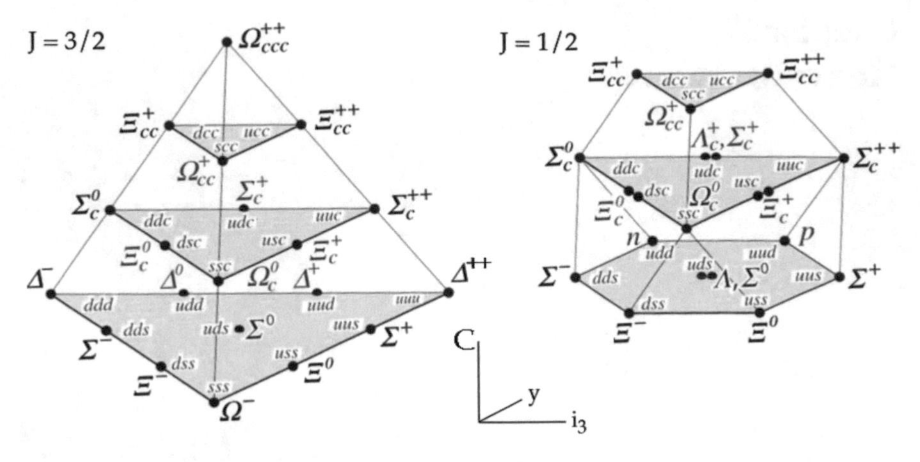

Fig. 17.1 SU(4)$_f$ multiplets of the ground state spin-$\frac{3}{2}$ and $\frac{1}{2}$ baryons including the c quark. The hypercharge is defined as $y = B + S - \frac{C}{3}$ [1]

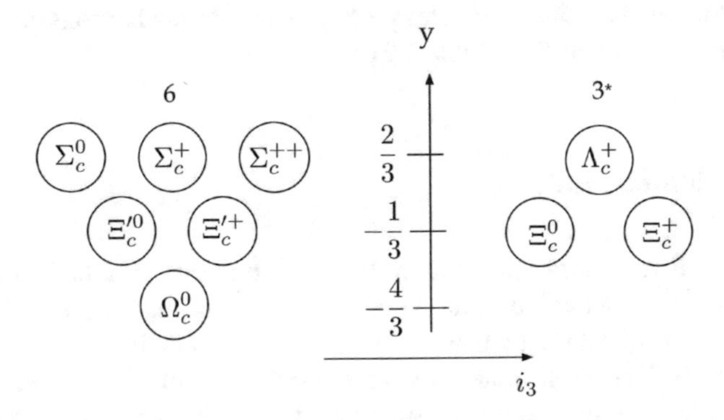

Fig. 17.2 The multiplet of the $C = 1$ spin-$\frac{1}{2}$ baryons in Fig. 17.1 splits into a 6 and a 3*

Table 16.1 to combine with a coloured charm quark. The diquark wavefunctions are given by (16.5) and (16.6) and the corresponding weight diagrams are found in Figs. 16.2a and 16.1a, respectively, to which a c quark is added with $i_3 = y = 0$. Figure 17.2 shows the resulting weight diagrams of the spin-$\frac{1}{2}$ singly charmed baryons. The quadruplet in (17.1) does not exist in the ground state but is realized for $L = 1$ excitations. Figure 17.3 shows the 4* weight diagram, conjugate to the fundamental one in Fig. 8.11, with the experimentally observed states.

The $C = 1$ baryons have all been observed. Table 17.1 lists those with established quantum numbers and their main decay modes. The parity of the Λ_c^+ is that of the c quark, defined as positive. Spin and parity have not been determined experimentally for most of the states. They follow the ordering and expectation from the quark model.

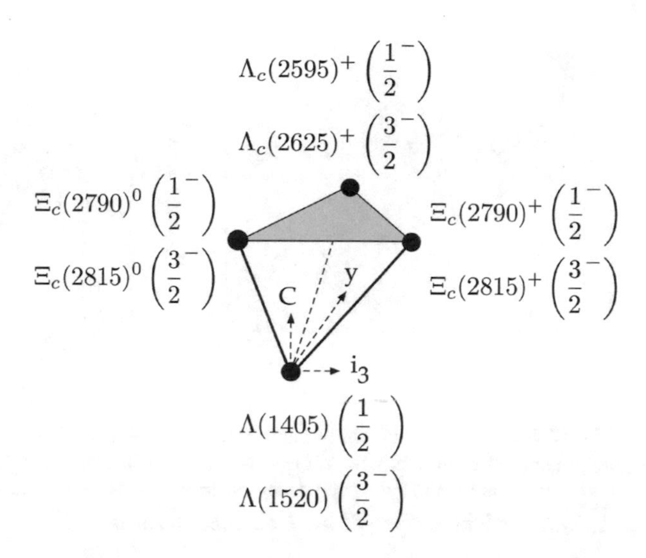

Fig. 17.3 The first orbital excitations contain a quadruplet of $s = \frac{1}{2}$ baryons with four $J^P = \frac{1}{2}^-$ and four $\frac{3}{2}^-$ baryons. The charmed ones are the partners of the $L = 1$ SU(3)$_f$ singlets listed in Table 15.3

Table 17.1 $C = 1$ baryons with established J^P and main decay modes [2]

$\Lambda_c[2286]$	$\frac{1}{2}^+$	$c \to s$, e.g. $p\overline{K}\pi(\pi)$, $\Sigma\pi(\pi)$, $\Lambda\pi(\pi)$, $\Lambda\ell\nu_\ell\ldots$	+
$\Sigma_c(2455)$	$\frac{1}{2}^+$	$\Lambda_c^+\pi$	++, +, 0
$\Xi_c[2469]$	$\frac{1}{2}^+$	$c \to s$, e.g. $\Xi + \pi's$	+, 0
$\Xi_c'(2578)$	$\frac{1}{2}^+$	$\Xi_c\gamma$	+, 0
$\Omega_c[2695]$	$\frac{1}{2}^+$	$c \to s$, e.g. $\Xi\overline{K}\pi$, Ω^- + pions ...	0
$\Sigma_c(2520)$	$\frac{3}{2}^+$	$\Lambda_c^+\pi$	++, +, 0
$\Xi_c(2645)$	$\frac{3}{2}^+$	$\Xi_c\pi$	+, 0
$\Omega_c(2770)$	$\frac{3}{2}^+$	$\Omega_c^0\gamma$	0
$\Lambda_c(2595)$	$\frac{1}{2}^-$	$\Sigma_c\pi(\pi)$	+
$\Lambda_c(2625)$	$\frac{3}{2}^-$	$\Lambda_c^+\pi^+\pi^-$	+
$\Xi_c(2790)$	$\frac{1}{2}^-$	$\Xi_c\pi$, $\Xi_c'\pi$	+, 0
$\Xi_c(2815)$	$\frac{3}{2}^-$	$\Xi_c\pi$, $\Xi_c\pi\pi$	+, 0

The electric charges are given in the last column. There are 9 spin-$\frac{1}{2}$ and 6 spin-$\frac{3}{2}$ baryons. The masses of the "stable" (weakly decaying) baryons is given in square brackets. A stable Σ_c does not exist. Note the spin-$\frac{1}{2}$ Ω_c^0. The bottom part lists the first excited quadruplet states

Figure 17.4 (left) shows the first experimental hint of a charmed baryon. The neutrino induced reaction

$$\nu_\mu p \to \mu^-\Lambda(\to \pi^- p)\pi^+\pi^+\pi^-\pi^+ \tag{17.2}$$

was observed at BNL in a bubble chamber exposed to a muon-neutrino beam [3]. The hadronic electric charge increases from +1 to +2, while the strangeness

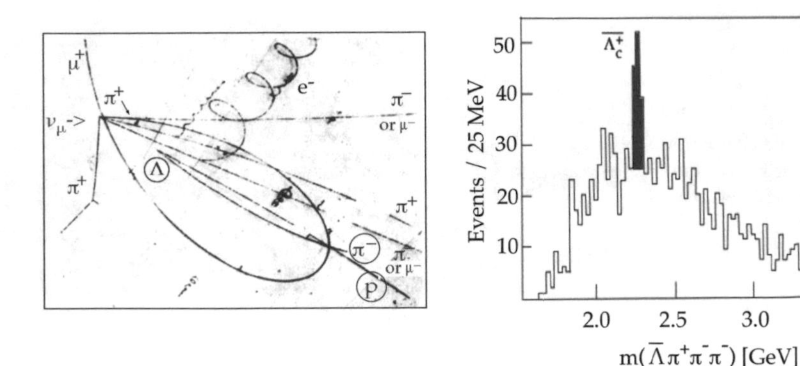

Fig. 17.4 Left: bubble chamber picture of the reaction $\nu_\mu p \to \mu^- \Lambda\pi^+\pi^+\pi^-\pi^+$ with $\Lambda \to \pi^- p$ (adapted from [3]). One of the two tracks labelled π^- or μ^- is the negative muon induced by the ν_μ. The loop is due to the decay chain $\pi^+ \to \mu^+ \to e^+$. Right: observation of the $(\overline{udc})$ $\overline{\Lambda_c^+}$ decaying into $\overline{\Lambda}\pi^+\pi^-\pi^-$ in a broadband photon beam [4]

decreases by one unit, thus seemingly violating the $\Delta S = \Delta Q$ rule for reactions involving leptons. The conundrum was solved by the observation at Fermilab of the Λ_c^+ (in fact, this was the charge conjugate $\overline{\Lambda_c^+}$ shown in Fig. 17.4, right). The (cuu) $\Sigma_c(2455)^{++}$ which was discovered later decays into $\pi^+\Lambda_c^+$, hence reaction (17.2) proceeds through the intermediate state

$$\nu_\mu p \to \mu^- \Sigma_c(2455)^{++} \tag{17.3}$$

which does not involve any strange quark and fulfils the $\Delta C = \Delta Q$ rule, followed by

$$\Sigma_c(2455)^{++} \to \Lambda_c^+\pi^+ \to \Lambda\pi^+\pi^+\pi^-\pi^+. \tag{17.4}$$

Figure 17.5a shows the $Q = +1$ (cud) partner, the $\Sigma_c(2455)^+$ and the first observation of the $\Sigma_c(2520)^+$ by the CLEO collaboration (a sketch of the CLEO III detector is shown in Fig. 9.7). The two states were produced in e^+e^- collisions at or below the $\Upsilon(4S)$. They decay almost exclusively into $\Lambda_c^+\pi^0$. Up to 15 different Λ_c^+ decay modes were included to increase the data sample. The spin $\frac{1}{2}$ has been assigned to the $\Sigma_c(2455)$ by the BaBar collaboration by studying the decay $B^- \to \Sigma_c(2455)^0\overline{p}$ with $\Sigma_c(2455)^0 \to \Lambda_c^+\pi^-$ (Fig. 17.5b).

The spin-$\frac{1}{2}$ Ω_c^0 is our next example of a charmed hyperon. Figure 17.6a shows the result of a high statistics sample collected by Belle running at or around the $\Upsilon(4S)$. The Ω^- was first reconstructed in its ΛK^- decay (inset). The Ω_c^0 appears as a strong peak when the Ω^- is combined with a positive pion, while the wrong charge combination (with a negative pion) does not show any signal (Fig. 17.6b). The $\Omega_c(2770)^0$ decaying radiatively into Ω_c^0 was also observed in this experiment [7].

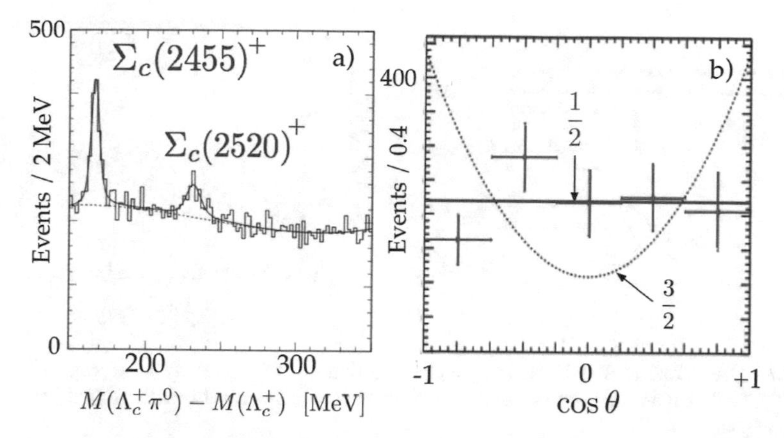

Fig. 17.5 (**a**) $\Sigma_c(2455)^+$ and $\Sigma_c(2520)^+$ decay into $\Lambda_c^+\pi^0$ [5]; (**b**) the angular distribution between the antiproton and the Λ_c^+ in $B^- \to (\Sigma_c(2455)^0 \to \Lambda_c^+\pi^-)\overline{p}$ favours spin $\frac{1}{2}^+$ for the $\Sigma_c(2455)$, in agreement with the quark model [6]

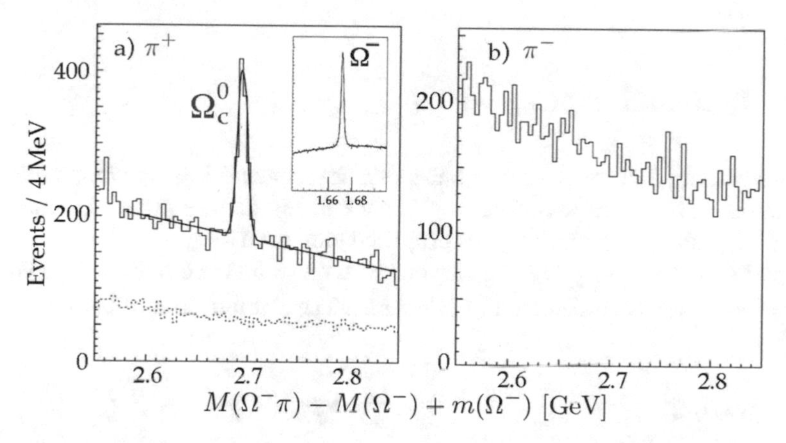

Fig. 17.6 (**a**) Measured invariant mass distribution $M(\Omega^-\pi^+)$ with 725 events. $M(\Omega^-\pi^+)$ refers to the reconstructed mass and $m(\Omega^-)$ to the tabulated Ω^- mass. The dotted histogram is the contribution from the side bands around the $\Omega^- \to \Lambda K^-$ signal shown in the inset (33,880 events); (**b**) the wrong charge distribution $M(\Omega^-\pi^-)$ does not show any signal (adapted from [7])

Two of the three spin-$\frac{1}{2}$ doubly charmed baryons in Fig. 17.1 have been reported, the Ξ_{cc}^+ [8] and the Ξ_{cc}^{++} [9]. The former was seen by SELEX at Fermilab in the $\Lambda_c^+K^-\pi^+$ mass spectrum, using a high energy hyperon beam, the latter by LHCb in the $\Lambda_c^+K^-\pi^+\pi^+$ mass distribution in pp collisions at 13 TeV. Figure 17.7 (left) shows a Feynman graph of the weakly decaying Ξ_{cc}^{++} and Fig. 17.7 (right) the mass distribution. The Λ_c^+ was reconstructed from charged particles emerging from a vertex clearly separated from the pp collision point. The SELEX and LHCb masses lie in the predicted 3500–3700 MeV mass range (see e.g. [10]). However, the Ξ_c^{++}

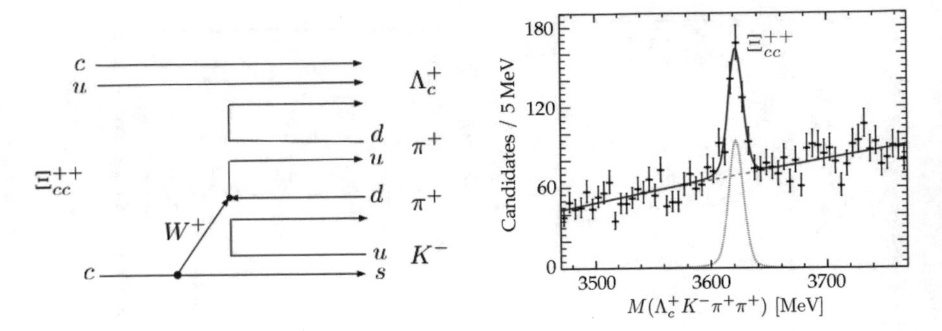

Fig. 17.7 Left: one of the Feynman graphs contributing to the decay of a *ccu* baryon into $\Lambda_c^+ K^- \pi^+ \pi^+$. Right: final state mass distribution and background substracted signal [9]

state lies about 100 MeV above the SELEX one, with a mass splitting far too large for *ucc* and *dcc* isospin partners. The two states need to be confirmed by other experiments before being firmly established.

17.2 Bottom Baryons

When replacing the *c* by a *b* quark one generates identical SU(4)$_f$ weight diagrams for bottom baryons, as was done for mesons in Fig. 8.12. Figure 17.8 shows the weight diagrams of bottom baryons projected onto the axis $i_3 - y$ plane.

Apart from the spin-$\frac{3}{2}$ Ω_b^- the baryons with one *b* and two light (u, d, s) quarks ($B' = -1$) have been detected (Table 17.2). The lightest bottom baryon, the Λ_b^0

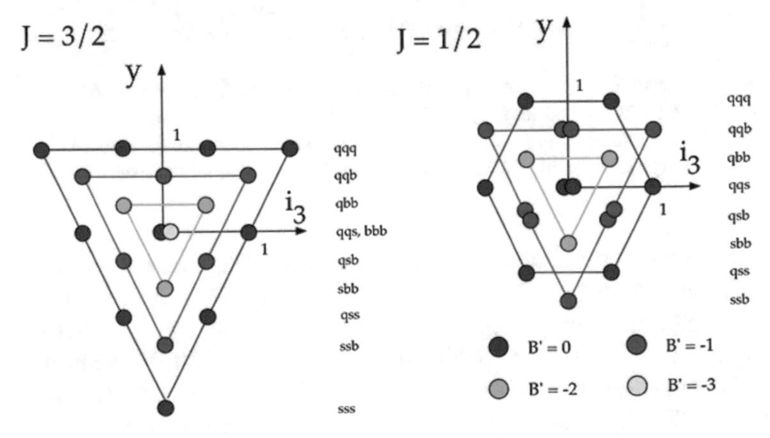

Fig. 17.8 SU(4)$_f$ multiplets of the ground state spin-$\frac{3}{2}$ and $\frac{1}{2}$ baryons including the *b* quark [11]. The hypercharge is defined as $y = B + S + \frac{B'}{3}$ (*q* stands for *u* or *d*)

Table 17.2 Established baryons with one b quark and main decay modes observed so far [2]

$\Lambda_b[5620]$	$\frac{1}{2}^+$	$b \to c$, e.g. $\Lambda_c^+ \ell \nu_\ell X$	0
$\Sigma_b(5811)$	$\frac{1}{2}^+$	$\Lambda_b^0 \pi$	+, 0, −
$\Xi_b(5793)$	$\frac{1}{2}^+$	$\Lambda_b^0 \pi$ and $b \to c$, e.g. $J/\psi\ \Xi$, $pD\overline{K}$	0, −
$\Xi_b'(5935)$	$\frac{1}{2}^+$	$\Xi_b \pi$	0, −
$\Omega_b^-[6046]$	$\frac{1}{2}^+$	$b \to c$, e.g. $J/\psi\ \Omega^- \ldots$	−
$\Sigma_b(5832)$	$\frac{3}{2}^+$	$\Lambda_b^0 \pi$	+, 0, −
$\Xi_b(5950)$	$\frac{3}{2}^+$	$\Xi_b \pi$	0, −
$\Lambda_b(5912)^0$	$\frac{1}{2}^-$	$\Lambda_b^0 \pi\pi$	0
$\Lambda_b(5920)^0$	$\frac{3}{2}^-$	$\Lambda_b^0 \pi\pi$	0

The parity of the b quark is defined to be positive and so is that of the Λ_b^0. The masses of weakly decaying baryons are given in square brackets. The spin-parity assignments follow the quark model and have not been established experimentally. The well-known first excited singlet states are also listed

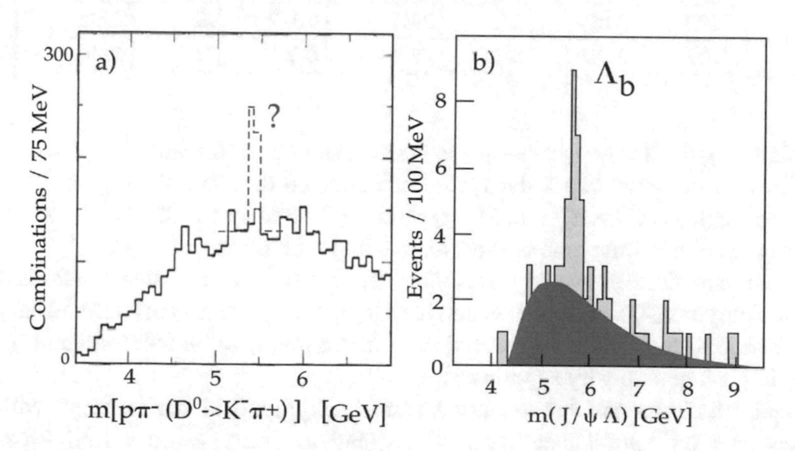

Fig. 17.9 (**a**) Search for the Λ_b in its $D^0 p\pi^-$ decay mode [12]. The dashed peak shows the signal that would be expected from a previous claim [13]; (**b**) observation in the $J/\psi\ \Lambda$ decay mode. The dark (orange) area is the background contribution (adapted from [14])

(udb) was first reported in 1981 at CERN's Intersecting Storage Rings (ISR), decaying into $D^0 p\pi^-$, but the discovery could not be reproduced by another ISR experiment (Fig. 17.9a) [12]. The Λ_b^0 was later established in its $J/\psi\,\Lambda$ decay mode by the UA1 Collaboration at the CERN $\overline{p}p$ collider [14], with 16 ± 5 events above background (Fig. 17.9b). The J/ψ was identified by $\mu^+\mu^-$ pairs and the Λ by its $\pi^- p$ decay mode.

The production of heavy baryons has now become almost routine at the LHC: Fig. 17.10a shows a high statistics sample of Λ_b^0 decays into $J/\psi\ \Lambda$ collected by LHCb (6870 events). Although final states involving the J/ψ are not dominant

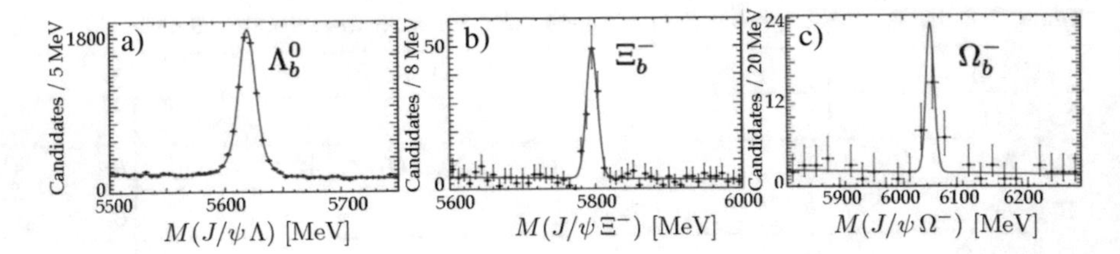

Fig. 17.10 The Λ_b^0, Ξ_b^- and Ω_b^- observed by LHCb in their J/ψ + hyperon decay modes [15]

Table 17.3 Mean lives of "stable" hyperons and heavy baryons [2]

	M [MeV]	τ [ps]		M [MeV]	τ [ps]		M [MeV]	τ [ps]
Λ	1116	263	Λ_c^+	2286	0.200	Λ_b^0	5620	1.47
Σ^+	1189	80	–			–		
Σ^-	1197	148	–			–		
Ξ^0	1315	290	Ξ_c^+	2468	0.442	Ξ_b^-	5793	1.56
Ξ^-	1322	164	Ξ_c^0	2471	0.112	Ξ_b^0	5792	1.46
Ω^-	1672	82	Ω_c^0	2695	0.268	Ω_b^-	6046	1.57

(Table 17.2), the decay $J/\psi \to \mu^+\mu^-$ is a clean and convenient trigger to enhance the fraction of heavy quark decays in the collected data sample. Figure 17.10b, c show the signals of the Ξ_b^- and Ω_b^- hyperons collected by LHCb, which have led to more precise measurements of their masses and widths [15].

The mean lives of weakly decaying heavy baryons are listed in Table 17.3. Bottom baryons (like bottom mesons) have longer mean lives than charmed baryons. This is due to the Cabibbo favoured $c \to s$ transition, while the dominant $b \to t$ coupling is kinematically suppressed.

Heavy hadrons can be separated from background in high energy colliders thanks to their long mean lives, which lead to decay vertices well separated from the beam-beam collision point. This is illustrated by the first observation by CMS of the $\Xi_b(5950)^0$, decaying into the $\Xi_b(5793)^-\pi^+$ [16]. The baryon cascade $\Xi_b(5793)^- \to \Xi^- \to \Lambda \to p$ can then be reconstructed by detecting the secondary vertices that are well separated from the primary collision point (Fig. 17.11a). High resolution position detectors located as close as possible to the beam-beam interaction region are used in collider experiments. For example, the CMS barrel pixel detector was the innermost device, consisting of three 53 cm long cylindrical layers at about 4, 7 and 10 cm from the beam axis. The barrel detector had 48 millions $150\times100\,\mu\text{m}^2$ silicon pixels with which the CMS tracker could reconstruct particle trajectories with a resolution of $\sigma \simeq 12\,\mu\text{m}$. Figure 17.11b shows a photograph of a half shell pixel detector.

The $\Xi_b(5793)^-$ signal was enhanced by triggering on the $J/\psi \to \mu^+\mu^-$ decay products. Figure 17.12a shows the $\Xi_b(5793)^-$ peak in the $J/\psi\ \Xi^-$ invariant mass and Fig. 17.12b the signal from the $\Xi_b(5950)^0$ (21 events).

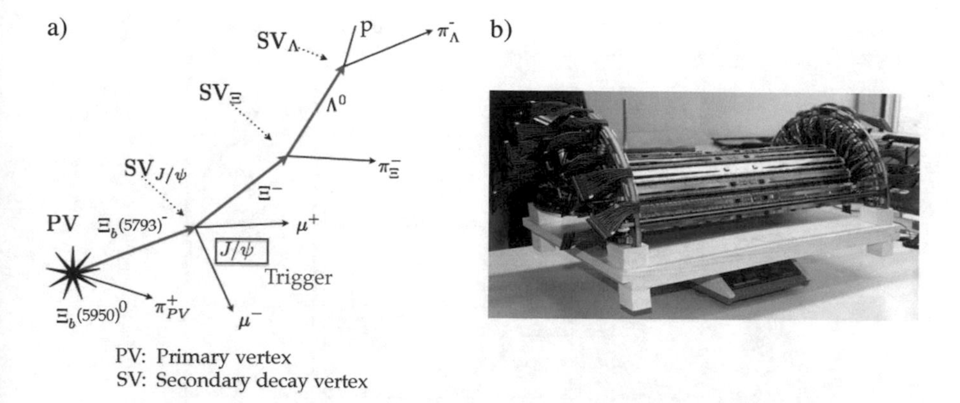

Fig. 17.11 (**a**) Decay cascade $\Xi_b(5950)^0 \rightarrow p\pi^+2\pi^-J/\psi(\rightarrow \mu^+\mu^-)$; (**b**) the lightweight (2.6 kg) half-shell CMS pixel detector (see the text) [17] (©SISSA Medialab Srl. Reproduced with permission of IOP publishing. All rights reserved)

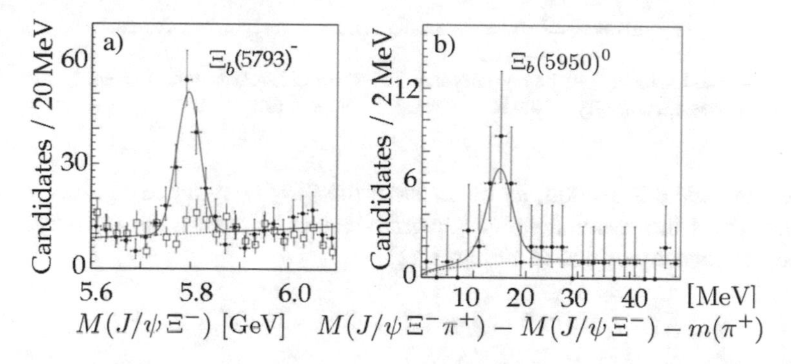

Fig. 17.12 (**a**) $J/\psi(\rightarrow \mu^+\mu^-)\Xi^-$ invariant mass distribution in pp collisions at 7 TeV with the $\Xi_b(5793)^-$ signal. The open (red) squares show the background from events associated with a wrong charge (positive) pion from Ξ^- decay; (**b**) observation of the $\Xi_b(5950)^0$ decaying into $\Xi_b^-\pi^+$. The dotted curve shows the background [16]

It is interesting to compare the mass splittings between strange baryons ($S = -1, -2, -3$) with those in which one s quark is replaced by a c or b quark (Fig. 17.13). It appears that the potential is only weakly flavour dependent, a feature that we have already encountered for mesons bound by the gluon exchange potential (Fig. 9.10). For example, the mass difference between the Ξ_b and the Λ_b is roughly the same as that between the Ξ_c and the Λ_c, the mass splitting between the spin-$\frac{1}{2}$ Ω_b and the Λ_b close to that between the spin-$\frac{1}{2}$ Ω_c and the Λ_c. The spin-$\frac{3}{2}$ states are heavier than the spin-$\frac{1}{2}$ ones, in agreement with expectations from the spin-spin force.

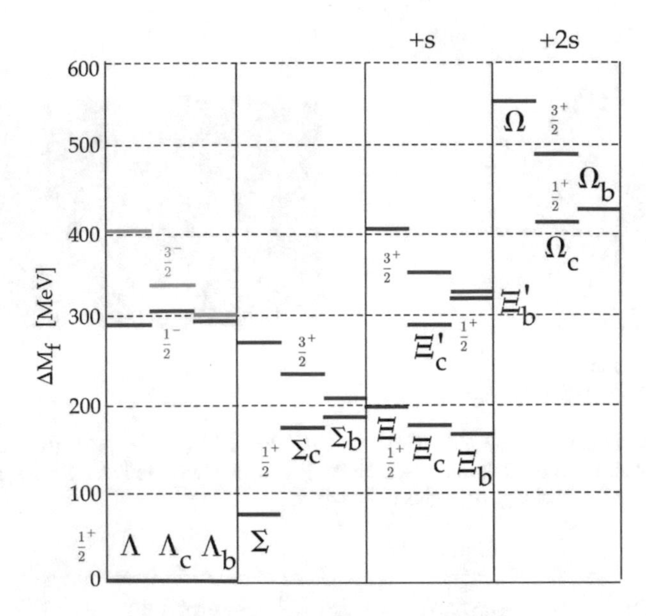

Fig. 17.13 Mass splittings of heavy baryons. The mass differences with respect to the ground states do not depend strongly on the heavy quark flavour s, c or b

To conclude this section, let us consider the $SU(5)_f$ symmetry group with five quarks. The 4-dimensional weight diagram cannot be drawn, but for baryons one expects the decomposition (Problem 17.1)

$$5 \times 5 \times 5 = 10^* + 40 + 40 + 35. \qquad (17.5)$$

The decuplet is not realized in the ground state. The two 40-plets have mixed symmetry ($s = \frac{1}{2}$) and the 35-plet is symmetric ($s = \frac{3}{2}$). The charmed and bottom $SU(4)_f$ multiplets contain 32 spin-$\frac{1}{2}$ states (8 + 2 × 9 + 2 × 3) and 30 spin-$\frac{3}{2}$ ones (10 + 2 × 6 + 2 × 3 + 2). Thus 13 ground state baryons containing both b and c quarks are predicted (8 spin-$\frac{1}{2}$ and 5 spin-$\frac{3}{2}$), yet to be discovered.

References

1. Amsler, C., DeGrand, T., Krusche, B.: In: Tanabashi, M., et al. (Particle Data Group): Phys. Rev. D 98, 030001 (2018), p. 287
2. Patrignani, C., et al. (Particle Data Group): Chin. Phys. C 40 100001 (2016); 2017 update
3. Cazzoli, E.G., et al.: Phys. Rev. Lett. 34, 1125 (1975)
4. Knapp, B., et al.: Phys. Rev. Lett. 37, 882 (1976)
5. Ammar, R., et al.: Phys. Rev. Lett. 86, 1167 (2001)
6. Aubert, B., et al.: Phys. Rev. D 78, 112003 (2008)
7. Solovieva, E., et al.: Phys. Lett. B 672, 1 (2009)

8. Mattson, M., et al.: Phys. Rev. Lett. 89, 112001 (2002)
9. Aaij, R., et al.: Phys. Rev. Lett. 119, 112001 (2017)
10. Karliner, M., Rosner, J.L.: Phys. Rev. D 90, 094007 (2014)
11. Amsler, C.: Nuclear and Particle Physics. IOP Publishing, Bristol (2015)
12. Drijard, D., et al.: Phys. Lett. B 108, 361 (1982)
13. Basile, M., et al.: Lett. Nuovo Cimento 31, 97 (1981)
14. Albajar, C., et al.: Phys. Lett. B 273, 540 (1991)
15. Aaij, R., et al.: Phys. Rev. Lett. 110, 182001 (2013)
16. Chatrchyan, S., et al.: Phys. Rev. Lett. 108, 252002 (2012)
17. Amsler, C., et al.: J. Instrum. 4, P05003 (2009)

Chapter 18
Decay Angular Distribution and Spin

A general method will be described to derive the angular distribution in the two-body decay of a particle with spin. Rotations of systems with angular momenta have already been introduced in Sect. 6.1. The operator

$$U = \mathrm{e}^{+i\vec{J}\cdot\vec{\chi}} \tag{18.1}$$

transforms the wavefunction of a system with spin $\vec{J}$ into the wavefunction expressed in a new coordinate system rotated by the angle χ about the direction $\vec{\chi}$. This transformation is referred to as a passive rotation, in contrast to active rotations in which the physical system itself is rotated in a fixed reference frame, in which case the operator reads

$$U = \mathrm{e}^{-i\vec{J}\cdot\vec{\chi}}. \tag{18.2}$$

The difference between passive and active rotations is illustrated in Fig. 18.1. Let us consider a passive rotation with the three Euler angles α, β and γ from the initial coordinate system Σ_1 into the final system Σ_4 (Fig. 18.2). Every rotation in the 3-dimensional space can be expressed in terms of three successive rotations:

1. a rotation about z_1 by the angle α ($\Sigma_1 \to \Sigma_2$, i.e. $x_1 \to x_2$, $y_1 \to y_2$, $z_1 = z_2$);
2. a rotation about y_2 by the angle β ($\Sigma_2 \to \Sigma_3$, $z_2 \to z_3$);
3. a rotation about z_3 by the angle γ ($\Sigma_3 \to \Sigma_4$).

These rotations are represented by the operator

$$U = \mathrm{e}^{iJ_{z_3}\gamma}\mathrm{e}^{iJ_{y_2}\beta}\mathrm{e}^{iJ_{z_1}\alpha} \tag{18.3}$$

© The Author(s) 2018

C. Amsler, *The Quark Structure of Hadrons*, Lecture Notes in Physics 949,
https://doi.org/10.1007/978-3-319-98527-5_18

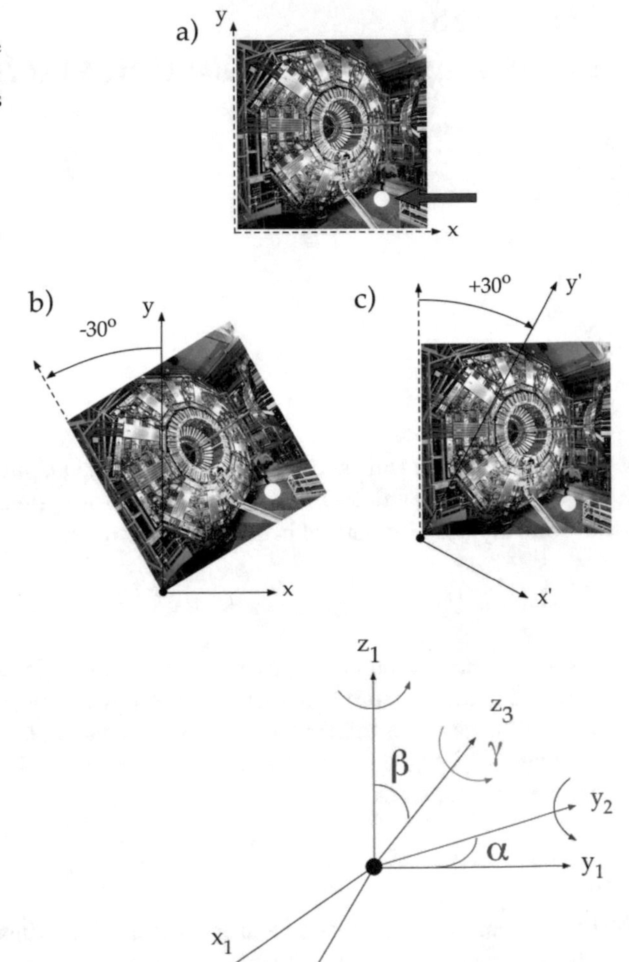

Fig. 18.1 Rotation of the CMS detector (**a**) by an active rotation of $-30°$ (**b**) in which the original reference frame is kept, or by a passive rotation by $+30°$ (**c**). The coordinates x' and y' refer to the rotated coordinate system. Note that the white dot (red arrow) gets the same coordinates after either rotation (image credit CERN)

Fig. 18.2 Passive rotations of the coordinate system by the three Euler angles α, β and γ

acting on the angular momentum wavefunction. However, each of the three terms operates in a different coordinate system. They can be expressed in Σ_1 with the usual transformation rule for operators, namely:

$$
\begin{aligned}
e^{iJ_{z_3}\gamma} &= e^{iJ_{y_2}\beta}\, e^{iJ_{z_2}\gamma}\, e^{-iJ_{y_2}\beta}\ (\Sigma_2 \to \Sigma_3),\\
e^{iJ_{y_2}\beta} &= e^{iJ_{z_1}\alpha}\, e^{iJ_{y_1}\beta}\, e^{-iJ_{z_1}\alpha}\ (\Sigma_1 \to \Sigma_2),\\
e^{iJ_{z_2}\gamma} &= e^{iJ_{z_1}\alpha}\, e^{iJ_{z_1}\gamma}\, e^{-iJ_{z_1}\alpha}\ (\Sigma_1 \to \Sigma_2).
\end{aligned}
\tag{18.4}
$$

Substituting into (18.3) gives with the help of Fig. 18.3

$$
U = e^{iJ_{z_1}\alpha} e^{iJ_{y_1}\beta} e^{iJ_{z_1}\gamma}. \tag{18.5}
$$

Fig. 18.3 Transformation of the exponents of operator (18.3) into (18.5)

$$(y_2\beta)(z_2\gamma)(-y_2\beta)\,|\,(z_1\alpha)(y_1\beta)(-z_1\alpha)\,|\,(z_1\alpha)$$

$$(z_1\alpha)(-y_1\beta)(-z_1\alpha)$$

$$(z_1\alpha)(z_1\gamma)(-z_1\alpha)$$

$$(z_1\alpha)(y_1\beta)(-z_1\alpha)$$

$$= (z_1\alpha)(y_1\beta)(z_1\gamma)$$

Let us replace α by ϕ, the azimuthal angle of the new quantization axis and β by θ its polar angle. The angle γ, which stands for a rotation around the quantization, axis can be ignored as it is not observable. We also drop the subscript "1" and express $U = \mathrm{e}^{iJ_z\phi}\mathrm{e}^{iJ_y\theta}$ as a matrix in the $2j+1$ dimensional space. As already discussed in Sect. 6.1 this is done by expanding the exponential functions as Taylor series and computing the matrix elements of the operators J_z and J_y.

The formalism most frequently used [1] is based on active rotations which are represented by the operator $\mathrm{e}^{-iJ_{z_1}\alpha}\mathrm{e}^{-iJ_{y_1}\beta}\mathrm{e}^{-iJ_{z_1}\gamma}$. The conversion between passive and active rotations is achieved by flipping the signs of the three Euler angles, hence $\theta \to -\theta$ and $\phi \to -\phi$. For instance, by using the Pauli matrices in (6.5) for $j = \frac{1}{2}$, one gets for the matrix representation of $\mathrm{e}^{-iJ_y\theta}$, see (6.7),

$$\langle m|\mathrm{e}^{-iJ_y\theta}|m'\rangle = \begin{pmatrix} \cos\frac{\theta}{2} & -\sin\frac{\theta}{2} \\ \sin\frac{\theta}{2} & \cos\frac{\theta}{2} \end{pmatrix} \equiv d^{\frac{1}{2}}_{mm'}(\theta). \tag{18.6}$$

The operator $\mathrm{e}^{-iJ_z\phi}\mathrm{e}^{-iJ_y\theta}$ is represented by a unitary matrix:

$$\mathcal{D}^j_{mm'}(\theta,\phi) \equiv \sum_{n=1}^{2j+1} \langle m|\mathrm{e}^{-iJ_z\phi}|n\rangle\langle n|\mathrm{e}^{-iJ_y\theta}|m'\rangle = \mathrm{e}^{-im\phi}\, d^j_{mm'}(\theta)\,, \tag{18.7}$$

since the first matrix term is diagonal. Note that the first row and first column in $d^j_{mm'}$ refer to the highest spin projection $m = j$, a convention that is not unique in the literature. The d-matrices for $j \leq 2$ are listed in Table 6.3. A passive rotation is represented by the conjugate transposed matrix:

$$[\mathcal{D}^j_{mm'}(\theta,\phi)]^{*T} = \mathcal{D}^{j*}_{m'm}(\theta,\phi) = \mathrm{e}^{im\phi}\,[d^j_{mm'}(\theta)]^T = \mathrm{e}^{im\phi}\, d^j_{m'm}(\theta) = \mathcal{D}^j_{mm'}(-\theta,-\phi). \tag{18.8}$$

Indeed, $d^j_{mm'}(-\theta) = d^j_{m'm}(\theta)$, as can be verified directly from the explicit d-matrices expressions or with formula (6.9).

Consider now the decay of particle A with spin j into two particles B and C with spins s_1 and s_2, respectively. We choose the quantization axis z along the flight direction of one of the daughters, say B (Fig. 18.4). Let M be the spin projection of A along z. Such a state is said to be pure, corresponding to an ensemble of

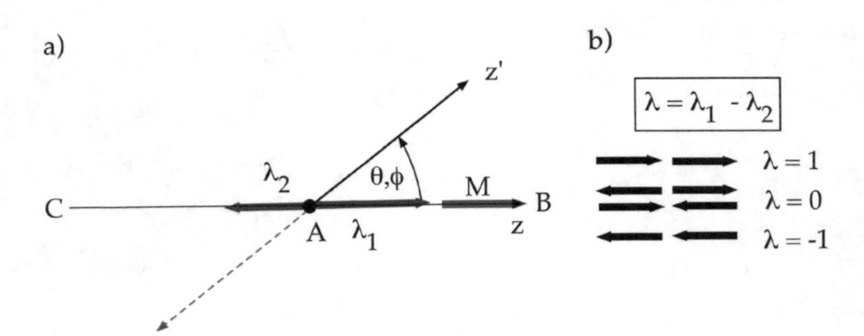

Fig. 18.4 (**a**) Decay of a particle A with spin projection M into B and C. The z-axis is along the flight direction of B, while the z'-axis is along the flight direction of A in the laboratory. The dashed pink arrow shows the flight direction of the laboratory observed in the rest frame of A. The parameter λ is defined as the difference between the two helicities; (**b**) decay into two spin-$\frac{1}{2}$ particles

$P = 100\%$ polarized particles. The projections m_ℓ of the angular momenta vanish along z for both B and C. Since total angular momentum is conserved, M is equal to the sum of the spin projections m_1+m_2, which is equal to the difference $\lambda = \lambda_1-\lambda_2$ of their helicities. (Recall that the helicity is the projection of the spin of a particle on its flight direction.) Hence $j \geq M = \lambda \geq -j$ and the system is described by the $2j+1$ dimensional unit spinor. The spins of B and C couple to the total spin s with $|s_1 - s_2| \leq s \leq s_1 + s_2$, which in turn couples with the angular momentum ℓ to j with $|\ell - s| \leq j \leq \ell + s$ and $m_\ell = 0$. The transition is described by the helicity amplitude

$$\boxed{T_{\lambda_1,\lambda_2} = \sum_{\ell,s} \alpha_{\ell s} \langle j\lambda|\ell s 0 \lambda\rangle \langle s\lambda | s_1 s_2 \lambda_1, -\lambda_2\rangle}\,, \tag{18.9}$$

where the parameters $\alpha_{\ell s}$ depend on the physics process. They can in principle be predicted for electromagnetic and weak decays, while for strong decays they are obtained from fits to experimental distributions.

Let us perform a passive rotation of the coordinate system by the angles θ and ϕ so that the new z-axis (denoted by z' in Fig. 18.4) coincides with the flight direction of A in some reference frame, such as the laboratory (that is A observes the laboratory flying in the $-z'$ direction). Particle B is emitted under the angles $-\theta$ and $-\phi$ in the rest frame of A. The transition amplitude is represented by the conjugate transposed rotation matrix $\mathcal{D}^{j*}_{m'm}(\theta, \phi) = \mathrm{e}^{im\phi}\, d^j_{m'm}(\theta)$, see (18.8), hence

$$\boxed{\mathcal{M}_{\lambda_1,\lambda_2;M} = \mathcal{D}^{j*}_{\lambda M}(\theta, \phi)\, T_{\lambda_1,\lambda_2} = \mathrm{e}^{iM\phi} d^j_{\lambda M}(\theta)\, T_{\lambda_1,\lambda_2}}\,. \tag{18.10}$$

The matrix $\mathcal{M}$ has $(2s_1 + 1)(2s_2 + 1)$ rows for the final state helicities of the daughters B and C, and $2j + 1$ columns for the spin projections M of particle A.

To compute the angular distribution one needs to specify the initial state polarization, while we have assumed so far the state A to be 100% polarized. The polarization state of A is described by a $(2j+1) \times (2j+1)$ initial spin density matrix ρ_i, which we take as diagonal:

$$\rho_i = \begin{pmatrix} p_j & 0 & \dots & 0 \\ 0 & p_{j-1} & \dots & 0 \\ \dots & \dots & \dots & \dots \\ 0 & \dots & ..0 & p_{-j} \end{pmatrix} \quad \text{with} \quad \sum_{M=-j}^{j} p_M = 1 \tag{18.11}$$

The system is said to be mixed.[1] The diagonal elements ρ_{MM} are the probabilities p_M to find the spin projection M of A, with to $j > M > -j$. The density matrix of the final state is given by

$$\boxed{\rho_f = \mathcal{M}\rho_i\mathcal{M}^\dagger} . \tag{18.12}$$

where $\mathcal{M}^\dagger$ is the conjugate transpose matrix of (18.10). The angular distribution of particle B in the rest frame of A is described by the function

$$\boxed{w(\theta, \phi) = \mathrm{Tr}\, \rho_f = \mathrm{Tr}\ (\mathcal{M}\rho_{\mathrm{i}}\mathcal{M}^\dagger)} , \tag{18.13}$$

up to a normalisation constant so that $\int_{4\pi} w(\theta, \phi) d\Omega = 1$. In the simple case of a pure state M the ρ_i is diagonal with all elements vanishing, but $\rho_{MM} = 1$. As shown in the examples below, often only one value of $\alpha_{\ell s}$ in (18.9) is possible in strong interactions, owing to P, C and total angular momentum conservations, in which case $\alpha_{\ell s}$ is absorbed into the normalisation constant. The expectation value of an operator O—such as the spin $\vec{s}$—is given by:

$$\boxed{\langle O \rangle = \frac{\mathrm{Tr}\,(\rho_f O)}{\mathrm{Tr}\, \rho_f}} . \tag{18.14}$$

Before illustrating the formalism with concrete examples, let us describe the procedure to derive the angular distributions for decay chains. The total transition matrix is obtained by multiplying the transition matrices for each decay. For example, for the decay chain shown in Fig. 18.5a the total transition matrix is

$$\mathcal{M} = \mathcal{M}(D)\mathcal{M}(B)\mathcal{M}(A). \tag{18.15}$$

There are four final state particle, C, E, F, G, hence $\mathcal{M}$ is a matrix with $(2s_C + 1)(2s_E+1)(2s_F+1)(2s_G+1)$ rows for the final state helicities and $(2j+1)$ columns

[1]For a discussion on pure and mixed states and on the density matrix, see the Appendix F.

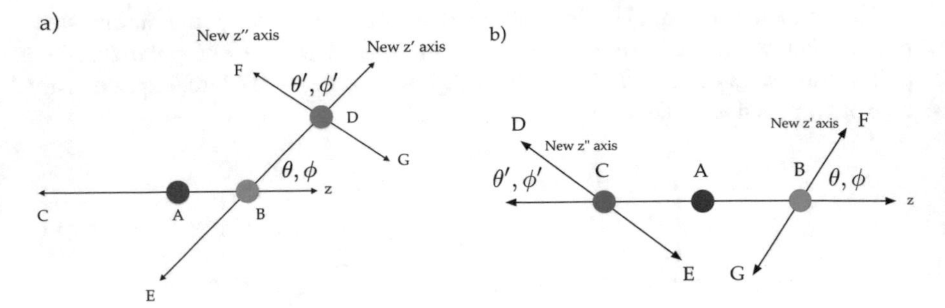

Fig. 18.5 (**a**) Decay cascade $A \to BC$, $B \to DE$, $D \to FG$; (**b**) decay $A \to BC$, $B \to FG$, $C \to DE$

for the spin multiplicity of A. The first coordinate system is the one in which A is at rest, with known polarization described by the density matrix ρ_i. One then lies the z-axis in the direction of one of the daughters, say B, and calculates $\mathcal{M}(A)$, which is the helicity amplitude (18.9). The axes are then rotated by the angles θ and ϕ and a Lorentz boost is performed into the rest frame of B, where one chooses the z-axis in the flight direction of D. One calculates the helicity amplitudes for the decay $B \to D + E$, before applying a rotation into the flight direction of B and computing $\mathcal{M}(B)$ with (18.10). The procedure is repeated for the decay $D \to F + G$. Each matrix $\mathcal{M}(k)$ in the chain depends on angles θ_k and ϕ_k, which are calculated in the rest frame of the corresponding two-body decay. The probability to observe the final state is then proportional to the weight $w(\theta_k, \phi_k) = \mathrm{Tr}\, \rho_f = \mathrm{Tr}\ (\mathcal{M}\rho_{\mathrm{i}}\mathcal{M}^{\dagger})$.[2]

For the decay chain shown in Fig. 18.5b the transition matrix is

$$\mathcal{M} = [\mathcal{M}(B) \otimes \mathcal{M}(C)]\mathcal{M}(A), \tag{18.16}$$

which is a matrix with $(2s_F + 1)(2s_G + 1)(2s_D + 1)(2s_E + 1)$ rows and $(2j + 1)$ columns. The symbol $\otimes$ denotes a tensor (Kronecker) product in which the matrix elements are multiplied sequentially one by one. For complicated decays chains the arithmetic becomes tedious and the weight should be calculated with a computer program [2].

Let us now illustrate the helicity formalism with a few examples (see also Problems 18.1 and 18.2).

[2]The weight w describes the angular distributions only. The final state probability is obtained by multiplying with phase space factors and Breit-Wigner amplitudes for all decay steps.

18.1 Angular Distribution in $\Lambda \to p\pi^-$

Let us derive the angular distribution (14.22) with the helicity formalism. The helicity amplitudes of the proton are given by (14.18) and (14.19),

$$T_{\frac{1}{2}} = \alpha_0 - \frac{1}{\sqrt{3}}\alpha_1 \text{ and } T_{-\frac{1}{2}} = \alpha_0 + \frac{1}{\sqrt{3}}\alpha_1, \tag{18.17}$$

where we have dropped the subscript $\lambda_2 = 0$ for the pion helicity. According to (18.10) the transition matrix is

$$\mathcal{M}_{\lambda_1;M} = e^{iM\phi} d^j_{\lambda_1 M}(\theta)\, T_{\lambda_1} = \begin{pmatrix} e^{i\frac{\phi}{2}} \cos\frac{\theta}{2}[\alpha_0 - \frac{1}{\sqrt{3}}\alpha_1], & -e^{-i\frac{\phi}{2}} \sin\frac{\theta}{2}[\alpha_0 - \frac{1}{\sqrt{3}}\alpha_1] \\ e^{i\frac{\phi}{2}} \sin\frac{\theta}{2}[\alpha_0 + \frac{1}{\sqrt{3}}\alpha_1], & e^{-i\frac{\phi}{2}} \cos\frac{\theta}{2}[\alpha_0 + \frac{1}{\sqrt{3}}\alpha_1] \end{pmatrix}. \tag{18.18}$$

For a 100% polarized Λ along the z-axis the elements of the initial density matrix are $\rho_{11} = 1$, $\rho_{12} = \rho_{21} = \rho_{-1-1} = 0$. The calculation of the final state density matrix (18.12) is then straightforward. Equation (14.20) is obtained by building the trace of ρ_f, leading to the angular distribution (14.22) of the proton, $w(\theta) \propto 1 + \alpha_\Lambda \cos\theta$.

On the other hand, for unpolarized Λ the initial density matrix is proportional to the unit matrix. The final state density matrix is then given by

$$\rho_f = \mathcal{M}_{\lambda_1;M}\mathcal{M}^\dagger_{\lambda_1;M} = \begin{pmatrix} |\alpha_0 - \frac{1}{\sqrt{3}}\alpha_1|^2 & 0 \\ 0 & |\alpha_0 + \frac{1}{\sqrt{3}}\alpha_1|^2 \end{pmatrix}. \tag{18.19}$$

Let us verify that the proton is longitudinally polarized, as advertised by (14.24), and choose the quantization axis along the flight direction of the proton in the Λ rest frame. The expectation value of s_z is, using the Pauli matrix (σ_3) and the parameters $S \equiv \alpha_0$, $P \equiv -\frac{\alpha_1}{\sqrt{3}}$, and α_Λ defined in (14.21):

$$\begin{aligned} \langle s_z \rangle &= \mathrm{Tr}\,\frac{1}{2}\begin{pmatrix} 1 & 0 \\ 0 & -1 \end{pmatrix}\begin{pmatrix} |\alpha_0 - \frac{1}{\sqrt{3}}\alpha_1|^2 & 0 \\ 0 & |\alpha_0 + \frac{1}{\sqrt{3}}\alpha_1|^2 \end{pmatrix}\left[\frac{1}{\mathrm{Tr}\,\rho_f}\right] \\ &= -\mathrm{Re}\,\frac{2\alpha_0^*\alpha_1}{\sqrt{3}}\left[\frac{1}{2(|\alpha_0|^2 + \frac{|\alpha_1|^2}{3})}\right] = \mathrm{Re}\left[\frac{S^*P}{|S|^2 + |P|^2}\right] \\ &= \frac{\alpha_\Lambda}{2}. \end{aligned} \tag{18.20}$$

It is easy to verify that $\langle s_x \rangle = \langle s_y \rangle = 0$ (Problem 18.1), hence the proton is longitudinally polarized with polarization $P_p = \alpha_\Lambda$.

The proton polarization for a Λ hyperon with polarization P_Λ can be computed in a similar way by replacing the diagonal elements in the initial spin density matrix by $\rho_{11} = \frac{1+P_\Lambda}{2}$ and $\rho_{-1-1} = \frac{1-P_\Lambda}{2}$, choosing the quantization axis along the Λ polarization and calculating the expectation value $\langle \vec{s} \cdot \vec{k} \rangle$, where $\vec{k}$ is a unit vector in the direction of the proton momentum. The full formulae are given in [3].

18.2 Proton-Antiproton Annihilation at Rest into $\rho\pi$

Stopping antiprotons are captured in the target to form antiprotonic atoms. As described in Sect. 2.3, antiprotonic hydrogen (protonium) is formed by ejecting the 1s electron, which then de-excites by X-ray emission to the ground state, or by annihilation from one of the atomic nS levels, a process dominant in liquid hydrogen due to Stark collision mixing [4]. In liquid hydrogen proton-antiproton annihilation into mesons occurs from the two hyperfine states n^3S_1 and n^1S_0 with quantum numbers $J^{PC} = 1^{--}$ and 0^{-+}, respectively. Let us now consider proton-antiproton annihilation into $\rho\pi$, followed by $\rho \to \pi\pi$. Annihilation into $\rho^\pm\pi^\mp$ ($\rho^\pm \to \pi^\pm\pi^0$) occurs from the 1^{--} and 0^{-+} states. On the other hand, annihilation into $\rho^0\pi^0(\rho^0 \to \pi^+\pi^-)$ occurs from 1^{--} but not from 0^{-+}, due to C parity conservation in strong interactions.

Let us work out the angular distribution of the π^+ in $\bar{p}p \to \rho^+\pi^-, \rho^+ \to \pi^+\pi^0$. For 1^{--} decay we have $s = 1$ (spin of the ρ^+) and $\ell = 1$ from parity conservation. With $\lambda_2 = 0$ for the π^- one obtains for the helicity amplitude (18.9)

$$T_{\lambda_1} = \alpha_1\langle 1\lambda_1|110\lambda_1\rangle\langle 1\lambda_1|10\lambda_1 0\rangle = -\frac{1}{\sqrt{2}}\alpha_1\lambda_1 \propto \lambda_1. \tag{18.21}$$

The transition matrix has three rows and three columns and reads, when choosing the quantization axis along the momentum vector of the ρ^+:

$$\mathcal{M}_{\lambda_1;M} = \begin{pmatrix} \mathcal{D}^{1*}_{11} & \mathcal{D}^{1*}_{10} & \mathcal{D}^{1*}_{1-1} \\ 0 & 0 & 0 \\ -\mathcal{D}^{1*}_{-11} & -\mathcal{D}^{1*}_{-10} & -\mathcal{D}^{1*}_{-1-1} \end{pmatrix}_{\theta=\phi=0} = \begin{pmatrix} 1\,0\,0 \\ 0\,0\,0 \\ 0\,0\,1 \end{pmatrix}. \tag{18.22}$$

For the subsequent $\rho^+ \to \pi^+\pi^0$ decay we have $j = 1$ ($\ell = 1$ and $s = 0$) and both helicities are equal to 0. The helicity amplitude is a constant. The transition matrix has three columns and one row:

$$\mathcal{M}_M = (\mathcal{D}^{1'*}_{01}, \mathcal{D}^{'1*}_{00}, \mathcal{D}^{1'*}_{0-1}) = \left(\frac{e^{i\phi'}}{\sqrt{2}}\sin\theta', \cos\theta', -\frac{e^{-i\phi'}}{\sqrt{2}}\sin\theta'\right), \tag{18.23}$$

where the apostrophe refers to the angles θ' and ϕ' of the π^+ in the rest frame of the ρ^+. The total transition matrix is

$$\mathcal{M} = \mathcal{M}_M \mathcal{M}_{\lambda_1;M} = \left(\frac{e^{i\phi'}}{\sqrt{2}} \sin\theta', 0, -\frac{e^{-i\phi'}}{\sqrt{2}} \sin\theta' \right). \tag{18.24}$$

A reasonable assumption is that the $^3S_1$ protonium state is not polarized. Therefore the trace of the final density matrix gives simply

$$\mathrm{Tr}\, \mathcal{M}\mathcal{M}^\dagger = \sin^2\theta' \Rightarrow \boxed{w(\theta') = \tfrac{3}{8\pi} \sin^2\theta'} \tag{18.25}$$

for the normalized angular distribution. This means that in the rest frame of the ρ the pions are emitted preferably in the direction perpendicular to the momentum vector of the ρ.

Let us repeat the calculation for the $^1S_0$ protonium state. Here again $\ell = 1$ and $s = 1$ but we have only one initial magnetic substate with $M = j = 0$, hence $\lambda_1 = 0$ (helicity of the ρ). Therefore with $\mathcal{D}^{j=0} \equiv 1$ the transition matrix is

$$\mathcal{M}_{\lambda_1;M} = T_{\lambda_1} \propto \begin{pmatrix} 0 \\ 1 \\ 0 \end{pmatrix}. \tag{18.26}$$

The full transition matrix is obtained by multiplying (18.23) with this last expression:

$$\mathcal{M} = \cos\theta' \Rightarrow \boxed{w(\theta') = \tfrac{3}{4\pi} \cos^2\theta'}. \tag{18.27}$$

The pions are emitted preferably in the direction parallel to the momentum vector of the ρ meson. This example illustrates clearly that the angular distribution depends on how the decaying state has been produced.

18.3 Proton-Antiproton Annihilation at Rest into $\rho^0\rho^0$ ($\rho^0 \to \pi^+\pi^-$)

C parity conservation implies that only the $^1S_0 = 0^{-+}$ protonium state contributes to this decay. On the other hand, *P* parity conservation requires ℓ to be odd. Since $j = 0$ the spins of the ρ do not couple to $s = 0$ or 2 but only to $s = 1$ with $\ell = 1$. We have again only one initial magnetic substate with $M = 0$, therefore $\lambda = 0$, hence $\lambda_1 = \lambda_2$. The helicity amplitude is equal to

$$T_{\lambda_1,\lambda_2} = \alpha_{11} \langle 00|1100\rangle \langle 10|11\lambda_1, -\lambda_1\rangle = -\alpha_{11} \frac{1}{\sqrt{2}} \frac{1}{\sqrt{3}} \lambda_1 \delta_{\lambda_1,\lambda_2} \propto \lambda_1 \delta_{\lambda_1,\lambda_2}. \tag{18.28}$$

Since $\mathcal{D}^{j=0} \equiv 1$ the transition matrix to $\rho^0\rho^0$ is simply

$$\mathcal{M}_{\lambda_1,\lambda_2};0 = \begin{pmatrix} 1 \\ 0 \\ 0 \\ 0 \\ 0 \\ 0 \\ 0 \\ 0 \\ -1 \end{pmatrix}. \tag{18.29}$$

We now deal with the ρ decays. The transition matrix is obtained by building the tensor product of two matrices of the form (18.23):

$$\begin{aligned}\mathcal{M}_M &= (\mathcal{D}_{01}^{1*}, \mathcal{D}_{00}^{1*}, \mathcal{D}_{0-1}^{1*})_{\rho_1} \otimes (\mathcal{D}_{01}^{1*}, \mathcal{D}_{00}^{1*}, \mathcal{D}_{0-1}^{1*})_{\rho_2} \\ &= (\mathcal{D}_{01}^{1*}(\rho_1)\mathcal{D}_{01}^{1*}(\rho_2), \mathcal{D}_{01}^{1*}(\rho_1)\mathcal{D}_{00}^{1*}(\rho_2), \ldots, \mathcal{D}_{0-1}^{1*}(\rho_1)\mathcal{D}_{0-1}^{1*}(\rho_2)). \end{aligned} \tag{18.30}$$

Multiplying with (18.29) gives

$$\begin{aligned}\mathcal{M} &= \mathcal{D}_{01}^{1*}(\rho_1)\mathcal{D}_{01}^{1*}(\rho_2) - \mathcal{D}_{0-1}^{1*}(\rho_1)\mathcal{D}_{0-1}^{1*}(\rho_2) \\ &= \left(\frac{e^{i[\phi_1+\phi_2]}}{2} \sin\theta_1 \sin\theta_2 - \frac{e^{-i[\phi_1+\phi_2]}}{2} \sin\theta_1 \sin\theta_2\right). \\ &= i \sin(\phi_1 + \phi_2) \sin\theta_1 \sin\theta_2, \end{aligned} \tag{18.31}$$

hence the weight

$$w(\theta_1, \phi_1, \theta_2, \phi_2) = \sin^2(\phi_1 + \phi_2) \sin^2\theta_1 \sin^2\theta_2. \tag{18.32}$$

In the rest frames of the ρ mesons the pions fly preferably in the direction perpendicular to the flight direction of the laboratory. Note that the pair of angles refer to different coordinate systems, the rest frames of the two ρ mesons. Thus the preferred azimuthal angle between the flight directions of the meson pairs is $\phi_1 + \phi_2$ = 90°, as illustrated in Fig. 18.6. We have already encountered this situation in the similar π^0 decay into two virtual photons, leading to the final state $2(e^+e^-)$, with which the spin-parity assignment 0^- was established for the neutral pion (Fig. 2.11).

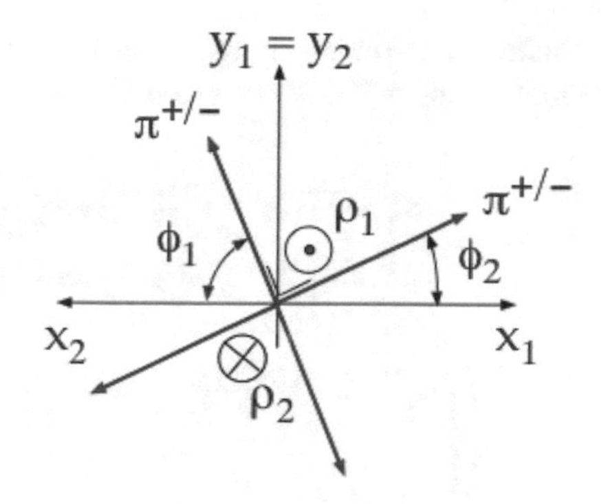

Fig. 18.6 Decay $0^{-+} \to \rho^0\rho^0 \to 2(\pi^+\pi^-)$. Shown is the most likely configuration projected on the plane transverse to the flight directions of the two ρ mesons

18.4 The Spin of the Ω^- Hyperon

Our last example shows how the spin $\frac{3}{2}$ of the Ω^- was established. The Ω^- decays dominantly into ΛK^- ($f = 68\%$). This is a strangeness changing and hence parity violating weak decay. Let us assume that $J_{\Omega^-} = \frac{3}{2}$ and work out the angular distribution of the Λ for a 100% polarized Ω^- in the magnetic state $M = \frac{3}{2}$. Angular momentum conservation allows $\ell = 1$ and 2 (relative P and D waves between the Λ and the K^-). The helicity amplitude of the Λ is

$$T_{\lambda_1} = \alpha_1 \left\langle \frac{3}{2}\lambda_1 \middle| 1\frac{1}{2}0\lambda_1 \right\rangle + \alpha_2 \left\langle \frac{3}{2}\lambda_1 \middle| 2\frac{1}{2}0\lambda_1 \right\rangle, \tag{18.33}$$

hence

$$T_{\frac{1}{2}} = \sqrt{\frac{2}{3}}\alpha_1 - \sqrt{\frac{2}{5}}\alpha_2 \quad \text{and} \quad T_{-\frac{1}{2}} = \sqrt{\frac{2}{3}}\alpha_1 + \sqrt{\frac{2}{5}}\alpha_2. \tag{18.34}$$

The transition matrix is

$$\mathcal{M}_{\lambda_1;M} = \sqrt{2} \times$$
$$\begin{pmatrix} \mathcal{D}^{\frac{3}{2}*}_{\frac{1}{2}\frac{3}{2}}\left[\frac{\alpha_1}{\sqrt{3}} - \frac{\alpha_2}{\sqrt{5}}\right] & \mathcal{D}^{\frac{3}{2}*}_{\frac{1}{2}\frac{1}{2}}\left[\frac{\alpha_1}{\sqrt{3}} - \frac{\alpha_2}{\sqrt{5}}\right] & \mathcal{D}^{\frac{3}{2}*}_{\frac{1}{2}-\frac{1}{2}}\left[\frac{\alpha_1}{\sqrt{3}} - \frac{\alpha_2}{\sqrt{5}}\right] & \mathcal{D}^{\frac{3}{2}*}_{\frac{1}{2}-\frac{3}{2}}\left[\frac{\alpha_1}{\sqrt{3}} - \frac{\alpha_2}{\sqrt{5}}\right] \\ \mathcal{D}^{\frac{3}{2}*}_{-\frac{1}{2}\frac{3}{2}}\left[\frac{\alpha_1}{\sqrt{3}} + \frac{\alpha_2}{\sqrt{5}}\right] & \mathcal{D}^{\frac{3}{2}*}_{-\frac{1}{2}\frac{1}{2}}\left[\frac{\alpha_1}{\sqrt{3}} + \frac{\alpha_2}{\sqrt{5}}\right] & \mathcal{D}^{\frac{3}{2}*}_{-\frac{1}{2}-\frac{1}{2}}\left[\frac{\alpha_1}{\sqrt{3}} + \frac{\alpha_2}{\sqrt{5}}\right] & \mathcal{D}^{\frac{3}{2}*}_{-\frac{1}{2}-\frac{3}{2}}\left[\frac{\alpha_1}{\sqrt{3}} + \frac{\alpha_2}{\sqrt{5}}\right] \end{pmatrix}. \tag{18.35}$$

With the d-functions listed in Table 6.3 the trace of the ΛK^- final state density matrix and a 100% polarized Ω^- is

$$\mathrm{Tr}\,\rho_f = \mathrm{Tr}\begin{pmatrix} \sqrt{3\times 2}\mathrm{e}^{i\frac{3}{2}\phi}\left(\frac{1+\cos\theta}{2}\right)\sin\frac{\theta}{2}\left[\frac{\alpha_1}{\sqrt{3}}-\frac{\alpha_2}{\sqrt{5}}\right] \cdots\cdots \\ \sqrt{3\times 2}\mathrm{e}^{i\frac{3}{2}\phi}\left(\frac{1-\cos\theta}{2}\right)\cos\frac{\theta}{2}\left[\frac{\alpha_1}{\sqrt{3}}+\frac{\alpha_2}{\sqrt{5}}\right] \cdots\cdots \end{pmatrix}\begin{pmatrix} 1\,0\,0\,0 \\ 0\,0\,0\,0 \\ 0\,0\,0\,0 \\ 0\,0\,0\,0 \end{pmatrix}$$

$$\times\begin{pmatrix} \sqrt{3\times 2}\mathrm{e}^{-i\frac{3}{2}\phi}\left(\frac{1+\cos\theta}{2}\right)\sin\frac{\theta}{2}\left[\frac{\alpha_1^*}{\sqrt{3}}-\frac{\alpha_2^*}{\sqrt{5}}\right] & \sqrt{3\times 2}\mathrm{e}^{-i\frac{3}{2}\phi}\left(\frac{1-\cos\theta}{2}\right)\cos\frac{\theta}{2}\left[\frac{\alpha_1^*}{\sqrt{3}}+\frac{\alpha_2^*}{\sqrt{5}}\right] \\ \cdot\cdot & \cdot\cdot \\ \cdot\cdot & \cdot\cdot \\ \cdot\cdot & \cdot\cdot \end{pmatrix}, \tag{18.36}$$

where only the explicit expressions for the non-vanishing terms have been written, hence

$$\mathrm{Tr}\,\rho_f = 6\left(\frac{1+\cos\theta}{2}\right)^2\sin^2\frac{\theta}{2}\left|\frac{\alpha_1}{\sqrt{3}}-\frac{\alpha_2}{\sqrt{5}}\right|^2 +6\left(\frac{1-\cos\theta}{2}\right)^2\cos^2\frac{\theta}{2}\left|\frac{\alpha_1}{\sqrt{3}}+\frac{\alpha_2}{\sqrt{5}}\right|^2. \tag{18.37}$$

After lengthy but straightforward algebraic manipulations one finally obtains the angular distribution

$$w(\theta) = \mathrm{Tr}\,\rho_f \propto (1+\alpha_\Omega\cos\theta)\sin^2\theta, \tag{18.38}$$

where

$$\alpha_\Omega \equiv \frac{2\mathrm{Re}\,P^*D}{|P|^2+|D|^2}, \tag{18.39}$$

with $P \equiv -\frac{\alpha_1}{\sqrt{3}}$ and $D \equiv \frac{\alpha_2}{\sqrt{5}}$. The asymmetry parameter α_Ω violates parity conservation and arises from the interference between P and D waves. Figure 18.7 shows the angular distribution of the Λ for $\alpha_\Omega = 0$ and 1.

We have demonstrated with (14.24) that in unpolarized Λ decay the proton is longitudinally polarized with polarization α_Λ. This would apply equally to the Λ in $\Omega^- \to \Lambda K^-$ decay if the Ω^- had spin-$\frac{1}{2}$ and were unpolarized, hence the Λ polarization would be longitudinal and equal to α_Ω. Let us show that this statement is also true for a spin-$\frac{3}{2}$ Ω^- by replacing the initial density matrix by a unit matrix to describe the unpolarized Ω^-. The rotations matrices are unitary, that is $\sum_M |\mathcal{D}^j_{\lambda M}|^2 = 1$. The density matrix ρ_f then simply reads

$$\rho_f = 2\begin{pmatrix} |\frac{\alpha_1}{\sqrt{3}}-\frac{\alpha_2}{\sqrt{5}}|^2 & 0 \\ 0 & |\frac{\alpha_1}{\sqrt{3}}+\frac{\alpha_2}{\sqrt{5}}|^2 \end{pmatrix}, \tag{18.40}$$

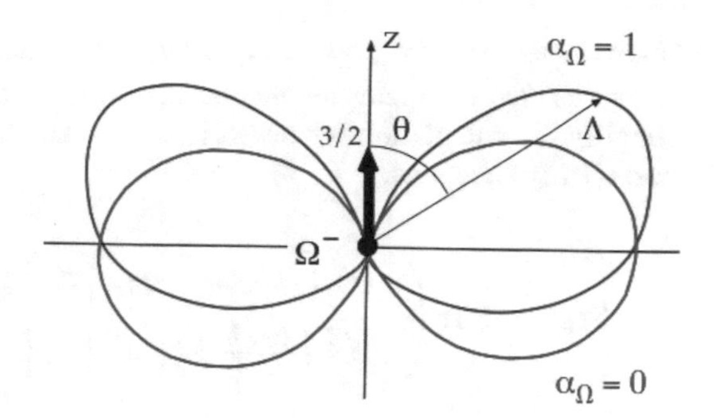

Fig. 18.7 Decay of a spin-$\frac{3}{2}$ hyperon into ΛK^-. The diagram shows the angular distribution of the Λ when the initial hyperon is produced in the $M = +\frac{3}{2}$ state

with trace

$$\mathrm{Tr}\,\rho_f = \frac{4}{3}|\alpha_1|^2 + \frac{4}{5}|\alpha_2|^2. \tag{18.41}$$

By cloning the calculation (18.20) we obtain for the longitudinal spin projection of the Λ, when choosing the quantization axis along its flight direction,

$$\begin{aligned}\langle s_z\rangle &= \mathrm{Tr}\begin{pmatrix}1 & 0\\ 0 & -1\end{pmatrix}\begin{pmatrix}|\frac{\alpha_1}{\sqrt{3}} - \frac{\alpha_2}{\sqrt{5}}|^2 & 0\\ 0 & |\frac{\alpha_1}{\sqrt{3}} + \frac{\alpha_2}{\sqrt{5}}|^2\end{pmatrix}\left[\frac{1}{\mathrm{Tr}\,\rho_f}\right]\\ &= -\mathrm{Re}\,\frac{4\alpha_1^*\alpha_2}{\sqrt{15}}\left[\frac{1}{\frac{4}{3}|\alpha_1|^2 + \frac{4}{5}|\alpha_2|^2}\right] = \frac{\mathrm{Re}\,P^*D}{|P|^2 + |D|^2}\\ &= \frac{\alpha_\Omega}{2}.\end{aligned} \tag{18.42}$$

The polarization of the Λ is indeed equal to α_Ω. The angular distribution of the proton in the rest frame of the Λ is then given by (14.23):

$$w(\theta) = \frac{1}{4\pi}(1 + \alpha_\Omega\alpha_\Lambda\cos\theta), \tag{18.43}$$

from which the asymmetry parameter α_Ω can be determined. The measurement was performed at Fermilab with 800 GeV protons impinging on a copper target [5]. The Ω^- hyperons were produced in the very forward direction to ensure a vanishing polarization. The polarization of the Λ was found to be $\alpha_\Omega = (1.78 \pm 0.25)\%$, small but still consistent with parity violation. The chief reason for this small value is the limited phase space for the decay ($Q = 71$ MeV) which suppresses the D wave contribution. Following (18.38) the decay of the Ω^- is thus nearly parity conserving and the angular distribution reads in good approximation

$$\boxed{w(\theta) = \frac{3}{8\pi}\sin^2\theta} \tag{18.44}$$

(blue curve labelled $\alpha_\Omega = 0$ in Fig. 18.7). This is also true for $M = -\frac{3}{2}$.

Let us also compute the angular distribution for spin-$\frac{3}{2}$ but with $M = \frac{1}{2}$. To simplify the calculation we neglect the very small D wave and set $\alpha_2 \equiv 0$. The trace (18.36) becomes

$$\mathrm{Tr}\,\rho_f = 2\,\mathrm{Tr}\begin{pmatrix} ..\ e^{i\frac{\phi}{2}}\left(\frac{3\cos\theta-1}{2}\right)\cos\frac{\theta}{2}\left[\frac{\alpha_1}{\sqrt{3}}\right] \cdots \\ ..\ e^{i\frac{\phi}{2}}\left(\frac{3\cos\theta+1}{2}\right)\sin\frac{\theta}{2}\left[\frac{\alpha_1}{\sqrt{3}}\right] \cdots \end{pmatrix}\begin{pmatrix} 0\,0\,0\,0 \\ 0\,1\,0\,0 \\ 0\,0\,0\,0 \\ 0\,0\,0\,0 \end{pmatrix}$$

$$\times\begin{pmatrix} .. & .. \\ e^{-i\frac{\phi}{2}}\left(\frac{3\cos\theta-1}{2}\right)\cos\frac{\theta}{2}\left[\frac{\alpha_1^*}{\sqrt{3}}\right] & e^{-i\frac{\phi}{2}}\left(\frac{3\cos\theta+1}{2}\right)\sin\frac{\theta}{2}\left[\frac{\alpha_1^*}{\sqrt{3}}\right] \\ .. & .. \\ .. & .. \end{pmatrix}$$

$$\propto \left[\left(\frac{3\cos\theta-1}{2}\right)^2\cos^2\frac{\theta}{2} + \left(\frac{3\cos\theta+1}{2}\right)^2\sin^2\frac{\theta}{2}\right]$$

$$\propto 1 + 3\cos^2\theta. \tag{18.45}$$

Hence the normalized angular distribution is

$$\boxed{w(\theta) = \frac{1}{8\pi}(1 + 3\cos^2\theta)} \tag{18.46}$$

for both $M = \pm\frac{1}{2}$.

Now assume that $j = \frac{1}{2}$ for the Ω^-. We have seen with (14.23) that the angular distribution of the proton in $\Lambda \to p\pi^-$ is governed by the parity violating parameter α_Λ, multiplied by the Λ polarization. By analogy, for an unpolarized spin-$\frac{1}{2}$ Ω^- decaying into ΛK^- (spins $\frac{1}{2} \to \frac{1}{2} + 0$) the angular distribution of the Λ would be isotropic. The Ω^- would be produced polarized perpendicularly to the scattering plane spanned by its flight direction and that of the incident particle (as required by parity conservation). However, it can be made on average unpolarized when ignoring the orientations of the scattering plane.[3] Figure 18.8 shows the angular distribution of the Λ from an early experiment at the CERN PS. The Ω^- hyperons were produced by a 8.25 GeV/c K^- beam interacting with protons in a 2 m long hydrogen bubble chamber. The average decay length was around 1 cm. The 40 decays plotted in Fig. 18.8 show that the angular distribution is compatible with a spin-$\frac{3}{2}$ Ω^-, mainly produced in the magnetic state $\pm\frac{1}{2}$, and exclude a constant

[3]We now know that the asymmetry parameter α_Ω in $\Omega^- \to \Lambda K^-$ is rather small $(1.78 \pm 0.25)\%$ [5]. Hence the angular distribution of the Λ would be nearly isotropic, even if the Ω^- hyperons were fully polarized.

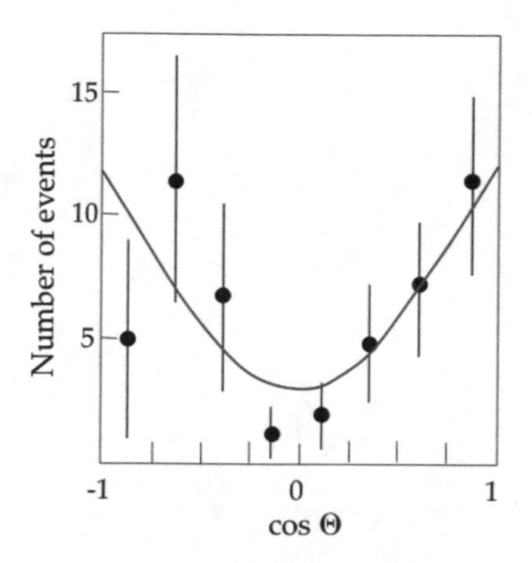

Fig. 18.8 Angular distribution of the Λ in the decay $\Omega^- \to \Lambda K^-$. The blue curve shows a fit to the data with the function $1 + 3\cos^2\theta$ (adapted from [6])

distribution with a probability of 99.7% [6]. The spin of the Ω^- is therefore (at least) equal to $\frac{3}{2}$.

References

1. Tanabashi, M., et al. (Particle Data Group): Phys. Rev. D 98, 030001 (2018)
2. Amsler, C., Bizot, J.C.: Comput. Phys. Commun. 30, 21 (1983)
3. Williams, W.S.C.: Nuclear and Particle Physics, p.155. Oxford University Press, Oxford (1991)
4. Batty, C.: J. Rep. Prog. Phys. 52, 1165 (1989)
5. Lu, L.C., et al.: Phys. Lett. B 617, 11 (2005)
6. Baubillier, M., et al.: Phys. Lett. 78B, 342 (1978)

Correction to: Resonance Analysis

Correction to:
Chapter 12 in: C. Amsler, *The Quark Structure of Hadrons*, Lecture Notes in Physics 949, https://doi.org/10.1007/978-3-319-98527-5_12

The original version of the book was inadvertently published without incorporating the author's proof correction for figure caption 12.2.

The corrected caption now reads as:

Fig. 12.2 Left: predicted $\eta\pi$ and $K\overline{K}$ mass distributions (in arbitrary units, full red curves) for $a_0(980)$ resonance production in $\overline{p}p$ annihilation into $a_0\pi$, according to the Flatté formalism with $g_1 = 324\,\text{MeV}$, $g_2 = 329\,\text{MeV}$ ($\Gamma_0' = 462\,\text{MeV}$). The observed width is about 54 MeV. The dashed blue line shows the $\eta\pi$ mass distribution for the same width Γ_0' in the absence of $K\overline{K}$ decay ($g_2 = 0$). The observed width increases to 300 MeV. Right: the $a_0(980) \to \eta\pi^0$ resonance in proton-antiproton annihilation at rest into $\eta\eta\pi^0$ [4]

The updated version of the chapter can be found at
https://doi.org/10.1007/978-3-319-98527-5_12

© The Author(s) 2018
C. Amsler, *The Quark Structure of Hadrons*, Lecture Notes in Physics 949,
https://doi.org/10.1007/978-3-319-98527-5_19

Appendix A
Landau-Yang Theorem

The theorem states that states with quantum numbers $1^{\pm}, 3^{-}, 5^{-}, 7^{-} \ldots$ do not decay into two spin-1 massless bosons (such as photons or gluons). We reproduce here an elegant proof from Ref. [1]. Consider the decay of a particle with parity P and spin j decaying into two massless spin-1 particles, say photons. The photons are emitted back-to-back and their spin projections (helicities) lie parallel or antiparallel to their flight directions. The photons are right-handed (R) or left-handed (L). The four possible configurations are shown in Fig. A.1. The projection of the total angular momentum onto the z axis is $M = 0$ for RR and LL (the contribution from any orbital angular momentum vanishes along the flight direction) and $M = +2$ and -2 for RL and LR, respectively. The parity eigenstates are

$$\begin{aligned} \psi_+ &= RR + LL & & P = +1, \\ \psi_- &= RR - LL \ \text{ with parity} & & P = -1, \\ \psi' &= RL + LR & & P = +1, \end{aligned} \tag{A.1}$$

since parity reverses the momenta but not the spins: $R \leftrightarrow L$ under P transformation. The superposition $RL - LR$ is antisymmetric under permutations and is forbidden by Bose-Einstein statistics.

The states ψ_+ and ψ_- are invariant under rotations about the z-axis, as well as under rotation of 180° about any axis perpendicular to z (say x). They are rotation eigenfunctions with eigenvalues +1. Therefore the wavefunction of the 2γ pair is given by the spherical harmonics $Y_j^{M=0}(\theta, \phi)$ with j even, since for odd j the eigenvalue under rotations by 180° around the x-axis is -1, $Y_j^{M=0}$ flipping its sign ($\cos\theta \to -\cos\theta$). Therefore, the wavefunction for odd j would have to be of the ψ' type, which has positive parity according to (A.1). Consequently, negative parity

© The Author(s) 2018

C. Amsler, *The Quark Structure of Hadrons*, Lecture Notes in Physics 949,
https://doi.org/10.1007/978-3-319-98527-5

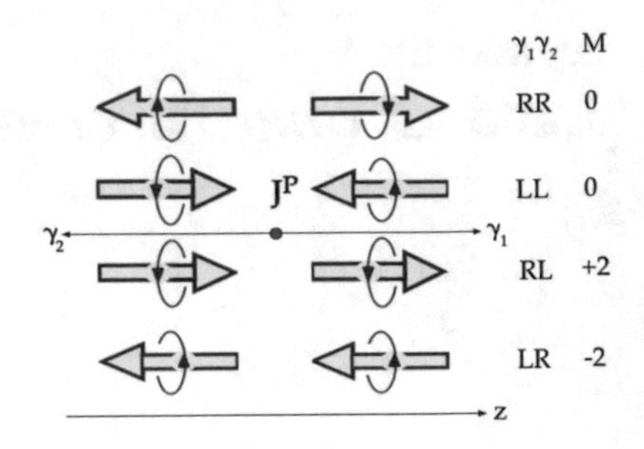

Fig. A.1 Decay of a state with parity P and spin (j, M) into two photons. Photon 1 flies along $+z$, photon 2 along $-z$

odd spin states cannot decay into two photons. In addition, spin-1 states cannot be described by ψ' since $M = \pm 2$ is impossible. Thus positive and negative $j = 1$ states do not decay into two photons.

Appendix B
Invariant Mass Distribution

Consider a three-body system with fixed total energy, e.g. a resonance with mass M decaying into three particles with masses m_1, m_2 and m_3. We have mentioned in Sect. 2.5.1 that phase space distributed events are homogeneously distributed when expressed as a function of two kinetic energies (e.g. T_1 and T_2). The Dalitz plot density is given by [2]

$$d^2\rho = \pi^2 dT_1 dT_2. \tag{B.1}$$

Alternatively, the squared invariant masses (say m_{12}^2 and m_{13}^2) can be used as independent variables. The two-body invariant masses are defined as

$$m_{12}^2 \equiv (P_1 + P_2)^2, \;\; m_{13}^2 \equiv (P_1 + P_3)^2, \;\; m_{23}^2 \equiv (P_2 + P_3)^2, \tag{B.2}$$

where the momenta P_i are 4-vectors, $P_i = (E_i, \vec{p}_i)$. The invariant masses satisfy the relation (Problem 2.1)

$$m_{12}^2 + m_{13}^2 + m_{23}^2 = m_1^2 + m_2^2 + m_3^2 + M^2, \tag{B.3}$$

so that only two of the three invariant masses can be chosen independently.

Let us now verify that the Dalitz plot m_{12}^2 vs. m_{13}^2 is uniformly populated. In the center-of-mass system of the three-body final state (the rest frame of M) the invariant masses read

$$\begin{aligned} m_{12}^2 &= (E_1 + E_2)^2 - (\overbrace{\vec{p}_1 + \vec{p}_2}^{-\vec{p}_3})^2 = (E_1 + E_2)^2 - (\overbrace{\;E_3\;}^{M-E_1-E_2})^2 + m_3^2, \\ m_{13}^2 &= (\underbrace{E_1 + E_3}_{M-E_2})^2 - (\underbrace{\vec{p}_1 + \vec{p}_3}_{-\vec{p}_2})^2 = (M - E_2)^2 - E_2^2 + m_2^2. \end{aligned} \tag{B.4}$$

© The Author(s) 2018
C. Amsler, *The Quark Structure of Hadrons*, Lecture Notes in Physics 949,
https://doi.org/10.1007/978-3-319-98527-5

The density (B.1) becomes in the new variables

$$d^2\rho = \pi^2 dT_1 dT_2 = \pi^2 dE_1 dE_2 = \pi^2 \left| \frac{\partial(m_{12}^2, m_{13}^2)}{\partial(E_1, E_2)} \right|^{-1} dm_{12}^2 dm_{13}^2, \tag{B.5}$$

with

$$\begin{aligned} \frac{\partial(m_{12}^2, m_{13}^2)}{\partial(E_1, E_2)} &= \begin{vmatrix} 2(E_1 + E_2) + 2(M - E_1 - E_2) & \dots \\ 0 & -2(M - E_2) - 2E_2 \end{vmatrix} \\ &= -4M^2, \end{aligned} \tag{B.6}$$

so that finally

$$\boxed{d^2\rho = \frac{\pi^2}{4M^2} dm_{12}^2 dm_{13}^2}\,, \tag{B.7}$$

which proves that the Dalitz plot density expressed in m_{12}^2 vs. m_{13}^2 is homogeneous.

Appendix C
Condon-Shortley-Wigner Convention

The matrix elements of the SU(2) generators (Table 6.2) follow from the commutation relation

$$[J_i, J_j] = i\epsilon_{ijk} J_k. \tag{C.1}$$

A reminder of the proof might be useful. The ladder operators increment or decrement the projection of the angular momentum:

$$\begin{aligned} J_+|jm\rangle &= a|jm+1\rangle, \\ J_-|jm\rangle &= b|jm-1\rangle. \end{aligned} \tag{C.2}$$

The complex constants a and b are obtained from the matrix elements

$$\begin{aligned} \langle jm|J_+^\dagger J_+|jm\rangle &= |a|^2 = \langle jm|J_- J_+|jm\rangle, \\ \langle jm|J_-^\dagger J_-|jm\rangle &= |b|^2 = \langle jm|J_+ J_-|jm\rangle. \end{aligned} \tag{C.3}$$

Noting that

$$\begin{aligned} J_- J_+ &= J_1^2 + J_2^2 + i[J_1, J_2] = \vec{J}^{\,2} - J_3^2 - J_3, \\ J_+ J_- &= J_1^2 + J_2^2 - i[J_1, J_2] = \vec{J}^{\,2} - J_3^2 + J_3, \end{aligned} \tag{C.4}$$

one gets by taking the eigenvalues

$$\begin{aligned} |a|^2 &= j(j+1) - m^2 - m = (j-m)(j+m+1), \\ |b|^2 &= j(j+1) - m^2 + m = (j+m)(j-m+1). \end{aligned} \tag{C.5}$$

© The Author(s) 2018

C. Amsler, *The Quark Structure of Hadrons*, Lecture Notes in Physics 949,
https://doi.org/10.1007/978-3-319-98527-5

The phases being arbitrary the convention is to choose a and b real. Therefore

$$J_+|jm\rangle = \sqrt{(j-m)(j+m+1)}|jm+1\rangle,$$
$$J_-|jm\rangle = \sqrt{(j+m)(j-m+1)}|jm-1\rangle. \tag{C.6}$$

Appendix D
SU(3) Young Tableau and Weight Diagram

In SU(n) a set of $n - 1$ integer numbers is associated to a weight diagram and to its corresponding Young tableau. In SU(3) the two integer numbers p and q indicate the number of times the ladder operator I_+ has to be applied from the left boundary to the right boundary of the weight diagram, p at the top and q at the bottom boundary. SU(3) Young tableaux have at most three rows. The integer p is equal to the number of boxes in the top row exceeding those in the second row, while q refers to the number of boxes in the second row exceeding those in the third row. A few illustrating examples are shown in Fig. D.1.

© The Author(s) 2018

C. Amsler, *The Quark Structure of Hadrons*, Lecture Notes in Physics 949,
https://doi.org/10.1007/978-3-319-98527-5

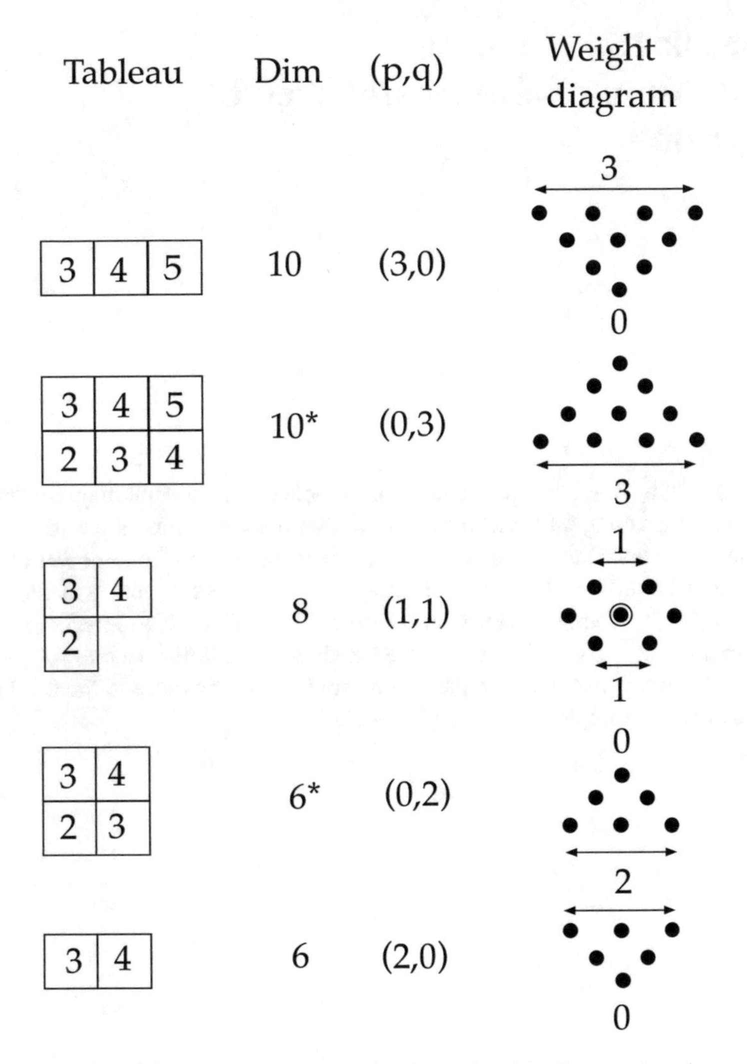

Fig. D.1 Examples of SU(3) tableaux and associated (i_3, y) weight diagrams (see the text)

Appendix E
Expectation Value of the Casimir Operator

The (Casimir) operator

$$\vec{G}^2 = \sum_{i=1}^{8} G_i^2, \tag{E.1}$$

commutes with the SU(3) generators G_i (Chap. 7) and therefore also with the ladder operators $I_\pm$ (7.8) and $U_\pm$, $V_\pm$ (7.9). Hence all members of a given SU(3) multiplet have the expectation value $\langle \vec{G}^2 \rangle$. We start by showing that

$$\vec{G}^2 = \frac{1}{2}\{I_+, I_-\} + \frac{1}{2}\{U_+, U_-\} + \frac{1}{2}\{V_+, V_-\} + G_3^2 + G_8^2, \tag{E.2}$$

where the curly brackets indicate anticommutators. For the first term

$$\frac{1}{2}\{I_+, I_-\} = \frac{1}{2}(G_1 + iG_2)(G_1 - iG_2) + \frac{1}{2}(G_1 - iG_2)(G_1 + iG_2) = G_1^2 + G_2^2 \tag{E.3}$$

and likewise for the other terms in curly brackets:

$$\frac{1}{2}\{U_+, U_-\} + \frac{1}{2}\{V_+, V_-\} = G_4^2 + G_5^2 + G_6^2 + G_7^2, \tag{E.4}$$

which completes the proof. Consider now the states $|\text{max}\rangle$ for which

$$I_+|\text{max}\rangle = V_+|\text{max}\rangle = U_-|\text{max}\rangle = 0. \tag{E.5}$$

© The Author(s) 2018

C. Amsler, *The Quark Structure of Hadrons*, Lecture Notes in Physics 949,
https://doi.org/10.1007/978-3-319-98527-5

Table E.1 SU(3) expectation values of I_3, Y for the states $|\text{max}\rangle$ and multiplet value $\langle \vec{G}^2 \rangle$

Dim	i_3	y	$\langle \vec{G}^2 \rangle$
1	0	0	0
3, 3*	$\frac{1}{2}$	$\pm\frac{1}{3}$	$\frac{4}{3}$
6, 6*	1	$\pm\frac{2}{3}$	$\frac{10}{3}$
8	1	0	3
10	$\frac{3}{2}$	1	6

These are the states on the rightmost edges of the weight diagrams, for instance the u-quark in the fundamental representation, or the Σ^+ in the octet one. Let us calculate with (E.2) the multiplet expectation value $\langle \vec{G}^2 \rangle$ for $|\text{max}\rangle$ states:

$$\begin{aligned} \langle \vec{G}^2 \rangle &= \frac{1}{2}\langle I_+ I_- \rangle + \frac{1}{2}\langle U_- U_+ \rangle + \frac{1}{2}\langle V_+ V_- \rangle + \langle \underbrace{G_3^2}_{I_3^2} \rangle + \langle \underbrace{G_8^2}_{\frac{3}{4}Y^2} \rangle \\ &= \frac{1}{2}\langle [I_+, I_-] \rangle + \frac{1}{2}\langle [U_-, U_+] \rangle + \frac{1}{2}\langle [V_+, V_-] \rangle + \langle I_3^2 \rangle + \frac{3}{4}\langle Y^2 \rangle \\ &= \frac{1}{2}\langle 2I_3 \rangle + \frac{1}{2}\left\langle I_3 - \frac{3}{2}Y \right\rangle + \frac{1}{2}\left\langle I_3 + \frac{3}{2}Y \right\rangle + \langle I_3^2 \rangle + \frac{3}{4}\langle Y^2 \rangle. \end{aligned} \tag{E.6}$$

We have used commutation properties from Table 7.2. Therefore

$$\boxed{\langle \vec{G}^2 \rangle = 2\langle I_3 \rangle + \langle I_3^2 \rangle + \tfrac{3}{4}\langle Y^2 \rangle}\,. \tag{E.7}$$

Table E.1 lists the values of $\langle \vec{G}^2 \rangle$ for various dimensions.

Appendix F
Density Matrix Formalism

F.1 Pure and Mixed States

The concept of pure and mixed states is best explained by considering as concrete example an ensemble of spin-$\frac{1}{2}$ polarized particles. To start with, let us assume that the particles are 100% polarized along the $+z$-direction. The system is said to be in a pure state and is represented by the spinor

$$|\psi_1\rangle = \begin{pmatrix} 1 \\ 0 \end{pmatrix} \tag{F.1}$$

in the 2-dimensional spin space. Let us now perform an active rotation of the physical system so that the spins all point along the direction determined by thc angles θ and ϕ. The full rotation matrix for the spin $j = \frac{1}{2}$ is, according to (18.7) and using the d-functions in Table 6.3,

$$\mathcal{D}^{\frac{1}{2}}_{mm'}(\theta, \phi) = \begin{pmatrix} e^{-i\frac{\phi}{2}} \cos\frac{\theta}{2} & -e^{-i\frac{\phi}{2}} \sin\frac{\theta}{2} \\ e^{i\frac{\phi}{2}} \sin\frac{\theta}{2} & e^{i\frac{\phi}{2}} \cos\frac{\theta}{2} \end{pmatrix}. \tag{F.2}$$

Hence the rotated wavefunction reads

$$\begin{aligned} |\Psi^1\rangle &= \mathcal{D}^{\frac{1}{2}}_{mm'}(\theta, \phi) \begin{pmatrix} 1 \\ 0 \end{pmatrix} = \begin{pmatrix} e^{-i\frac{\phi}{2}} \cos\frac{\theta}{2} \\ e^{i\frac{\phi}{2}} \sin\frac{\theta}{2} \end{pmatrix} \\ &= e^{-i\frac{\phi}{2}} \cos\frac{\theta}{2} \underbrace{|\psi_1\rangle}_{\begin{pmatrix} 1 \\ 0 \end{pmatrix}} + e^{i\frac{\phi}{2}} \sin\frac{\theta}{2} \underbrace{|\psi_2\rangle}_{\begin{pmatrix} 0 \\ 1 \end{pmatrix}}. \end{aligned} \tag{F.3}$$

© The Author(s) 2018

C. Amsler, *The Quark Structure of Hadrons*, Lecture Notes in Physics 949,
https://doi.org/10.1007/978-3-319-98527-5

The wavefunction $|\Psi^1\rangle$ is a superposition of the states $|\psi_1\rangle$ and $|\psi_1\rangle$ and still represents a pure state of $+100\%$ polarization. Similarly, the wavefunction $|\Psi^2\rangle$, representing a rotated pure state with –100% polarization, is

$$
\begin{aligned}
|\Psi^2\rangle = \mathcal{D}^{\frac{1}{2}}_{mm'}(\theta,\phi)\begin{pmatrix}0\\1\end{pmatrix} &= \begin{pmatrix}-\mathrm{e}^{-i\frac{\phi}{2}}\sin\frac{\theta}{2}\\ \mathrm{e}^{i\frac{\phi}{2}}\cos\frac{\theta}{2}\end{pmatrix}\\
&= -\mathrm{e}^{-i\frac{\phi}{2}}\sin\frac{\theta}{2}|\psi_1\rangle + \mathrm{e}^{i\frac{\phi}{2}}\cos\frac{\theta}{2}|\psi_2\rangle. \qquad (F.4)
\end{aligned}
$$

Let us calculate the expectation value of an operator in the $(|\psi_1\rangle, |\psi_2\rangle)$ basis, for example the average value of the spin component s_z in the state Ψ^1:

$$
\begin{aligned}
\langle s_z\rangle = \langle\Psi^1|s_z|\Psi^1\rangle &= \frac{1}{2}\langle\Psi^1|\sigma_z|\Psi^1\rangle\\
&= \frac{1}{2}\left(\mathrm{e}^{i\frac{\phi}{2}}\cos\frac{\theta}{2}, \mathrm{e}^{-i\frac{\phi}{2}}\sin\frac{\theta}{2}\right)\begin{pmatrix}1 & 0\\ 0 & -1\end{pmatrix}\begin{pmatrix}\mathrm{e}^{-i\frac{\phi}{2}}\cos\frac{\theta}{2}\\ \mathrm{e}^{i\frac{\phi}{2}}\sin\frac{\theta}{2}\end{pmatrix}\\
&= \frac{1}{2}\left(\cos^2\frac{\theta}{2} - \sin^2\frac{\theta}{2}\right) = \frac{\cos\theta}{2}. \qquad (F.5)
\end{aligned}
$$

Similarly, for the state $|\Psi^2\rangle$ the average value of $\langle s_z\rangle$ is $\langle\Psi^2|s_z|\Psi^2\rangle = -\frac{\cos\theta}{2}$.

Consider now a polarized spin-$\frac{1}{2}$ system Ψ with incomplete polarization ($P_T < 1$) and denote the probability to find a spin pointing along $+z$ by p_1 and along $-z$ by p_2. The polarization is then equal to $P_T = \frac{p_1 - p_2}{p_1 + p_2}$ with $p_1 + p_2 = 1$. Such a system is said to be in a mixed state $|\Psi\rangle$. The expectation value of s_z is equal to

$$
\langle s_z\rangle = \langle\Psi|s_z|\Psi\rangle = P_T\left(\frac{\cos\theta}{2}\right). \qquad (F.6)
$$

The derivation is most conveniently obtained by introducing the density matrix formalism which is applied here to the particular case of systems with spins [3].

Let us rotate the system with spin j by the angle θ and ϕ. The $2j+1$ pure states $|\Psi^i\rangle$ ($i = 1\ldots 2j+1$) can be written as linear superpositions of the $2j+1$ basis spinors $|\psi_n\rangle$:

$$
|\Psi^i\rangle = \sum_n^{2j+1} a_n^i|\psi_n\rangle, \qquad (F.7)
$$

as in (F.3) or (F.4) for $j = \frac{1}{2}$. For each pure state $|\Psi^i\rangle$ the expectation value of an operator O in the $(|\psi_1\rangle\ldots|\psi_n\rangle)$ basis is then given by

$$
\langle O\rangle^i = \langle\Psi^i|O|\Psi^i\rangle = \sum_{n,m}^{2j+1} a_n^{i*}a_m^i\langle\psi_n|O|\psi_m\rangle. \qquad (F.8)
$$

If the states $|\Psi^i\rangle$ are populated each with the probability p_i, the overall expectation value is equal to

$$\langle O\rangle = \sum_{i=1}^{2j+1} p_i \langle O\rangle^i = \sum_{n,m}^{2j+1}\sum_{i}^{2j+1} p_i a_m^i a_n^{i*} \langle\psi_n|O|\psi_m\rangle = \sum_{n,m}^{2j+1} \rho_{mn} O_{nm}, \tag{F.9}$$

= Tr $(O\rho)$, where we have defined the $(2j+1)\times(2j+1)$ spin density matrix as

$$\rho_{mn} = \sum_{i=1}^{2j+1} p_i a_m^i a_n^{i*}. \tag{F.10}$$

The last equation assumes that the probabilities p_i are normalized so that $\mathrm{Tr}\,\rho = \sum_n \rho_{nn} = 1$. The normalization is guaranteed by writing the expectation value as

$$\langle O\rangle = \frac{\mathrm{Tr}\,(\rho O)}{\mathrm{Tr}\,\rho}. \tag{F.11}$$

The density matrix is Hermitian and positive definite,

$$\rho^\dagger = \rho \Rightarrow \rho_{mn}^* = \rho_{nm}, \quad \rho_{nn} \ge 0, \tag{F.12}$$

and $\rho^2 = \rho$ for a pure state. The proofs are left as simple exercises.

As an application, consider a spin-$\frac{1}{2}$ system. The density matrix is

$$\rho = \begin{pmatrix} p_1|a_1^1|^2 + p_2|a_1^2|^2 & p_1 a_1^1 a_2^{1*} + p_2 a_1^2 a_2^{2*} \\ p_1 a_1^{1*} a_2^1 + p_2 a_1^{2*} a_2^2 & p_1|a_2^1|^2 + p_2|a_2^2|^2 \end{pmatrix}, \tag{F.13}$$

with the parameters a_i from (F.3) and (F.4):

$$a_1^1 = \mathrm{e}^{-i\frac{\phi}{2}} \cos\frac{\theta}{2}, \quad a_2^1 = \mathrm{e}^{i\frac{\phi}{2}} \sin\frac{\theta}{2}, \quad a_1^2 = -\mathrm{e}^{-i\frac{\phi}{2}} \sin\frac{\theta}{2}, \quad a_2^2 = \mathrm{e}^{i\frac{\phi}{2}} \cos\frac{\theta}{2}. \tag{F.14}$$

Thus, the average spin projection $\langle s_z\rangle$ is with the Pauli matrix (σ_3):

$$\begin{aligned}\langle s_z\rangle = \mathrm{Tr}\,(\rho s_z) &= \mathrm{Tr}\begin{pmatrix} \rho_{11} & \rho_{12} \\ \rho_{21} & \rho_{22}\end{pmatrix} \frac{1}{2}\begin{pmatrix} 1 & 0 \\ 0 & -1\end{pmatrix} \\ &= \frac{1}{2}\mathrm{Tr}\begin{pmatrix} \rho_{11} & -\rho_{12} \\ \rho_{21} & -\rho_{22}\end{pmatrix} = \frac{1}{2}(\rho_{11} - \rho_{22}),\end{aligned} \tag{F.15}$$

or with $\rho_{11} + \rho_{22} = p_1 + p_2 = 1$:

$$\langle s_z\rangle = \frac{1}{2}(p_1 - p_2)\cos^2\frac{\theta}{2} + \frac{1}{2}(p_2 - p_1)\sin^2\frac{\theta}{2} = (p_1 - p_2)\frac{\cos\theta}{2} = P_T\left(\frac{\cos\theta}{2}\right), \tag{F.16}$$

which completes the proof of (F.6). For example, the density matrix for a $P_T = 100\%$ polarized proton beam along the z-axis ($p_1 = 1$, $p_2 = 0$, $\theta = \phi = 0$) is diagonal with $\rho_{11} = 1$ and $\rho_{22} = 0$. The average spin projection is $\langle s_z \rangle = \frac{1}{2}$. On the other hand, for a 100% polarized proton beam along the y-axis ($p_1 = 1$, $p_2 = 0$, $\theta = \phi = 90°$), the expectation value of the spin component s_y is with (F.14):

$$\langle s_y \rangle = \mathrm{Tr}\,\frac{1}{2}\begin{pmatrix} 1 & -i \\ i & 1 \end{pmatrix}\frac{1}{2}\begin{pmatrix} 0 & -i \\ i & 0 \end{pmatrix} = \frac{1}{4}\mathrm{Tr}\begin{pmatrix} 1 & -i \\ i & 1 \end{pmatrix} = \frac{1}{2}, \tag{F.17}$$

as expected.

F.2 Angular Distributions in Two-Body Decays

The density matrix (F.10) can be rewritten as

$$\rho = \sum_i p_i \Psi^i \Psi^{i\dagger}. \tag{F.18}$$

The products in the sum are tensor (Kronecker) products in which each element a^i_m of the column vector Ψ^i is multiplied by each element a^{i*}_n of the row vector $\Psi^{i\dagger}$. Let us now consider a transition between the initial state $|i\rangle$ with density matrix ρ_i and a final state $|f\rangle$ with density matrix ρ_f, such as the decay of an initial state A into two particles B and C. We orient the coordinate system in the rest frame of A so that B is emitted under the angles θ and ϕ. The final state wavefunctions are obtained with the transition matrix $\mathcal{M}$:

$$\Psi^i(f) = \mathcal{M}\Psi^i(i), \tag{F.19}$$

each pure state $\Psi^i(i)$ being populated with the probability p_i. The density matrix in the final state reads

$$\rho_f = \sum_i p_i \Psi^i(f)\Psi^{i\dagger}(f) = \sum_i p_i \mathcal{M}\Psi^i(i)[\mathcal{M}\Psi^i(i)]^\dagger = \sum_i p_i \mathcal{M}\Psi^i(i)\Psi^{i\dagger}(i)\mathcal{M}^\dagger$$

$$= \mathcal{M}\left[\sum_i p_i \Psi^i(i)\Psi^{i\dagger}(i)\right]\mathcal{M}^\dagger \;\Rightarrow\; \rho_f = \mathcal{M}\rho_i\mathcal{M}^\dagger. \tag{F.20}$$

The angular distribution of B in the rest frame of A, is described by the weight function

$$w(\theta, \phi) = \mathrm{Tr}\,\rho_f = \mathrm{Tr}\,(\mathcal{M}\rho_i\mathcal{M}^\dagger), \tag{F.21}$$

apart from a normalisation constant, so that $\int_{4\pi} w(\theta, \phi)d\Omega = 1$.

Problems

2.1 Invariant Masses

A particle of mass M decays into three daughters with masses m_1, m_2 and m_3. Show that

$$m_{12}^2 + m_{13}^2 + m_{23}^2 = m_1^2 + m_2^2 + m_3^2 + M^2. \quad \text{(F.22)}$$

3.1 Isospin and Clebsch-Gordan Coefficients

(a) The $f_2(1270)$ meson is an isoscalar meson ($i = 0$). Predict the ratio of partial widths

$$\frac{\Gamma(f_2 \to \pi^+\pi^-)}{\Gamma(f_2 \to \pi^0\pi^0)}. \quad \text{(F.23)}$$

(b) Which isospin values are possible in the reactions $K^- p \to \Sigma^0\pi^0$ and $K^- p \to \Sigma^+\pi^-$?

(c) Predict the ratio of cross sections.

Ignore mass differences within isospin multiplets.

3.2 Pion-Nucleon Scattering

The pion-nucleon system couples to $i = \frac{1}{2}$ and $i = \frac{3}{2}$. With the help of the Clebsch-Gordan coefficients (Fig. 2.8) determine the isospin decomposition of the

© The Author(s) 2018

C. Amsler, *The Quark Structure of Hadrons*, Lecture Notes in Physics 949,
https://doi.org/10.1007/978-3-319-98527-5

three states $|\pi^+ p\rangle$, $|\pi^- p\rangle$ and $|\pi^0 n\rangle$ into $i = \frac{1}{2}$ and $\frac{3}{2}$. Use isospin conservation and charge independence to predict the ratios of cross sections

$$\frac{\sigma_1}{\sigma_2} \equiv \frac{\sigma(\pi^+ p \to \pi^+ p)}{\sigma(\pi^- p \to \pi^- p)},$$
$$\frac{\sigma_1}{\sigma_3} \equiv \frac{\sigma(\pi^+ p \to \pi^+ p)}{\sigma(\pi^- p \to \pi^0 n)}. \tag{F.24}$$

The low energy cross sections are dominated by the excitation of the $(i = \frac{3}{2})$ $\Delta(1232)$ resonance. Predict the ratios of cross sections

$$\frac{\sigma_1}{\sigma_2} \equiv \frac{\sigma(\pi^+ p \to \Delta^{++} \to \pi^+ p)}{\sigma(\pi^- p \to \Delta^0 \to \pi^- p)},$$
$$\frac{\sigma_1}{\sigma_3} \equiv \frac{\sigma(\pi^+ p \to \Delta^{++} \to \pi^+ p)}{\sigma(\pi^- p \to \Delta^0 \to \pi^0 n)}. \tag{F.25}$$

Compare your results to the cross sections measured with a pion beam of $\simeq$190 MeV:

$$\sigma_1 : \sigma_2 : \sigma_3 \simeq 204 : 23 : 47 \text{ mb}. \tag{F.26}$$

4.1 Exotic Quantum Numbers

Show that the following quantum numbers are impossible for quark-antiquark states:

$$J^{PC} = 0^{--}, 0^{+-}, 1^{-+}, 2^{+-}, 3^{-+}.$$

4.2 Nomenclature and Quantum Numbers

Derive from the properties (4.1)–(4.4) the quantum numbers $J^{PC}(i^G)$ and the quark content of the following mesons:

$$a_2^+(1320),\ h_1(1380),\ f_2'(1525),\ \pi_1^0(1400),\ K_1^+(1400),$$
$$\overline{K_0}^*(1430),\ D^-,\ B_{s2}^*(5840)^0,\ \chi_{c1}(1P).$$

5.1 Pseudoscalar Mixing Angle

The following branching ratios have been measured with stopping antiprotons in liquid hydrogen (Z. Phys. C58, 175 (1993)):

$$f(p\bar{p} \to \pi^0\eta) = (2.12 \pm 0.12) \times 10^{-4},$$
$$f(p\bar{p} \to \pi^0\eta') = (1.23 \pm 0.13) \times 10^{-4}.$$

In hydrogen the stopping antiprotons are captured by the proton and build antiprotonic atoms (Fig. 2.7). The $\bar{p}p$ atom then annihilates from one of the atomic states with orbital momentum ℓ.

(a) Consider $\ell = 0, 1$ and 2. Which atomic states ${}^{2s+1}\ell_J$ contribute to $\pi^0\eta$ and $\pi^0\eta'$?
(b) Assume that the dominant relative angular momentum in $\pi^0\eta$ and $\pi^0\eta'$ is $\ell = 0$. Estimate the ratio of intensities $|\langle p\bar{p}|A|\eta'\rangle|^2/|\langle p\bar{p}|A|\eta\rangle|^2$ (where A is the transition operator) by using the wavefunctions of the η and η' as a function of pseudoscalar mixing angle. Then apply the OZI rule and estimate the pseudoscalar mixing angle. Assume that the phase space factor for $\ell = 0$ is proportional to the momentum p in two-body decays.

6.1 3-Dimensional SU(2) Generators

Find the matrix representations of the SU(2) generators for isovector mesons and verify the commutation relations (6.2).

6.2 Kaon Doublets

1. Use the G parity transformations for the three light quarks and Table 6.4 to show that the kaon states transform according to (6.41).
2. Derive the isospin decomposition (6.45).
3. Symmetrize the states (6.45) for $i = 0$ and $i = 1$ for angular momenta $\ell = 0$ (or even) and $\ell = 1$ (or odd).

7.1 Commutation Rules for SU(3) Ladder Operators

The ladder operators obey the rules listed in Table 7.2. Show for example that $[U_-, I_+] = 0$ and $[U_+, U_-] = \frac{3}{2}Y - I_3$.

7.2 Ladder Operators on Antiquarks

Show that the operators $I_\pm$, $V_\pm$ and $U_\pm$ reverse the signs when applied on $\overline{u}$, $\overline{d}$, and $\overline{s}$ antiquarks.

7.3 Radiative Kaon Decay

Predict the ratio

$$R = \frac{\Gamma(K^{*0} \to K^0\gamma)}{\Gamma(K^{*+} \to K^+\gamma)}. \tag{F.27}$$

The experimental value is $R = 2.5 \pm 0.3$. Calculate the ratio $\frac{m_s}{m_u}$ of strange to light quark masses.

8.1 Width of the $\Upsilon(1S)$ Resonance

The width of the $\Upsilon(1S)$ observed with colliding e^+e^- beams is dominated by the energy uncertainty in the beam energies. Figure F.1 shows the cross section for the $\Upsilon(1S)$ decaying into hadrons, measured by the Crystal Ball (Fig. 7.8) at DORIS. The branching ratio for $\Upsilon(1S)$ decay into lepton pairs is $\simeq 7.5\%$.

(a) What are the branching ratios for $\Upsilon(1S)$ decay into e^+e^-, $\mu^+\mu^-$ and in $\tau^+\tau^-$ (ignoring mass differences)?

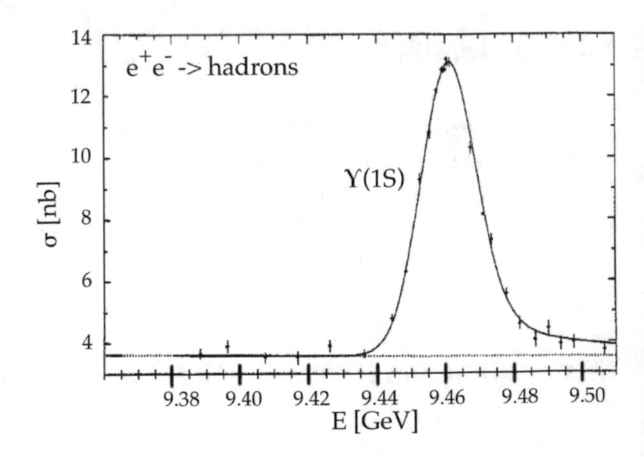

Fig. F.1 The $\Upsilon(1S)$ in $e^+e^- \to \Upsilon(1S) \to$ hadrons, after Kobel, M., et al.: Z. Phys. C 53, 193 (1992)

(b) What is the branching ratio for the decay into hadrons?
(c) What kind of interactions are responsible for the decays into leptons and hadrons?
(d) Estimate from the data in Fig. F.1 the natural width of the $\Upsilon(1S)$ and compare your result with the table value (8.6).

8.2 Van Royen-Weisskopf Formula

Neutral vector mesons V decay into e^+e^- (or $\mu^+\mu^-$) pairs with the partial widths

$$\Gamma(V \rightarrow e^+e^-) = \frac{16\pi\alpha^2}{M_V^2} Q^2 |\Psi(0)|^2, \tag{F.28}$$

where Q is the average charge of the quark flavour and $\Psi(0)$ is the $q\bar{q}$ orbital wavefunction at the interquark distance $r = 0$. The factor $\alpha^2 Q^2$ arises from the couplings of the photon to the $q\overline{q}$ and e^+e^- pairs (Fig. F.2). Predict the ratios of partial widths between the mesons ρ^0, ω, ϕ, $J/\psi(1S)$ and $\Upsilon(1S)$and compare with the experimental data in Table F.1. Assume ideal mixing and also that the ratio $\frac{|\Psi(0)|^2}{M_V^2}$ is constant.

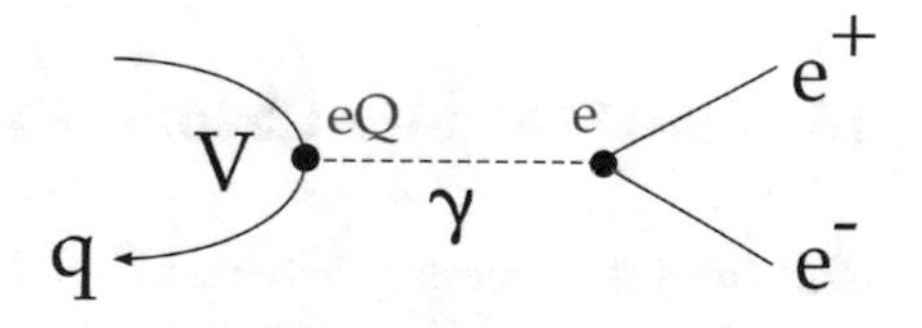

Fig. F.2 Drell-Yan pair production in vector meson decays

Table F.1 Measured partial widths into electron-positron pairs

	$\Gamma(V \rightarrow e^+e^-)$ [keV]
ρ^0	7.04 ± 0.06
ω	0.60 ± 0.02
ϕ	1.26 ± 0.02
$J/\psi(1S)$	5.55 ± 0.14
$\Upsilon(1S)$	1.340 ± 0.018

9.1 Logarithmic and Harmonic Potentials

Show that the logarithmic potentials $\kappa \ln \frac{r}{r_0}$ and br^2 predict the P levels to lie below the S levels, e.g. the $1P$ lies below the $2S$.

10.1 6 x 6*, 3 x 6

Decompose the products of SU(3) representations $6 \times 6^*$ and 3×6.

10.2 Two-Gluon Coupling

Gluons are associated to the 8-dimensional representation of $SU(3)_c$. Decompose the 8×8 representation of two gluons and show the occurrence of the glueball colour singlet.

11.1 $f' \rightarrow \eta\eta$ Decay Amplitude

Derive the decay amplitude for $\gamma(f' \rightarrow \eta\eta)$ listed in Table 11.2.

12.1 Flatté Coupled Channel Formula

Start from the production vector $T = (1 - iK\rho)^{-1}P$ (12.19) and prove that the amplitude to decay into channel 1 is given by the Flatté formula (12.28).

13.1 Gell-Mann-Nishijima Formula

Check the Gell-Mann-Nishijima formula for the following particles:
D_s^+, K^0, $\overline{\Sigma^+}$, Λ_c^+, Ξ_b^0, t-quark.

13.2 Λ and Σ^0 Flavour Wavefunctions

Derive the mixed antisymmetric SU(3)$_f$ wavefunctions of the Λ and Σ^0 listed in Table 13.4.

14.1 Magnetic Dipole Moment of the Nucleon

Predict the ratio of neutron to proton magnetic moments, assuming the flavour $\times$ spin wavefunctions of the ground state baryons to be antisymmetric, i.e. given by the combination $\frac{1}{\sqrt{2}}|\phi_{MS}\chi_{MA} - \phi_{MA}\chi_{MS}\rangle$.

14.2 Magnetic Dipole Moment of the Σ^-

Derive the magnetic moment of the Σ^-:

$$\mu_{\Sigma^-} = -\frac{4}{9}\left(\frac{e}{2m}\right) + \frac{1}{9}\left(\frac{e}{2m_s}\right). \tag{F.29}$$

14.3 Magnetic Moment of the Δ^{++}

Predict the magnetic moment of the Δ^{++} from the quark model (in units of nuclear magnetons) and compare your result with the experiment, Bosshard, A., et al.: Phys. Rev. 44 1962 (1991).

17.1 $SU(4)_f$ and $SU(5)_f$

(a) For baryons made of three quarks (u, d, s or c) prove the decomposition of SU(4) $\times$ SU(4) $\times$ SU(4):
$4 \times 4 \times 4 = 4^* + 20 + 20 + 20$.
(b) Show that by adding b quarks:
$5 \times 5 \times 5 = 10^* + 40 + 40 + 35$.

18.1 Polarization of the Proton in Unpolarized $\Lambda \to \pi^- p$ decay

We have shown that the polarization of the proton is equal to the asymmetry parameter α_Λ for unpolarized Λ. Show that the transverse polarization of the proton vanishes.

18.2 Angular Distribution in $J/\psi \rightarrow \mu^+\mu^-$ Decay

(a) This is a parity conserving electromagnetic decay. What are the possible angular momenta between the two muons?
(b) Let the J/ψ fly in the z-direction and predict the angular distribution of the muons in the rest frame of the J/ψ with respect to its flight direction for 100% polarization along the flight direction of the J/ψ,
(c) perpendicular to its flight direction.

Note that the muons cannot be emitted with the same helicites since spin flips are not allowed for the muon current which absorbs the γ.

Solutions

2.1 Invariant Masses

Let P_1, P_2 and P_3 be the 4-momenta, then

$$\begin{aligned} m_{12}^2 + m_{13}^2 + m_{23}^2 &= (P_1 + P_2)^2 + (P_1 + P_3)^2 + (P_2 + P_3)^2 \\ &= 2P_1^2 + 2P_2^2 + 2P_3^2 + 2P_1P_2 + 2P_1P_3 + 2P_2P_3 \\ &= 2m_1^2 + 2m_2^2 + 2m_3^2 + (P_1 + P_2 + P_3)^2 - m_1^2 - m_2^2 - m_3^2 \\ &= m_1^2 + m_2^2 + m_3^2 + M^2, \end{aligned} \tag{F.30}$$

where we have used energy and momentum conservation: $P^2 = (P_1 + P_2 + P_3)^2 = M^2$.

3.1 Isospin and Clebsch-Gordan Coefficients

(a) The Clebsch-Gordan decomposition is

$$|00\rangle = -\frac{1}{\sqrt{3}}|\pi^0\pi^0\rangle + \frac{1}{\sqrt{3}}|\pi^+\pi^-\rangle + \frac{1}{\sqrt{3}}|\pi^-\pi^+\rangle. \tag{F.31}$$

Hence, neglecting the mass difference between neutral and charged pions,

$$\frac{\Gamma(f_2 \to \pi^+\pi^-)}{\Gamma(f_2 \to \pi^0\pi^0)} \simeq 2. \tag{F.32}$$

© The Author(s) 2018

C. Amsler, *The Quark Structure of Hadrons*, Lecture Notes in Physics 949,
https://doi.org/10.1007/978-3-319-98527-5

(b) The π and the Σ have $i = 1$ and therefore combine to $i = 0, 1$ or 2, while K and p combine to $i = 0$ or 1. Due to isospin conservation in strong interactions the transition to $i = 2$ is excluded. The π^0 and the Σ^0 have both $i_3 = 0$ and the $i = 1$ Clebsch-Gordan coefficient $\langle 10|1100\rangle$ for $\Sigma^0\pi^0$ vanishes, therefore the transition $K^- p \to \Sigma^0\pi^0$ is pure $i = 0$, again due to isospin conservation. For $K^- p \to \Sigma^+\pi^-$ both $i = 0$ and 1 contribute. The isospin decompositions are

$$\begin{aligned} |K^- p\rangle &= -\frac{1}{\sqrt{2}}|0\rangle + \frac{1}{\sqrt{2}}|1\rangle, \\ |\Sigma^0\pi^0\rangle &= -\frac{1}{\sqrt{3}}|0\rangle + \sqrt{\frac{2}{3}}|2\rangle, \\ |\Sigma^+\pi^-\rangle &= \frac{1}{\sqrt{3}}|0\rangle + \frac{1}{\sqrt{2}}|1\rangle + \frac{1}{\sqrt{6}}|2\rangle. \end{aligned} \tag{F.33}$$

With isospin conservation the ratio of cross sections is therefore given by (ignoring mass differences in the final state)

$$\frac{\sigma(K^- p \to \Sigma^0\pi^0)}{\sigma(K^- p \to \Sigma^+\pi^-)} \simeq \left| \frac{\mathcal{M}_0}{\mathcal{M}_0 - \sqrt{\frac{3}{2}}\mathcal{M}_1} \right|^2, \tag{F.34}$$

where $\mathcal{M}_0$ and $\mathcal{M}_1$ are the transition amplitudes for $i = 0$ and 1.

3.2 Pion-Nucleon Scattering

The isospin decomposition is from the table in Fig. 2.8:

$$\begin{aligned} |\pi^+ p\rangle &= \left| \overbrace{\frac{3}{2}\frac{3}{2}}^{i\ i_3} \right\rangle, \\ |\pi^- p\rangle &= \sqrt{\frac{1}{3}}\left|\frac{3}{2} - \frac{1}{2}\right\rangle - \sqrt{\frac{2}{3}}\left|\frac{1}{2} - \frac{1}{2}\right\rangle, \\ |\pi^0 n\rangle &= \sqrt{\frac{2}{3}}\left|\frac{3}{2} - \frac{1}{2}\right\rangle + \sqrt{\frac{1}{3}}\left|\frac{1}{2} - \frac{1}{2}\right\rangle. \end{aligned} \tag{F.35}$$

Isospin conservation forbids transitions matrix elements $\frac{1}{2} \leftrightarrow \frac{3}{2}$. Furthermore the transition amplitudes do not depend on i_3 due to charge independence. Let us write the transition amplitudes as $\mathcal{M}_{\frac{1}{2}}$ and$\mathcal{M}_{\frac{3}{2}}$. The total cross sections for $\pi^+ p$ and $\pi^- p$ scattering are given by the amplitudes squared, times kinematical factors (phase space), which we assume to be equal by neglecting the mass differences between charged and neutral pions, and between proton and neutron. Hence

$$\sigma_1 \equiv \sigma(\pi^+ p \to \pi^+ p) \propto |\mathcal{M}_{\frac{3}{2}}|^2,$$

$$\sigma_2 \equiv \sigma(\pi^- p \to \pi^- p) \propto \left|\frac{1}{3}\mathcal{M}_{\frac{3}{2}} + \frac{2}{3}\mathcal{M}_{\frac{1}{2}}\right|^2,$$

$$\sigma_3 \equiv \sigma(\pi^- p \to \pi^0 n) \propto \left|\frac{\sqrt{2}}{3}\mathcal{M}_{\frac{3}{2}} - \frac{\sqrt{2}}{3}\mathcal{M}_{\frac{1}{2}}\right|^2. \quad \text{(F.36)}$$

Since the Δ resonance dominates low energy pion-nucleon scattering we neglect the $\mathcal{M}_{\frac{1}{2}}$ term and predict from (F.36) the ratios of total cross sections

$$\sigma_1 : \sigma_2 : \sigma_3 \propto 9 : 1 : 2, \quad \text{(F.37)}$$

which are in very good agreement with data.

4.1 Exotic Quantum Numbers

For $q\overline{q}$ mesons: $|\ell - s| < j < \ell + s, \quad P = -(-1)^\ell, \quad C = (-1)^{\ell+s}$.

The following quantum numbers are impossible:

$$\begin{aligned}
0^{--} &: j = 0,\ P = -1 \Rightarrow \ell = 0 \Rightarrow s = 0 \Rightarrow C = +1 \text{ not } -1,\\
0^{+-} &: j = 0,\ P = +1 \Rightarrow \ell = 1 \Rightarrow s = 1 \Rightarrow C = +1 \text{ not } -1,\\
1^{-+} &: j = 1,\ P = -1 \Rightarrow \ell = 0, 2 \Rightarrow s = 1 \Rightarrow C = -1 \text{ not } +1,\\
2^{+-} &: j = 2,\ P = +1 \Rightarrow \ell = 1, 3 \Rightarrow s = 1 \Rightarrow C = +1 \text{ not } -1,\\
3^{-+} &: j = 3,\ P = -1 \Rightarrow \ell = 2, 4 \Rightarrow s = 1 \Rightarrow C = -1 \text{ not } +1.
\end{aligned} \quad \text{(F.38)}$$

4.2 Nomenclature and Quantum Numbers

Notation: $J^{PC}(I^G)$ [quark content]. C and G are omitted when not defined.

$$a_2^+(1320) : 2^+(1^-)[u\overline{d}], \quad h_1(1380) : 1^{+-}(0^-)[u\overline{u}, d\overline{d}, s\overline{s}]^\dagger,$$

$$f_2'(1525) : 2^{++}(0^+)[s\overline{s}]^\dagger,$$

$$\pi_1^0(1400) : 1^{-+}(1^-)[\text{non} - q\overline{q}], \quad K_1^+(1400) : 1^+\left(\frac{1}{2}\right)[u\overline{s}],$$

$$\overline{K}_0^{*0}(1430) : 0^+ \left(\frac{1}{2}\right) [s\overline{d}],$$

$$D^- : 0^- \left(\frac{1}{2}\right) [\overline{c}d], \;\; B_{s2}^*(5840)^0 : 2^+(0)[\overline{b}s], \;\; \chi_{c1}(1P) : 1^{++}(0^+)[c\overline{c}]. \tag{F.39}$$

[†]The superposition of $u\overline{u}, d\overline{d}$ and $s\overline{s}$ pairs in isoscalar mesons is discussed in Sect. 5.1.

5.1 Pseudoscalar Mixing Angle

(a) The annihilation into two pseudoscalar mesons (0^{-+}) is possible with the angular momenta $\ell = 0$ and $\ell = 2$ from the 0^{++} (or $^3P_0$) and 2^{++} (or $^3P_2$) protonium levels, respectively. This follows from (2.5) and parity conservation, and from (3.17) and C parity conservation. For $\ell = 1$ the quantum numbers 1^{-+} are exotic and do not couple to $\overline{p}p$.

(b) We apply the OZI rule and drop the $s\overline{s}$ components in the octet and singlet wavefunctions. The η and η' mesons then contribute only their light quark components and we write the truncated states as

$$|\tilde{\eta}\rangle = \left[\frac{1}{\sqrt{6}}\cos\theta_P - \frac{1}{\sqrt{3}}\sin\theta_P\right]|u\overline{u} + d\overline{d}\rangle,$$

$$|\tilde{\eta}'\rangle = \left[\frac{1}{\sqrt{6}}\sin\theta_P + \frac{1}{\sqrt{3}}\cos\theta_P\right]|u\overline{u} + d\overline{d}\rangle. \tag{F.40}$$

The ratio of annihilation rates is then given by (neglecting $\ell = 2$)

$$\frac{I(\overline{p}p \to \pi^0\tilde{\eta}')}{I(\overline{p}p \to \pi^0\tilde{\eta})} = \left[\frac{\frac{1}{\sqrt{6}}\sin\theta_P + \frac{1}{\sqrt{3}}\cos\theta_P}{\frac{1}{\sqrt{6}}\cos\theta_P - \frac{1}{\sqrt{3}}\sin\theta_P}\right]^2 \cdot \frac{p_{\eta'}}{p_\eta} = \left[\frac{1 + \frac{1}{\sqrt{2}}\tan\theta_P}{\frac{1}{\sqrt{2}} - \tan\theta_P}\right]^2 \cdot \frac{p_{\eta'}}{p_\eta}. \tag{F.41}$$

For annihilation at rest $p_{\eta'} = 685\,\text{MeV/c}$ and $p_\eta = 852\,\text{MeV/c}$. Inserting the experimental ratio of intensities leads to $\theta_P = -20.4°$.

6.1 3-Dimensional SU(2) Generators

For the matrix representations of I_2 and I_3, see (6.8) and (6.12), respectively. With Table 6.2 one also obtains

$$(I_1) = \langle m|I_1|m'\rangle = \frac{1}{\sqrt{2}}\begin{pmatrix} 0 & 1 & 0 \\ 1 & 0 & 1 \\ 0 & 1 & 0 \end{pmatrix} . \tag{F.42}$$

The commutation relations (6.2) are then easily verified. For example,

$$(I_1)(I_3) = \frac{1}{\sqrt{2}}\begin{pmatrix} 0 & 0 & 0 \\ 1 & 0 & -1 \\ 0 & 0 & 0 \end{pmatrix} , \quad (I_3)(I_1) = \frac{1}{\sqrt{2}}\begin{pmatrix} 0 & 1 & 0 \\ 0 & 0 & 0 \\ 0 & -1 & 0 \end{pmatrix} , \tag{F.43}$$

hence

$$(I_1)(I_3) - (I_3)(I_1) = \frac{i}{\sqrt{2}}\begin{pmatrix} 0 & i & 0 \\ -i & 0 & +i \\ 0 & -i & 0 \end{pmatrix} = i \underbrace{\epsilon_{132}}_{-1} \underbrace{\frac{1}{\sqrt{2}}\begin{pmatrix} 0 & -i & 0 \\ i & 0 & -i \\ 0 & i & 0 \end{pmatrix}}_{(I_2)} . \tag{F.44}$$

6.2 Kaon Doublets

1. By using (6.29) and $Gs = \overline{s}$ one gets with Table 6.4

$$\begin{aligned} G|K^+\rangle &= \frac{1}{\sqrt{2}}G|u\overline{s} + \overline{s}u\rangle = +\frac{1}{\sqrt{2}}|\overline{d}s + s\overline{d}\rangle = -|\overline{K}^0\rangle, \\ G|K^0\rangle &= \frac{1}{\sqrt{2}}G|d\overline{s} + \overline{s}d\rangle = -\frac{1}{\sqrt{2}}|\overline{u}s + s\overline{u}\rangle = +|K^-\rangle, \\ G|\overline{K}^0\rangle &= -\frac{1}{\sqrt{2}}G|s\overline{d} + \overline{d}s\rangle = +\frac{1}{\sqrt{2}}|\overline{s}u + u\overline{s}\rangle = +|K^+\rangle, \\ G|K^-\rangle &= -\frac{1}{\sqrt{2}}G|s\overline{u} + \overline{u}s\rangle = -\frac{1}{\sqrt{2}}|\overline{s}d + d\overline{s}\rangle = -|K^0\rangle. \end{aligned} \tag{F.45}$$

2. The isospin decompositions of the $\overline{K}K$ states are derived from Fig. 2.8 or by copying (2.29) and (2.30),

$$|0\,0\rangle = \frac{1}{\sqrt{2}}\left(\left|\frac{1}{2} - \frac{1}{2}\right\rangle - \left| - \frac{1}{2}\frac{1}{2}\right\rangle\right) = \frac{1}{\sqrt{2}}(|\overline{K}^0 K^0\rangle + |K^- K^+\rangle),$$
$$|1\,0\rangle = \frac{1}{\sqrt{2}}\left(\left|\frac{1}{2} - \frac{1}{2}\right\rangle + \left| - \frac{1}{2}\frac{1}{2}\right\rangle\right) = \frac{1}{\sqrt{2}}(|\overline{K}^0 K^0\rangle - |K^- K^+\rangle), \quad \text{(F.46)}$$

where we have flipped the signs of the second terms to account for the minus sign in the K^- isospinor.

3. The symmetrized eigenstates of G are

$$\ell = 0, i = 0 : \frac{1}{2}[|\overline{K}^0 K^0\rangle + |K^- K^+\rangle + |K^0 \overline{K}^0\rangle + |K^+ K^-\rangle]\ (G = +1),$$
$$\ell = 0, i = 1 : \frac{1}{2}[|\overline{K}^0 K^0\rangle - |K^- K^+\rangle + |K^0 \overline{K}^0\rangle - |K^+ K^-\rangle]\ (G = -1),$$
$$\ell = 1, i = 0 : \frac{1}{2}[|\overline{K}^0 K^0\rangle + |K^- K^+\rangle - |K^0 \overline{K}^0\rangle - |K^+ K^-\rangle]\ (G = -1),$$
$$\ell = 1, i = 1 : \frac{1}{2}[|\overline{K}^0 K^0\rangle - |K^- K^+\rangle - |K^0 \overline{K}^0\rangle + |K^+ K^-\rangle]\ (G = +1),$$
$$\text{(F.47)}$$

as is easily verified by applying the transformations (F.45). The eigenstates fulfil the relation (4.4), $G = (-1)^{i+\ell}$.

7.1 Commutation Rules for SU(3) Ladder Operators

Let us show that $U_- I_+ = I_+ U_-$ follows from the structure constants in Table 7.1:

$$U_- I_+ = (G_6 - iG_7)(G_1 + iG_2) = G_6 G_1 + i(G_6 G_2 - G_7 G_1) + G_7 G_2, \quad \text{(F.48)}$$

and

$$I_+ U_- = (G_1 + iG_2)(G_6 - iG_7) = G_1 G_6 + i(G_2 G_6 - G_1 G_7) + G_2 G_7$$
$$= \frac{i}{2}G_5 + G_6 G_1 + i\left(-\frac{i}{2}G_4 + G_6 G_2 + \frac{i}{2}G_4 - G_7 G_1\right) - \frac{i}{2}G_5 + G_7 G_2$$
$$= G_6 G_1 + i(G_6 G_2 - G_7 G_1) + G_7 G_2. \quad \text{(F.49)}$$

Proof that $[U_+, U_-] = \frac{3}{2}Y - I_3$:

$$
\begin{aligned}
U_+U_- - U_-U_+ &= (G_6 + iG_7)(G_6 - iG_7) - (G_6 - iG_7)(G_6 + iG_7) \\
&= \cancel{G_6^2} + i[G_7, G_6] + \cancel{G_7^2} - (\cancel{G_6^2} + i[G_6, G_7] + \cancel{G_7^2}) \\
&= 2i[G_7, G_6] = -2\left(\frac{1}{2}G_3 - \frac{\sqrt{3}}{2}G_8\right) = \frac{3}{2}Y - I_3. \quad \text{(F.50)}
\end{aligned}
$$

7.2 Ladder Operators on Antiquarks

For antiquarks $V_\pm$ are represented by the matrices (7.21)

$$
(V_+)' = -(G_4) + i(G_5) = \begin{pmatrix} 0 & 0 & 0 \\ 0 & 0 & 0 \\ -1 & 0 & 0 \end{pmatrix},
$$

$$
(V_-)' = -(G_4) - i(G_5) = \begin{pmatrix} 0 & 0 & -1 \\ 0 & 0 & 0 \\ 0 & 0 & 0 \end{pmatrix}, \quad \text{(F.51)}
$$

where we have used the matrices (7.6). Therefore

$$
V_+|\overline{u}\rangle = (V_+)' \begin{pmatrix} 1 \\ 0 \\ 0 \end{pmatrix} = \begin{pmatrix} 0 \\ 0 \\ -1 \end{pmatrix} = -|\overline{s}\rangle\,,
$$

$$
V_-|\overline{s}\rangle = (V_-)' \begin{pmatrix} 0 \\ 0 \\ 1 \end{pmatrix} = \begin{pmatrix} -1 \\ 0 \\ 0 \end{pmatrix} = -|\overline{u}\rangle\,. \quad \text{(F.52)}
$$

Similarly for $(U_\pm)$:

$$
(U_+)' = -(G_6) + i(G_7) = \begin{pmatrix} 0 & 0 & 0 \\ 0 & 0 & 0 \\ 0 & -1 & 0 \end{pmatrix},
$$

$$
(U_-)' = -(G_6) - i(G_7) = \begin{pmatrix} 0 & 0 & 0 \\ 0 & 0 & -1 \\ 0 & 0 & 0 \end{pmatrix}, \quad \text{(F.53)}
$$

Thus

$$U_+|\overline{d}\rangle = (U_+)'\begin{pmatrix}0\\1\\0\end{pmatrix} = -\begin{pmatrix}0\\0\\1\end{pmatrix} = -|\overline{s}\rangle\ ,$$

$$U_-|\overline{s}\rangle = (U_-)'\begin{pmatrix}0\\0\\1\end{pmatrix} = -\begin{pmatrix}0\\1\\0\end{pmatrix} = -|\overline{d}\rangle\ , \tag{F.54}$$

and for $(I_\pm)$:

$$(I_+)' = -(G_1) + i(G_2) = \begin{pmatrix}0&0&0\\-1&0&0\\0&0&0\end{pmatrix},$$

$$(I_-)' = -(G_1) - i(G_2) = \begin{pmatrix}0&-1&0\\0&0&0\\0&0&0\end{pmatrix} \tag{F.55}$$

hence

$$I_+|\overline{u}\rangle = (I_+)'\begin{pmatrix}1\\0\\0\end{pmatrix} = -\begin{pmatrix}0\\1\\0\end{pmatrix} = -|\overline{d}\rangle\ ,$$

$$I_-|\overline{d}\rangle = (I_-)'\begin{pmatrix}0\\1\\0\end{pmatrix} = -\begin{pmatrix}1\\0\\0\end{pmatrix} = -|\overline{u}\rangle\ . \tag{F.56}$$

7.3 Radiative Kaon Decay

The wavefunctions are listed in Table 6.4. For $K^{*0} \to K^0\gamma$:

$$M_{K^{*0}} = \frac{1}{\sqrt{2}}\left\langle \frac{d\overline{s} - \overline{s}d}{\sqrt{2}}\,[\uparrow\uparrow]\,\middle|\, e_1\xi_1 s_{+1} + e_2\xi_2 s_{+2} \,\middle|\, \frac{d\overline{s} + \overline{s}d}{\sqrt{2}}\left[\frac{\uparrow\downarrow - \downarrow\uparrow}{\sqrt{2}}\right]\right\rangle. \tag{F.57}$$

The first and second terms give with $\xi = \frac{1}{m_u}$ and $\xi_s = \frac{1}{m_s}$

$$-\frac{e}{4}\left(-\frac{\xi}{3} - \frac{\xi_s}{3}\right) = \frac{e}{12}(\xi + \xi_s) \text{ and } \frac{e}{4}\left(\frac{\xi}{3} + \frac{\xi_s}{3}\right) = \frac{e}{12}(\xi + \xi_s), \tag{F.58}$$

respectively, hence

$$M_{K^{*0}} = \frac{e}{6}(\xi + \xi_s). \tag{F.59}$$

For $K^{*+} \to K^+\gamma$:

$$M_{K^{*+}} = \frac{1}{\sqrt{2}}\left\langle \frac{u\bar{s} - \bar{s}u}{\sqrt{2}} [\uparrow\uparrow] \middle| e_1\xi_1 s_{+1} + e_2\xi_2 s_{+2} \middle| \frac{u\bar{s} + \bar{s}u}{\sqrt{2}} \left[\frac{\uparrow\downarrow - \downarrow\uparrow}{\sqrt{2}}\right]\right\rangle. \tag{F.60}$$

The first and second terms give

$$-\frac{e}{4}\left(\frac{\xi}{3} - \frac{\xi_s}{6}\right) = -\frac{e}{12}\left(\xi + \frac{\xi_s}{2}\right) \text{ and } \frac{e}{4}\left(\frac{\xi_s}{6} - \frac{\xi}{3}\right) = \frac{e}{12}\left(\frac{\xi_s}{3} - \xi\right), \tag{F.61}$$

respectively, therefore

$$M_{K^{*+}} = \frac{e}{6}(\xi_s - 2\xi). \tag{F.62}$$

The ratio of partial widths is

$$R = \frac{\Gamma(K^{*0} \to K^0\gamma)}{\Gamma(K^{*+} \to K^+\gamma)} = \left(\frac{\xi + \xi_s}{2\xi - \xi_s}\right)^2 = \left(\frac{r+1}{2r-1}\right)^2 \tag{F.63}$$

with $r \equiv \frac{m_s}{m_u}$. The experimental value $R = 2.5 \pm 0.3$ leads to the mass ratio $r = 1.20 \pm 0.07$.

8.1 Width of the $\Upsilon(1S)$ Resonance

(a) The lepton masses are almost negligible compared to the $\Upsilon(1S)$ mass and therefore the decay branching ratios into the three lepton pairs should be almost equal by virtue of lepton universality. The measured values are $f(e^+e^-) = 2.38 \pm 0.11\%$, $f(\mu^+\mu^-) = 2.48 \pm 0.01\%$ and $f(\tau^+\tau^-) = 2.60 \pm 0.10\%$. Let us assume them equal to 2.5%.
(b) The decay branching ratio into hadrons is then $f(h) = 1 - 3 \times 0.025 = 92.5\%$. This includes radiative decays ($\sim$2%) into light mesons.
(c) The decay into (charged) leptons is mediated by the electromagnetic interaction, that into hadrons by the (OZI suppressed) strong interaction via gluons.

(d) The integrated cross section in Fig. F.1 is roughly equal to 240 nb·MeV. The natural width can be calculated from (8.1) with $M = 9460$ MeV:

$$\int_{-\infty}^{\infty} \sigma dE = \frac{3}{4} \cdot \frac{8\pi^2}{M^2} f(e^+e^-) f(h)\Gamma = 1.53 \times 10^{-8}\,\mathrm{MeV}^{-2}\,\Gamma. \tag{F.64}$$

With 1 nb = $2.57 \times 10^{-12}\,\mathrm{MeV}^{-2}$, see the transformation of units (1.7), one obtains the estimate $\Gamma \sim 40$ keV, to be compared to the Particle Data value of 54 keV.

8.2 Van Royen-Weisskopf Formula

The coupling to the photon is eQ, where Q is the charge of the emitting quark (Fig. F.2). Hence

$$|\rho^0\rangle = \frac{1}{\sqrt{2}}|d\overline{d} - u\overline{u}\rangle \Rightarrow Q^2 = \left(\frac{1}{\sqrt{2}}\left[-\frac{1}{3} - \frac{2}{3}\right]\right)^2 = \frac{1}{2},$$

$$|\omega\rangle = \frac{1}{\sqrt{2}}|d\overline{d} + u\overline{u}\rangle \Rightarrow Q^2 = \left(\frac{1}{\sqrt{2}}\left[-\frac{1}{3} + \frac{2}{3}\right]\right)^2 = \frac{1}{18},$$

$$|\phi\rangle = -|s\overline{s}\rangle \Rightarrow Q^2 = \left(-\frac{1}{3}\right)^2 = \frac{1}{9},$$

$$|J/\psi(1S)\rangle = |c\overline{c}\rangle \Rightarrow Q^2 = \left(\frac{2}{3}\right)^2 = \frac{4}{9},$$

$$|\Upsilon(1S)\rangle = |b\overline{b}\rangle \Rightarrow Q^2 = \left(-\frac{1}{3}\right)^2 = \frac{1}{9}. \tag{F.65}$$

Therefore one predicts the ratios of partial widths

$$\rho^0 : \omega : \phi : J/\psi(1S) : \Upsilon(1S) = 9 : 1 : 2 : 8 : 2, \tag{F.66}$$

in approximate agreement with the data in Table F.1:

$$9 : 0.77 \pm 0.03 : 1.61 \pm 0.03 : 7.09 \pm 0.18 : 1.71 \pm 0.03. \tag{F.67}$$

9.1 Logarithmic and Harmonic Potentials

$$V(r) = \kappa \ln \frac{r}{r_0} \Rightarrow \Delta V = \frac{d^2V}{dr^2} + \frac{2}{r}\frac{dV}{dr} = -\frac{\kappa}{r^2} + \frac{2\kappa}{r^2} = \frac{\kappa}{r^2} > 0,$$

$$V(r) = br^2 \Rightarrow \Delta V = 6b > 0, \tag{F.68}$$

hence from (9.14) $E(n, \ell) > E(n-1, \ell+1)$, e.g. the $1P$ lies below the $2S$ ($n = 1$ node).

10.1 6 x 6*, 3 x 6

The Young tableaux for 6* and 6 are given by (10.14) and (10.12) in Chap. 10, hence

$$\begin{array}{|c|c|}\hline 3&4\\\hline 2&3\\\hline\end{array} \times \begin{array}{|c|c|}\hline 3&4\\\hline\end{array} = \begin{array}{|c|c|}\hline 3&4\\\hline 2&3\\\hline 1&2\\\hline\end{array} + \begin{array}{|c|c|c|}\hline 3&4&5\\\hline 2&3\\\cline{1-2} 1\\\cline{1-1}\end{array} + \begin{array}{|c|c|c|c|}\hline 3&4&5&6\\\hline 2&3\\\cline{1-2}\end{array}$$

$$= 6^* \times 6 = \frac{144}{144} + \frac{360}{45} + \frac{360 \times 6}{5 \times 4 \times 2 \times 2} = 1 + 8 + 27. \tag{F.69}$$

For 3×6:

$$\begin{array}{|c|}\hline 3\\\hline\end{array} \times \begin{array}{|c|c|}\hline 3&4\\\hline\end{array} = \begin{array}{|c|c|}\hline 3&4\\\hline 2\\\cline{1-1}\end{array} + \begin{array}{|c|c|c|}\hline 3&4&5\\\hline\end{array}$$

$$= 3 \times 6 = 8 + 10. \tag{F.70}$$

10.2 Two-Gluon Coupling

Let us couple the two $SU(3)_c$ tableaux of dimension 8 (instructions in Sect. 6.3):

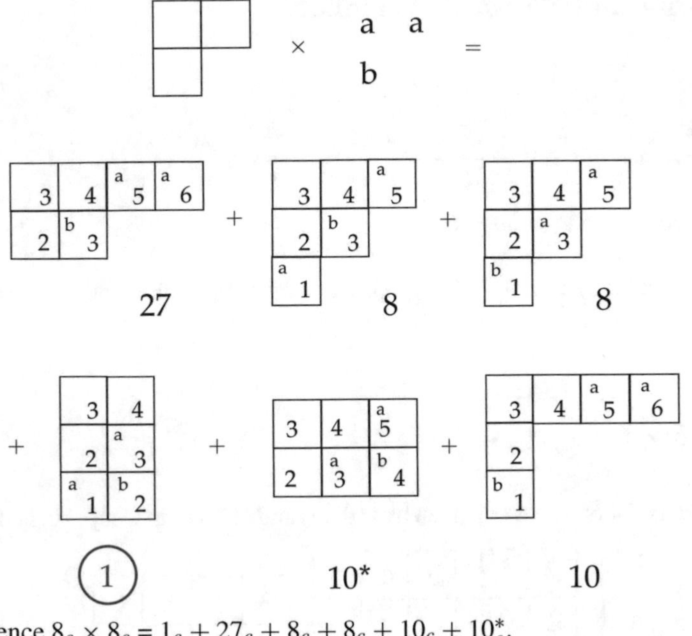

Hence $8_c \times 8_c = 1_c + 27_c + 8_c + 8_c + 10_c + 10^*_c$.

11.1 $f' \to \eta\eta$ Decay Amplitude

The octet and singlet couplings are given by, see (11.10, 11.11),

$$\begin{aligned} \gamma(f_8 \to \eta\eta) &= \gamma(f_8 \to 8 \times 8)\cos^2\theta - 2g_{18}\sin\theta\cos\theta, \\ \gamma(f_1 \to \eta\eta) &= \gamma(f_1 \to 8 \times 8)\cos^2\theta + g_{11}\sin^2\theta. \end{aligned} \tag{F.71}$$

One gets with the superposition $|f'\rangle = |f_8\rangle\sqrt{\frac{2}{3}} - |f_1\rangle\sqrt{\frac{1}{3}}$, copying from (11.13),

$$\begin{aligned} \gamma(f' \to \eta\eta) &= \sqrt{\frac{2}{3}}(-\sqrt{\frac{1}{5}}g_8\cos^2\theta - 2\underbrace{g_{18}}_{\sqrt{\frac{2}{5}}g_8}\sin\theta\cos\theta) \\ &\quad -\sqrt{\frac{1}{3}}(\underbrace{-\sqrt{\frac{1}{8}}g_1}_{\sqrt{\frac{2}{5}}g_8}\cos^2\theta + \underbrace{g_{11}}_{\sqrt{\frac{2}{5}}g_8}\sin^2\theta) \\ &= g_8\sqrt{\frac{2}{3}}\left(-\sqrt{\frac{1}{5}}\cos^2\theta - 2\sqrt{\frac{2}{5}}\sin\theta\cos\theta\right) \end{aligned}$$

$$
\begin{aligned}
&+g_8\sqrt{\frac{1}{3}}\left(-\sqrt{\frac{2}{5}}\cos^2\theta-\sqrt{\frac{2}{5}}\sin^2\theta\right)\\
&=\frac{g_8}{\sqrt{15}}(-2\sqrt{2}\cos^2\theta-4\sin\theta\cos\theta-\sqrt{2}\sin^2\theta)\\
&=-g_8\sqrt{\frac{2}{15}}(\sqrt{2}\cos\theta+\sin\theta)^2. \qquad \text{(F.72)}
\end{aligned}
$$

12.1 Flatté Coupled Channel Formula

According to (12.27) the production amplitude for channel 1 is

$$
T_1=\frac{(1-iK_{22}\rho_2)P_1+iK_{12}\rho_2P_2}{1-\rho_1\rho_2D-i(\rho_1K_{11}+\rho_2K_{22})}. \qquad \text{(F.73)}
$$

The K matrix components are

$$
K_{11}=\frac{\gamma_1^2m_0\Gamma_0'}{\Delta},\quad K_{22}=\frac{\gamma_2^2m_0\Gamma_0'}{\Delta},\quad K_{12}=K_{21}=\frac{\gamma_1\gamma_2m_0\Gamma_0'}{\Delta}, \qquad \text{(F.74)}
$$

where we have defined $\Delta\equiv m_0^2-m^2$. With the couplings (12.29)

$$
g_1\equiv\gamma_1\sqrt{m_0\Gamma_0'},\quad g_2\equiv\gamma_2\sqrt{m_0\Gamma_0'},\quad b\equiv\beta\sqrt{m_0\Gamma_0'}, \qquad \text{(F.75)}
$$

the components of the K-matrix and P-vector become

$$
\begin{aligned}
K_{11}&=\frac{g_1^2}{\Delta},\quad K_{22}=\frac{g_2^2}{\Delta},\quad K_{12}=\frac{g_1g_2}{\Delta},\\
P_1&=\frac{\beta\gamma_1m_0\Gamma_0'}{\Delta}=\frac{\beta g_1\sqrt{m_0\Gamma_0'}}{\Delta}=\frac{bg_1}{\Delta},\\
P_2&=\frac{\beta\gamma_2m_0\Gamma_0'}{\Delta}=\frac{\beta g_2\sqrt{m_0\Gamma_0'}}{\Delta}=\frac{bg_2}{\Delta}. \qquad \text{(F.76)}
\end{aligned}
$$

Therefore with (F.73) and $D=0$ one gets

$$
\begin{aligned}
T_1&=\frac{1}{\Delta}\left[\frac{(\Delta-ig_2^2\rho_2)bg_1+ig_1g_2\rho_2bg_2}{\Delta-i\rho_1g_1^2-i\rho_2g_2^2}\right]\\
&=\frac{bg_1}{\Delta-i(\rho_1g_1^2+\rho_2g_2^2)}=\frac{bg_1}{m_0^2-m^2-i(\rho_1g_1^2+\rho_2g_2^2)}. \qquad \text{(F.77)}
\end{aligned}
$$

13.1 Gell-Mann-Nishijima Formula

$$Q = i_3 + \frac{B + S + B' + C + T}{2} \tag{F.78}$$

hence (by default $B = S = B' = C = T = 0$):

$$\begin{aligned} D_s^+(c\overline{s}) &: Q = 1, i_3 = 0, S = 1, C = 1, \\ K^0(d\overline{s}) &: Q = 0, i_3 = -\frac{1}{2}, S = 1, \\ \overline{\Sigma^+}(\overline{uus}) &: Q = -1, i_3 = -1, B = -1, S = 1 \\ \Lambda_c^+(udc) &: Q = 1, i_3 = 0, B = 1, C = 1, \\ \Xi_b^0(usb) &: Q = 0, i_3 = \frac{1}{2}, B = 1, S = -1, B' = -1, \\ t\text{−quark} &: Q = \frac{2}{3}, i_3 = 0, B = \frac{1}{3}, T = 1. \end{aligned} \tag{F.79}$$

13.2 Λ and Σ^0 Flavour Wavefunctions

Applying V_- on the proton gives for the MA case

$$\begin{aligned} |\varphi\rangle &= V_- \left[\frac{1}{\sqrt{2}} |udu - duu\rangle \right] \\ &= \frac{1}{\sqrt{2}} |sdu + uds - dsu - dus\rangle. \end{aligned} \tag{F.80}$$

Applying I_- on Σ^+ leads to

$$\begin{aligned} |\varphi'\rangle &= I_- \left[\frac{1}{\sqrt{2}} |usu - suu\rangle \right] \\ &= \frac{1}{\sqrt{2}} |dsu + usd - sdu - sud\rangle. \end{aligned} \tag{F.81}$$

The state orthogonal to φ' is given by

$$|\Lambda\rangle = |\varphi'\rangle - \alpha|\varphi\rangle \tag{F.82}$$

with

$$\langle\Lambda|\varphi'\rangle = 0 = \langle\varphi'|\varphi'\rangle - \alpha\langle\varphi|\varphi'\rangle = 2 + \alpha \Rightarrow \alpha = -2, \tag{F.83}$$

hence

$$|\Lambda\rangle = |\varphi'\rangle + 2|\varphi\rangle = \frac{1}{\sqrt{2}}|sdu + 2uds - dsu - 2dus + usd - sud\rangle. \tag{F.84}$$

After normalizing one obtains for the Λ and Σ^0 in the MA case

$$|\Lambda\rangle = \frac{1}{2\sqrt{3}}|sdu - dsu + usd - sud + 2uds - 2dus\rangle, \tag{F.85}$$

$$|\Sigma^0\rangle = \frac{1}{2}|dsu - sdu + usd - sud\rangle. \tag{F.86}$$

14.1 Magnetic Dipole Moment of the Nucleon

For the antisymmetric case $\frac{1}{\sqrt{2}}|\phi_{MS}\chi_{MA} - \phi_{MA}\chi_{MS}\rangle$ the proton wavefunction would be given by

$$\begin{aligned} |p\uparrow\rangle &= \frac{1}{\sqrt{2}}\frac{1}{\sqrt{6}}\frac{1}{\sqrt{2}}[|udu + duu - 2uud\rangle|\uparrow\downarrow\uparrow - \downarrow\uparrow\uparrow\rangle \\ &\quad -|udu - duu\rangle|\uparrow\downarrow\uparrow + \downarrow\uparrow\uparrow - 2\uparrow\uparrow\downarrow\rangle] \\ &= \frac{1}{2}\frac{1}{\sqrt{6}} \\ &\quad \times 2|duu\uparrow\downarrow\uparrow\rangle - 2|udu\downarrow\uparrow\uparrow\rangle - 2|uud\uparrow\downarrow\uparrow\rangle \\ &\quad +2|uud\downarrow\uparrow\uparrow\rangle + 2|udu\uparrow\uparrow\downarrow\rangle - 2|duu\uparrow\uparrow\downarrow\rangle. \end{aligned} \tag{F.87}$$

The spins of the u quarks are always antiparallel, hence only μ_d contributes to $\sum\mu_i\sigma_{zi}$. Taking into account the orthonormality of the kets, one obtains by left-multiplying with $\langle p\uparrow|$

$$\mu_p = \left[\frac{1}{2}\frac{1}{\sqrt{6}}\right]^2 (4\times 6)\,\mu_d = \mu_d. \tag{F.88}$$

The wavefunction of the neutron would be given by

$$\begin{aligned} |n\uparrow\rangle &= -\frac{1}{\sqrt{2}}\frac{1}{\sqrt{6}}\frac{1}{\sqrt{2}}[|udd + dud - 2ddu\rangle|\uparrow\downarrow\uparrow - \downarrow\uparrow\uparrow\rangle \\ &\quad -|udd - dud\rangle|\uparrow\downarrow\uparrow + \downarrow\uparrow\uparrow - 2\uparrow\uparrow\downarrow\rangle] \end{aligned}$$

$$
\begin{aligned}
&= -\frac{1}{2}\frac{1}{\sqrt{6}} \\
&\quad \times 2|udd\uparrow\downarrow\uparrow\rangle - 2|ddu\uparrow\downarrow\uparrow\rangle + 2|ddu\downarrow\uparrow\uparrow\rangle \\
&\quad -2|dud\downarrow\uparrow\uparrow\rangle - 2|udd\uparrow\uparrow\downarrow\rangle + 2|dud\uparrow\uparrow\downarrow\rangle. \qquad \text{(F.89)}
\end{aligned}
$$

The spins of the d quarks are always antiparallel, hence only μ_u contributes to $\sum \mu_i \sigma_{zi}$. Hence we get again

$$\mu_n = \left[\frac{1}{2}\frac{1}{\sqrt{6}}\right]^2 (4 \times 6)\,\mu_u = \mu_u. \qquad \text{(F.90)}$$

The prediction

$$\frac{\mu_n}{\mu_p} = \frac{\mu_u}{\mu_d} = -2, \qquad \text{(F.91)}$$

which follows from (F.88) and (F.90), differs from the symmetric quark model prediction (14.15) and is in violent disagreement with data.

14.2 Magnetic Dipole Moment of the Σ^-

The Σ^- is a dds state. We therefore take the magnetic moment of the neutron (14.13) and replace the u by an s quark:

$$\mu_{\Sigma^-} = -\frac{1}{3}\mu_s + \frac{4}{3}\mu_d = \frac{1}{9}\left(\frac{e}{2m_s}\right) - \frac{4}{9}\left(\frac{e}{2m}\right). \qquad \text{(F.92)}$$

14.3 Magnetic Moment of the Δ^{++}

The magnetic moment of the Δ^{++} (uuu) is given by:

$$
\begin{aligned}
\mu_{\Delta^{++}} &= \sum_{i=1}^{3} \langle \Delta^{++}\uparrow |\mu_i \sigma_{zi}| \Delta^{++}\uparrow\rangle = 3 \times \frac{2}{3}\left(\frac{e}{2m}\right) \\
&= 2\mu_p = 5.58\,\mu_N = 2.00 \pm 0.06\,\mu_p. \qquad \text{(F.93)}
\end{aligned}
$$

The measurement (Bosshard, A., et al.: Phys. Rev. 44, 1962 (1991)) was performed by measuring the left-right asymmetry in the reaction $\pi^+ p \rightarrow \pi^+ p\gamma$ with 298 MeV pions striking a polarized proton target. The scattered pions were analyzed in a magnetic spectrometer, the proton detected in a scintillation counter array and the photons in a modular NaI detector. The result for the magnetic dipole moment of the Δ^{++} is $(1.62 \pm 0.18)\,\mu_p$.

17.1 SU(4)$_f$ and SU(5)$_f$

$$\left(\begin{array}{|c|}\hline 4\\ \hline 3\\ \hline\end{array} \times \begin{array}{|c|c|}\hline 4&5\\ \hline\end{array}\right) \times \begin{array}{|c|}\hline 4\\ \hline\end{array} = \begin{array}{|c|}\hline 4\\ \hline 3\\ \hline 2\\ \hline\end{array} + \begin{array}{|c|c|}\hline 4&5\\ \hline 3 \\ \cline{1-1}\end{array} + \begin{array}{|c|c|}\hline 4&5\\ \hline 3 \\ \cline{1-1}\end{array} + \begin{array}{|c|c|c|}\hline 4&5&6\\ \hline\end{array}$$

$$= 4 \times 4 \times 4 = 4^* + 20 + 20 + 20\,.$$

$$\left(\begin{array}{|c|}\hline 5\\ \hline 4\\ \hline\end{array} \times \begin{array}{|c|c|}\hline 5&4\\ \hline\end{array}\right) \times \begin{array}{|c|}\hline 5\\ \hline\end{array} = \begin{array}{|c|}\hline 5\\ \hline 4\\ \hline 3\\ \hline\end{array} + \begin{array}{|c|c|}\hline 5&6\\ \hline 4 \\ \cline{1-1}\end{array} + \begin{array}{|c|c|}\hline 5&6\\ \hline 4 \\ \cline{1-1}\end{array} + \begin{array}{|c|c|c|}\hline 5&6&7\\ \hline\end{array}$$

$$= 5 \times 5 \times 5 = 10^* + 40 + 40 + 35\,.$$

18.1 Polarization of the Proton in Unpolarized $\Lambda \to \pi^- p$ Decay

By analogy one gets from (18.20)

$$\langle s_x \rangle = \mathrm{Tr}\,\frac{1}{2}\begin{pmatrix} 0 & 1 \\ 1 & 0 \end{pmatrix} \begin{pmatrix} |\alpha_0 - \frac{1}{\sqrt{3}}\alpha_1|^2 & 0 \\ 0 & |\alpha_0 + \frac{1}{\sqrt{3}}\alpha_1|^2 \end{pmatrix} \left[\frac{1}{\mathrm{Tr}\,\rho_f}\right] = 0, \quad \text{(F.94)}$$

and similarly for $s_y = \frac{1}{2}\sigma_y$.

18.2 $J/\psi \to \mu^+\mu^-$

The decay $J/\psi \to \mu^+\mu^-$ is a parity conserving electromagnetic process with $P = -1$, hence the angular momentum between the muons is $\ell = 0$ or 2, the muons having opposite parities. Since the spin of the J/ψ is $j = 1$, the muon spins must add to $s = 1$. The helicity amplitude is given by

$$T_{\lambda_1,\lambda_2} = \alpha_0\left\langle 1\lambda \middle| 010\lambda \right\rangle\left\langle 1\lambda \middle| \frac{1}{2}\frac{1}{2}\lambda_1, -\lambda_2 \right\rangle + \alpha_2\left\langle 1\lambda \middle| 210\lambda \right\rangle\left\langle 1\lambda \middle| \frac{1}{2}\frac{1}{2}\lambda_1, -\lambda_2 \right\rangle, \quad \text{(F.95)}$$

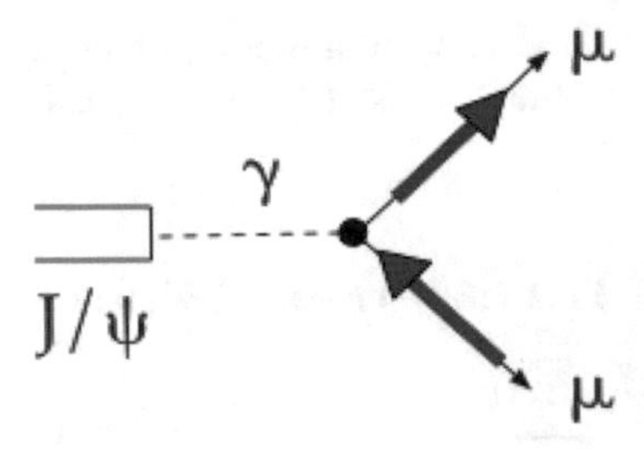

Fig. F.3 In the decay $J/\psi \to \mu^+\mu^-$ the muons have opposite helicites, hence $\lambda = \pm 1$

hence

$$T_{\frac{1}{2},-\frac{1}{2}} = \alpha_0 + \frac{1}{\sqrt{10}}\alpha_2 = T_{-\frac{1}{2},\frac{1}{2}} \equiv A \tag{F.96}$$

for the helicities $\lambda = \pm 1$. On the other hand, in the two helicity states $\lambda = 0$ the muon spins would point into opposite directions. This is not allowed by the electromagnetic interaction which absorbs the photon (Fig. F.3), as a consequence of the Dirac equation. The contributions from α_0 and α_2 are such that $T_{\frac{1}{2},\frac{1}{2}}$ vanishes, and we are left with one helicity amplitude only. The transition matrix is

$$\mathcal{M}_{\lambda_1,\lambda_2;M} = A \begin{pmatrix} \mathcal{D}^{1*}_{11} & \mathcal{D}^{1*}_{10} & \mathcal{D}^{1*}_{1-1} \\ 0 & 0 & 0 \\ 0 & 0 & 0 \\ \mathcal{D}^{1*}_{-11} & \mathcal{D}^{1*}_{-10} & \mathcal{D}^{1*}_{-1-1} \end{pmatrix}. \tag{F.97}$$

In analogy to the nomenclature used for the photon, the spin projection of a transversely polarized spin-1 particle is along (or opposite to) its flight direction. Hence with $\rho_{11} = 1$ and $\rho_{i\neq 1 \text{ or } j\neq 1} = 0$ the angular distribution is given by

$$w(\theta) = |A|^2(|\mathcal{D}^1_{11}|^2 + |\mathcal{D}^1_{-11}|^2) \propto \left[\left(\frac{1+\cos\theta}{2}\right)^2 + \left(\frac{1-\cos\theta}{2}\right)^2\right]$$

$$\Rightarrow \boxed{w(\theta) = \tfrac{3}{16\pi}(1+\cos^2\theta)}. \tag{F.98}$$

The same result is obtained with $\rho_{-1-1} = 1$ and hence (F.98) is valid for transversally polarized J/ψ, independent of the probabilities ρ_{11} and ρ_{-1-1}, in accord with parity conservation. For a longitudinally polarized J/ψ (that is with spin projection orthogonal to its flight direction) one gets with $\rho_{00} = 1$ and $\rho_{i\neq 0 \text{ or } j\neq 0} = 0$ the distribution

$$w(\theta) = |A|^2(|\mathcal{D}^1_{10}|^2 + |\mathcal{D}^1_{-10}|^2) \Rightarrow \boxed{w(\theta) = \tfrac{3}{8\pi}\sin^2\theta}. \tag{F.99}$$

The angular distribution for any spin density matrix can be found in Beneke, M., et al.: Phys. Rev. D 57, 4258 (1998).

References

1. Yang, C.N.: Phys. Rev. 77 242 (1950)
2. Amsler, C.: Nuclear and Particle Physics, p. 12-7. IOP Publishing, Bristol (2015)
3. Fano, U.: Rev. Mod. Phys. 29, 74 (1957)

Index

© The Author(s) 2018
C. Amsler, *The Quark Structure of Hadrons*, Lecture Notes in Physics 949,
https://doi.org/10.1007/978-3-319-98527-5

Printed in the United States
By Bookmasters